Pesticides

Theory and Application

GEORGE W. WARE
UNIVERSITY OF ARIZONA

W. H. FREEMAN AND COMPANY
New York

Project Editor: Patricia Brewer
Copy Editor: Amy Einsohn
Production Coordinator: Sarah Segal
Illustration Coordinator: Richard Quiñones
Artist: Donna Salmon
Compositor: Bi-Comp, Inc.
Printer and Binder: Malloy Lithographing

Library of Congress Cataloging in Publication Data

Ware, George Whitaker, 1927–
Pesticides, theory and application.

Updated ed. of: The pesticide book. c1978.
Bibliography: p.
Includes index.
1. Pesticides. I. Title.
SB951.W318 1983 628.9′6 82-7412
ISBN 0-7167-1416-7 (pbk.) AACR2

Printed in the United States of America

2 3 4 5 6 7 8 9 0 ML 0 8 9 8 7 6 5 4

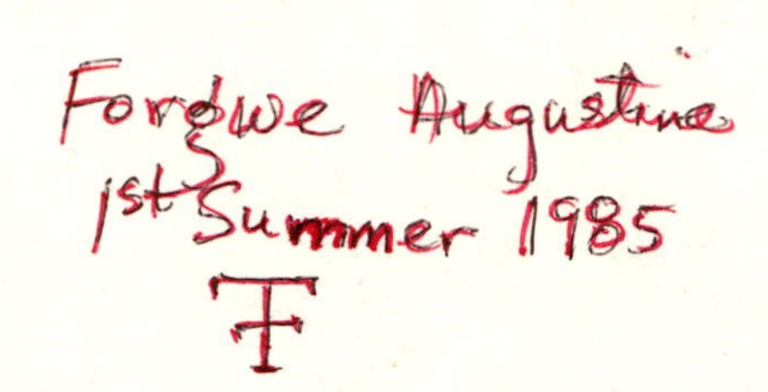

Pesticides

Theory and Application

To my wife,
Doris

DISCLAIMER CLAUSE
or
STATEMENT OF WARRANTY

Contents

Preface

Pesticides: Theory and Application is based on *The Pesticide Book*, which I wrote in 1978, but this book is not merely a revision. Each chapter has been rewritten, sharpened in focus, and updated. Part VI, "Biological Interactions with Pesticides," has been added—four new and much-needed chapters on modes of action and pesticide resistance. Abundant new information is presented on the application and use of pesticides, basic manufacturers are identified, and pesticides that have been restricted, suspended, withdrawn, or canceled are listed. The appendixes and glossary are more detailed, as are use and sales data. The product of these and many other changes is a new book, one that is useful both as a textbook and as a reference.

The book is not merely a catalogue of pesticides. Rather, individual pesticides are used to illustrate various points about chemical properties, usage and application, ecological and environmental interactions. (The inclusion of an individual pesticide in no way indicates an endorsement of the chemical or formulation; similarly, the omission of a pesticide does not indicate any rejection of it.)

The need for pesticides will not diminish. They are as indispensable to American agriculture as the tractor or mechanical harvester. Thus, we must know what they are, what they do, when to use them, and how to use them for the most effective yet safest results.

We must continue to weigh the advantages and disadvantages of the chemical control of pests and to determine and improve the risk-benefit ratio. Indeed, the times dictate the need for a source of reliable information, clearly and simply written in a factual and unemotional tone. I believe this book satisfies that need. Throughout the text I explore the difficult questions related to pesticide use and, based on current knowledge, suggest responsible and rational approaches to answering them.

In addition, the book presents practical advice on the handling and storage of pesticides. According to a recent nationwide survey of 8000 households, about 90 percent of American homes contain pesticidal products. Most householders make their pesticide purchases on impulse, without prior selection or authoritative information about method and site of application, safety, effectiveness, or storage. A chapter on the law and pesticides presents the history and current state of pesticide regulation and consumer and user protection.

No book on pesticides could be complete, because of the tremendous amount of material to be covered. This volume is intended to present a comprehensive picture of pesticides as a subject; it is not intended as an exhaustive study of any individual class of pesticides.

This book was written for students of agriculture, ecology, toxicology, pest management, and plant protection; for agricultural pest control advisors, structural pest control specialists, groundskeepers, gardeners, and landscape maintenance personnel; and for interested laypeople—whether urban householders or weekend gardeners. It will be of special value to those preparing for applicator certification and licensing in the field of pesticide usage.

A great many ideas and some data are presented in various sections of the book without direct citation of the sources. The bibliography lists the contributors whose work or writings were used, and it should be a guide for readers who wish to pursue specific topics.

During the writing of this book I received valuable advice, criticism, and assistance from many people. In particular, for reviewing the manuscript and making valuable suggestions, I want to express my sincere appreciation to Dr. Larry A. Crowder of the University of Arizona, Dr. Lena B. Brattsten of the University of Tennessee, and Dr. Larry P. Pedigo of Iowa State University. Special recognition is given to Mrs. Hazel Tinsley and Mrs. Grace Baker, who typed parts of the manuscript while under the pressures of carrying out their regular duties.

Finally, to Doris, Sam, Julie, Cindy, and Lynn, my wife and children, who sacrificed many days, evenings, and weekends of sharing time, thus enabling me to see this book to its completion, I owe a great debt of appreciation.

June 1982 George W. Ware

Theory and Application

PART I

BACKGROUND: NAMES AND PERSONALITIES OF PESTICIDES

mecarbam heptachlor bensulide thiabendazole
aldrin phoxim cyanazine paraquat dazomet
butazon carbaryl nicotine biphenyl fospirate
ioxynil ethazol malathion antu trichloronat
dimetilan acrolein chlordane terbacil endrin
ziram propazine disulfoton rotenone fonofos
sulfur molinate sulfoxide chlorazine captan
warfarin trifluralin dicumarol cycloheximide
naptalam barban dieldrin monuron fenac

CHAPTER 1

Pesticides: Chemical Tools

> Let's get our priorities in perspective....
> We must feed ourselves and protect ourselves
> against the health hazards of the world.
> To do that, we must have agricultural chemicals.
> Without them, the world population will starve.
>
> Norman E. Borlaug,
> 1970 Nobel Peace Prize

Pesticides are chemical substances used to kill or control pests. To the grower or farmer, pests could include insects and mites that damage crops; weeds that compete with field crops for nutrients and moisture; aquatic plants that clog irrigation and drainage ditches; diseases of plants caused by fungi, bacteria, and viruses; nematodes, snails, and slugs; rodents that feed on grain, young plants, and the bark of fruit trees; and birds that eat their weight every day in young plant seedlings and grain from fields and feedlots as well as from storage.

To the homeowner or apartment dweller, pests may include filthy, annoying, and disease-transmitting flies, mosquitoes, and cockroaches; moths that eat woolens; beetles that feed on leather goods and infest packaged foods; slugs, snails, aphids, mites, beetles, caterpillars, and bugs feeding on lawns, gardens, and ornamentals; termites that nibble away at wooden buildings; diseases that mar and destroy plants; algae growing on the walls or clouding the water of swimming pools; slimes and mildews that grow on shower curtains and stalls and under the rims of sinks; rats and mice that leave their fecal pellets scattered around in exchange for the food they eat; dogs that designate their territories by urinating on shrubs and favorite flowers; alley cats that yowl and screech at night; and annoying birds that defecate on window ledges, sidewalks, and statues of yesterday's forgotten heroes.

Pesticides are big business. The United States market is the world's largest, representing 34 percent of the total. In 1980 U.S. manufacturers produced 660 million kg of synthetic organic pesticides, valued at $4.2 billion (Figure 1-1), and the retail value of pesticide sales in the United States reached $5.8 billion (Aspelin and Ballard, 1980). More than 50 U.S. firms manufacture pesticides of one kind or another, with 14 firms accounting for about 85 percent of all pesticide sales. Most of the 50 basic producers prepare one or more ready-to-use forms of their products, and another 3250 formulators throughout the country prepare 35,000 different products for retail sales. In 1980 some 530 million kg of pesticides were used in

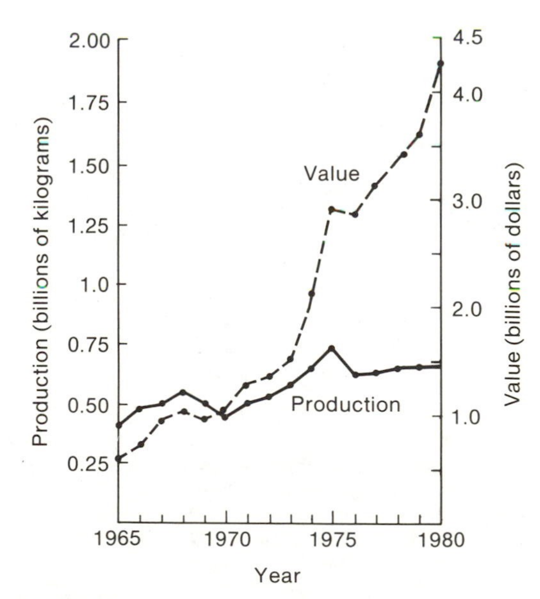

FIGURE 1-1
Production and sales of synthetic organic pesticides by the United States, 1965–1980. (*Source:* Fowler and Mahan, 1980, p. 3; U.S. Department of Agriculture, Agricultural Stabilization and Conservation Service; Storck, 1980.)

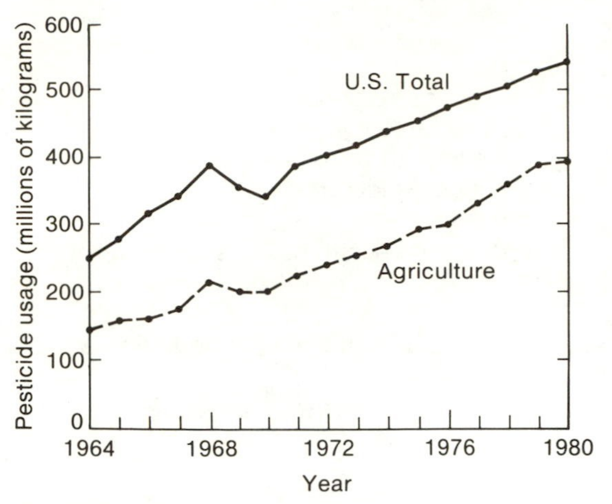

FIGURE 1-2
United States pesticide usage: total and estimated agricultural sector share, 1964–1980. (*Source:* Aspelin and Ballard, 1980.)

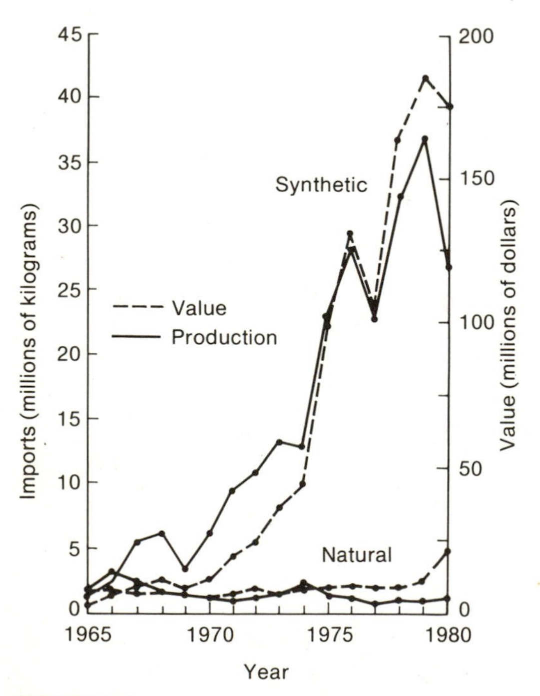

FIGURE 1-3
Volume and value of synthetic and natural organic pesticides imported by the United States, 1965–1980. (*Source:* Fowler and Mahan, 1980, p. 7; U.S. Department of Agriculture, Agricultural Stabilization and Conservation Service.)

the production of food, clothing, and durable goods for the more than 270 million persons living in the United States, that is, 2 kg (4.4 lb) per person.

Although our domestic consumption has grown substantially (Figure 1-2), the rest of the world is also using pesticides at an increasingly faster rate as more countries develop their agricultural and industrial economies. Thus, U.S. pesticide exports rose from 184 million kg in 1970 to 335 million kg in 1980 (Storck, 1980), while our imports of synthetic pesticides increased some sevenfold during that period (Figure 1-3).

The reader may be surprised to learn that the agricultural market consumes only 72 percent of the pesticides sold in the United States. Industry and government utilize 21 percent, while home and garden consumption amounts to only 7 percent (Aspelin and Ballard, 1980). Industrial and commercial use consists of pesticides used by pest control operators, turf and sod producers, floral and shrub nurseries, railroads, highways, utility rights-of-way, and industrial plant sites. Government use includes federal and state pest suppression and eradication programs, and municipal and state health protection efforts involving the control of disease vectors such as mosquitoes, flies, cockroaches, and rodents.

A survey of household, lawn, and garden use in 1976 and 1977 showed that 91 percent of American households used some form of pesticide during that period (Savage, Keefe, and Wheeler, 1979). Applications included the use of retail products, professional pest control treatments, and pet preparations. Specifically, 84 percent of the householders applied pesticides around the house (including garage, foundation, and house-dwelling pets), 21 percent in their flower and vegetable gardens, and 39 percent around their yards (including trees, shrubs, lawns, and outside pets).

Householders depend on pesticides more than they perhaps realize: for algae control in the swimming pool, mildew control in the shower and laundry area, and weed control in the yard. They use flea powder on pets, outdoor sprays to control a myriad of garden and lawn insects and diseases, indoor sprays for ants and roaches, aerosols for flies and mosquitoes. Soil and wood treatment offers termite protection, baits control mice and other rodents, woolen treatment provides moth protection, and repellents keep biting flies, chiggers, and mosquitoes off hikers and campers. Few homes in urban America are without some kind of pesticide spray, liquid, aerosol, paste, powder, or cleanser, disinfectant, or deodorizer.

As expected, the homeowner pays more for his pesticides than agriculture or industry. The difference in price is mainly due to the specialized formulations used and the small quantities purchased. The average price per pound in 1980 for pesticides was $9.88 for home and garden use, $5.63 for government and industry use, and $4.25 for agricultural use (Aspelin and Ballard, 1980).

The U.S. Environmental Protection Agency (EPA) has registered more than 2600 pesticide active ingredients: 575 herbicides, 610 insecticides, 670 fungicides and nematicides, 125 rodenticides, and 630 disinfectants. The EPA's list includes many solvents, stabilizers, surfactants, and the like; the number of physiologically active agents is about 1400. These include 335 herbicides, 425 insecticides, 250

TABLE 1-1
Pesticide usage by class and use sector for the United States in 1980 (percentage of total kilograms, active ingredients).

	Herbicides [a]	*Insecticides* [b]	*Fungicides* [c]	*Other* [d]
Agriculture	52.6	36.2	5.9	5.3
Industry and government	33.	19.	24.3	23.5
Home and garden	34.1	51.2	12.2	2.4

[a] Includes plant growth regulators.
[b] Includes miticides and contact nematicides.
[c] Does not include wood preservatives.
[d] Includes rodenticides, fumigants, and molluscicides.
Source: Aspelin and Ballard (1980).

fungicides and nematicides, 120 rodenticides, and 270 disinfectants. In 1981, some 1100 pesticide active ingredients were still in production. Of these only 200 are considered major active ingredients, and these are listed in Appendixes A through D. Table 1-1 shows the pesticide usage by type among the primary user groups in 1980.

IMPORTANCE OF PESTS AND THEIR DAMAGE

The world's main source of food is plants. They are susceptible to 80,000 to 100,000 diseases caused by viruses, bacteria, mycoplasma-like organisms, rickettsias, fungi, algae, and parasitic higher plants. They compete with 30,000 species of weeds the world over, of which approximately 1800 species cause serious economic losses. Some 3000 species of nematodes attack crop plants, and more than 1000 of these cause severe damage. Among the 800,000 species of insects, about 10,000 plant-eating species add to the devastating loss of crops throughout the world.

Approximately one-third of the world's food crops is destroyed by these pests during growth, harvesting, and storage. Losses are even higher in emerging countries: Latin America loses to pests approximately 40 percent of everything produced. Cocoa production in Ghana, the largest exporter in the world, has been trebled by the use of insecticides to control just one insect species. Pakistan sugar production was increased 33 percent through the use of insecticides. The Food and Agriculture Organization (FAO) has estimated that 50 percent of cotton production in developing countries would be destroyed without the use of insecticides.

In the United States alone crop losses due to pests are about 30 percent or $20 billion annually, despite the use of pesticides and other current control methods. What would the losses be without chemical controls?

Studies were conducted in 1976–1978 to answer that question. Insecticides were used to control insects in test plots, and the yields were compared with adjacent plots in which the insects were allowed to feed and multiply uncontrolled. From a single insect pest

TABLE 1-2
Comparison of losses caused by insects in plots treated by conventional use of insecticides and untreated plots.

Commodity	*Calculated losses (percentage)*		*Increased yield (percentage)*
	With treatment	*Without treatment*	
Corn			
Southwestern corn borer	9.9	34.3	24.4
Leafhopper on silage corn	38.3	76.7	38.4
Corn rootworm	5.0	15.7	10.7
Soybeans			
Mexican bean beetle	0.4	26.0	25.6
Stink bugs	8.5	15.0	6.5
Velvet bean caterpillar	2.4	16.6	14.2
Looper caterpillar	10.5	25.5	15.0
Wheat			
Brown wheat mite	21.0	100.0	79.0
Cutworms	7.7	54.7	47.0
White grubs	9.3	39.0	29.7
Cotton			
Boll weevil	19.0	30.9	11.9
Bollworm	12.1	90.8	78.7
Pink bollworm	10.0	25.5	25.5
Thrips	16.7	57.0	40.3
Potatoes			
Colorado potato beetle	1.0	46.6	45.6
European corn borer	1.5	54.3	52.8
Potato leafhopper	0.4	43.2	42.8

Source: *Washington Farmletter* (1979).

TABLE 1-3
Share of total agricultural herbicides and insecticides used in 1980 on corn, cotton, soybeans, and wheat.

Crop	*Percentage of pesticide used in agriculture (AI)*	
	Herbicides	*Insecticides*
Corn	53	20
Cotton	5	40
Soybeans	21	5
Wheat	6	5
Combined	85	70

Source: Eichers (1981).

species, under severe conditions, each of the major crops suffered substantial losses, as Table 1-2 shows. For example, 100 percent of the wheat in the untreated sample was destroyed by wheat mite.

Equally important are the agricultural losses from weeds, the principal agricultural pests on most farms. Weeds deprive crop plants of moisture and nutritive substances in the soil. They shade crop plants and hinder their normal growth. They contaminate harvested grain with seeds that may be poisonous to humans and animals. In some instances, complete loss of the crop results from disastrous competitive effects of weeds.

Herbicide use on all field crops is now a well-established practice. In 1980, 85 to 90 percent of the acreages of corn, cotton, soybeans, peanuts, and rice were treated with herbicides at an average rate of about 2.25 kg of active ingredient (AI) per hectare (Eichers, 1981). Of these crops, corn and soybeans together received 74 percent of all herbicides used in agriculture, while corn and cotton received 60 percent of all insecticides used in agriculture (Table 1-3).

In a study by the University of Illinois, the use of herbicides to control weeds in corn, soybeans, and wheat was found to be economically essential. In this 10-year study (Table 1-4), the herbicide treatments increased average yields for corn and soybeans by roughly 20 percent. Wheat yields were not significantly affected. An

TABLE 1-4
Increased yields of corn, soybeans, and wheat in herbicide and crop sequence experiments from 1966 through 1975.

Crop sequence and treatment	*Average yield (bushels)*	*Percentage yield increase with herbicides*
Corn		
Continuous corn		
Conventional herbicide rotation	128.5	26.4
No herbicide treatment	101.7	
Corn/soybeans/wheat sequence		
Conventional herbicide rotation	138.9	21.9
No herbicide treatment	113.9	
Soybeans		
Corn/corn/soybeans sequence		
Conventional herbicide rotation	53.5	25.6
No herbicide treatment	42.6	
Corn/soybeans/wheat sequence		
Conventional herbicide rotation	54.9	23.6
No herbicide treatment	44.4	
Wheat		
Corn/soybeans/wheat sequence		
Conventional herbicide rotation	50.8	3.0
No herbicide treatment	49.3	

Source: Hawkins, Slife, and Swanson (1977).

economic analysis of these results indicated a rate of return on herbicide expenditures of about four dollars for each dollar spent.

Were it not for herbicides, we would still have 10 to 12 percent of our population working on farms, instead of the present 3 percent. Today's farms would quickly become perpetuating weed fields that would require tremendous levels of our human energy. Indeed, it has been estimated that more energy is expended on the weeding of crops than on any other single human task (Holm, 1971). Of the world's ten worst weeds, presented in Table 1-5, eight are grasses and five are perennials. All can be readily controlled with herbicides.

TABLE 1-5
The world's ten worst weeds.

Purple nutsedge (*Cyperus rotundus* L.)
Bermudagrass (*Cynodon dactylon* (L.) Pers.)
Barnyardgrass (*Echinochloa crusgalli* (L.) Beauv.)
Junglerice (*Echinochloa colonum* L.) Link
Goosegrass (*Eleusine indica* (L.) Gaertn.)
Johnsongrass (*Sorghum halepense* (L.) Pers.)
Guineagrass (*Panicum maximum* Jacq)
Waterhyacinth (*Eichhornia crassipes* Mart.) Solms
Cogongrass (*Imperata cylindrica* (L.) Beauv.)
Lantana (*Lantana camara* L.)

Source: Holm (1969).

Minimum Tillage

Reduced tillage, or minimum tillage, farming is a new farming practice that promotes energy savings and soil conservation by reducing plowing and cultivation. Farmers once believed that fields required thorough plowing, and in some instances cross-plowing, before new rows were made and the crop planted. With some crops, particularly corn, farmers now till only enough to plant the new crop, leaving most of last year's weed and crop debris remaining on the soil surface. Insecticides and herbicides play an important role in this method because the weeds and crop stubble provide a new source of weed seed and harborage for overwintering insects. The chemical control of the weeds and insects requires about 80 percent less energy than mechanical control by cultivation. Pesticides thus form the cornerstone of minimum tillage, a new agricultural concept for energy and soil conservation.

Organic Farming

Organic and conventional farming are similar in many respects, but differ in their use of the products of modern chemical technology. Conventional farmers use synthetic chemical products while organic farmers prefer to avoid them. Organic farmers prefer to use naturally occurring chemicals such as rock phosphate and limestone, manures produced by domestic animals, nitrogen derived from the atmosphere by leguminous (pea family) plants, and substances with pesticidal properties that are produced by certain plants. Conventional farmers use these chemicals as well as using commercial fertilizers, synthetic pesticides, nutritional additives in animal feeds, and animal drugs.

Some people commonly use the word *organic* as a synonym for *natural* and regard organically grown food as nutritionally superior to conventionally grown food. Scientifically, natural substances are not necessarily organic, and organic substances are not necessarily natural. As far as is known, conventionally grown plants are just as organic and just as nutritious as are organically grown plants, and both absorb virtually all their supply of nutrients from the soil in inorganic forms.

Conventional and organic farming also differ in their use of energy. Both conventional and organic farming are largely powered by fossil fuels, but conventional farmers use more energy per hectare than do organic farmers because the production of synthetic fertilizers and pesticides requires fossil fuels. Fertilizers accounted for an estimated 33 percent of the total energy used in agricultural production in the United States in 1974, and pesticides accounted for 5 percent. Since agricultural production consumed 3 percent of the total amount of energy used in the United States, fertilizers and pesticides accounted for about 1 percent of the total energy used in this country.

However, adopting organic farming methods would not decrease national energy consumption by 1 percent. Additional land with lower yielding capability would have to be farmed to compensate for the lower yields obtained with organic farming. For example, the adoption of organic methods by farms now using a mixed grain-livestock system would result in decreased crop yields estimated at 15 to 25 percent per hectare if there were little or no change in cropping pattern. If non-livestock farms were to adopt organic methods there would be a considerably greater decrease in total yield of the high-value crops because the acreage of these would be reduced by introduction of legumes, which supply nitrogen, into the crop rotation. To offset such a 15 percent decline in yield, farmers would need to use 18 percent more of the same kind of land. But since the same kind of land is not available, any additional land would be less productive, and more of it would be needed. Finally, if legumes were to become part of the cropping sequence on intensively cropped, nearly level land, more row crops would have to be planted on the sloping land to maintain the output; this would mean increased erosion (Council for Agricultural Science and Technology, 1980).

PESTS IN HISTORY

History offers innumerable examples of the mass destruction of crops by diseases and insects. In the period from 1845 to 1851, the potato famine in Ireland occurred, as a result of a massive infection of potatoes by a fungus, *Phytophthora infestans*, now commonly referred to as *late blight*. (Either of two common fungicides, maneb or anilazine, would now control that disease handily with two or three applications.) This resulted in the loss of about a million lives and mass migrations from Ireland. Surprisingly, the infected potatoes were edible and nutritious, but the superstitious population refused to use diseased tubers. In 1930, 30 percent of the U.S. wheat crop was lost to stem rust, the same disease that destroyed 3 million tons of wheat in western Canada in 1954.

Many kinds of animal and human disease are caused by organisms carried by insects. In 1971, Venezuelan equine encephalitis appeared in southern Texas, moving in from Mexico. Through a very concerted suppression effort, involving horse vaccination, a quarantine on horse movement, and extensive spraying for mosquito control, the reported cases were limited to 88 humans and 192 horses. With the other arthropod-borne encephalitides, there were an average of 205 human cases in the United States annually between 1964 and 1973.

Since the first recorded epidemic of the Black Death, or bubonic plague, more than 65 million persons have died from this disease that is transmitted by the rat flea (*Nosopsyllus fasciatus*). In the nineteenth century, the Panama Canal was abandoned by the French because more than 30,000 laborers died from yellow fever.

As late as 1955, malaria (transmitted from person to person only by female mosquitoes belonging to the genus *Anopheles*) infected more than 200 million persons throughout the world. The annual death rate from this debilitating disease has been reduced from 6 million in 1939 to 2.5 million in 1965 to less than 1 million today. Through the use of insecticides, similar progress has been made in controlling other important tropical diseases, such as yellow fever (transmitted by *Aedes aegypti* mosquitoes), sleeping sickness (transmitted by tsetse flies in the genus *Glossina*), and Chagas' disease (transmitted by "kissing bugs" of the genus *Triatoma*).

The number of deaths resulting from all wars appears paltry beside the toll taken by insect-borne diseases. Currently there is the ever-lurking danger to humans from such diseases as encephalitis, typhus, relapsing fever, and sleeping sickness (Table 1-6).

TABLE 1-6
Some of the most common diseases known to be transmitted to humans by insects, ticks, or mites.

Disease	*Vector*
African sleeping sickness	Tsetse flies
Anthrax	Horse flies
Bubonic plague	A rat flea
Chagas' disease	Assassin bugs
Dengue fever	Two mosquitoes
Dysenteries	Several flies
Encephalitides	Several mosquitoes
Endemic typhus	Oriental rat flea
Epidemic typhus	Human louse
Filariasis	Several mosquitoes
Hemorrhagic fevers	Several mites and ticks
Leishmaniases	Psychodid flies
Louping ill	Castor bean tick
Malaria	*Anopheles* mosquitoes
Onchocerciasis	Several black flies
Pappataci fever	A psychodid fly
Q fever	Ticks
Relapsing fevers	Several ticks
Rocky Mountain spotted fever	Two ticks
Scrub typhus	Chigger mites
Trypanosomiasis	Several flies
Tularemia	Several flies, fleas, lice, ticks
Yaws	Several flies
Yellow fever	Several mosquitoes

PESTICIDES IN HISTORY

The earliest record of any material being used as a pesticide is by Homer, the Greek poet, who referred to the burning of sulfur for fumigation of homes in about 1000 BC. Pliny the Elder's *Natural History*, written in AD 70, includes a summary of pest control practices extracted from the Greek literature of the preceding 200–300 years. Most of the materials employed were useless, based on superstition and folklore.

Not until the mid-nineteenth century were pests controlled to any degree of success with chemicals. Pyrethrum, lime and sulfur combination, arsenic, sulfur, mercuric chloride, and soaps were the materials found effective between 1800 and 1825. Between 1825 and 1850 quassia, phosphorous paste, and rotenone were employed. With the use of the arsenical Paris green and kerosene emulsions as dormant sprays for deciduous fruit trees (1867–1868), the scientific use of pesticides had begun. The application of pesticides at the specific time when the pests to be controlled are most vulnerable is the most precise use of a pesticide, a subject continuously being reexamined by researchers.

Table 1-7 presents a chronology of the important events in the development and use of pesticides. Such events include not only the introduction of new chemicals and advances in formulation technology but also the demise of certain pesticides. Only in very recent years, particularly following the creation of the EPA in 1970, have data requirements for pesticide registration or reregistration been detailed enough to reveal some of the subtle long-range effects of certain pesticidal chemicals on our environment and on nontarget organisms including human beings.

BENEFITS AND RISKS OF USING PESTICIDES

What are the costs in terms of the effects on the environment, human health, wildlife, beneficial plants and insects, and the safety of our food and feed crops if we continue to use pesticides at the current rate? We have seen the costs in the past three decades, and will continue to see these penalties unless we reduce our dependence on pesticides as the single answer to pest control. As a society we must advocate specific, carefully planned utilization of pesticides integrated with other control measures. Integrated pest management (IPM) is exactly that: the thinking man's pest control. IPM devises a workable combination of the best parts of all control methods applicable to a pest problem. IPM is the practical manipulation of pest populations using sound ecological principles to keep pests below economic injury levels.

Pesticides are, however, an integral and indispensable part of IPM. They remain the first line of defense in pest control when crop injuries and losses become economic, and they are the *only* answer to a severe pest outbreak or emergency.

The effects of pesticides on nontarget organisms and the environment have been a source of worldwide contention for more than a decade and are the basis for most legislation intended to control or prohibit the use of specific pesticides. The most readily identified pesticide nontarget consequences were those of the persistent organochlorine insecticides—such as DDT—and their metabolites or conversion products on certain species of birds and fish. Consequences less readily identifiable include the effects of pesticide residues in food and the environment on humans and domestic animals. Ranging between these extremes are the unintentional effects of pesticides on plants, arthropod predators and parasites, soil microorganisms, wildlife, pollinating insects, and soil- and water-inhabiting invertebrates.

TABLE 1-7
History of pesticides.

BC	
1200	Biblical armies salt and ash the fields of the conquered; first reported use of nonselective herbicides.
1000	Homer refers to sulfur used in fumigation and other forms of pest control.
100	The Romans apply hellebore for the control of rats, mice, and insects.
25	Virgil reports seed treatment with "nitre and amurca."
AD	
70	Pliny the Elder reports pest control practices from Greek literature of the preceding three centuries; most practices based on folklore and superstition.
900	Chinese use arsenic to control garden insects.
1300	Marco Polo writes of mineral oil being used against mange of camels.
1649	Rotenone used to paralyze fish in South America.
1669	Earliest mention of arsenic as insecticide in Western world, used with honey as an ant bait.
1690	Tobacco extracts used as contact insecticide.
1773	Nicotine fumigation by heating tobacco and blowing smoke on infested plants.
1787	Soap mentioned as insecticide.
	Turpentine emulsion recommended to kill and repel insects.
1800	Persian louse powder (pyrethrum) known to the Caucasus.
	Sprays of lime and sulfur recommended in insect control.
	Whale oil prescribed as scalecide.
1810	Dip containing arsenic suggested for sheep scab control.
1820	Fish oil advocated as insecticide.
1821	Sulfur reported as fungicide for mildew by John Robertson in England.
1822	Mixture of mercuric chloride and alcohol recommended for bedbug control.
1825	Quassia used as insecticide in fly baits.
1842	Whale-oil soap mentioned as insecticide.
1845	Phosphorus paste declared as official rodenticide for rats by Prussia; by 1859 it was used in cockroach control.
1848	Derris (rotenone) reported being used in insect control in Asia.
1851	Boiled lime-sulfur employed at Versailles by Grison.
1854	Carbon disulfide tested experimentally as grain fumigant.
1858	Pyrethrum first used in the United States.
1860	Mercuric chloride solutions applied to destroy soil-inhabiting forms such as earthworms.
1867	Paris green used as an insecticide.
1868	Kerosene emulsions employed as dormant sprays for deciduous fruit trees.
1877	Hydrogen cyanide (HCN) first used as fumigant, to fumigate museum cases.
1878	London purple reported as a substitute for Paris green (both are arsenicals).
1880	Lime-sulfur used in California against San Jose scale.
1882	Naphthalene cakes used to protect insect collections.
1883	Millardet discovers the value of Bordeaux mixture in France.
1886	Hydrogen cyanide used for citrus tree fumigation in California.
	Resin fish-oil soap used as scalecide in California.
1890	Carbolineum, a coal-tar fraction, used in Germany on dormant fruit trees.
1892	Lead arsenate first prepared and used to control gypsy moth in Massachusetts.
	First use of a dinitrophenol compound, the potassium salt of 4-6-dinitro-*o*-cresol, as insecticide.
1896	Copper sulfate used selectively to kill weeds in grain fields.
	British patent refers to inorganic fluorine compounds as insecticides.
1897	Oil of citronella used as a mosquito repellent.
1902	The value of lime-sulfur as apple scab control discovered in New York.
1906	Passage of Federal Food, Drug, and Cosmetic Act (Pure Food Law).
	Lubricating oil emulsions first applied to citrus trees.
1907	Calcium arsenate in experimental use as an insecticide.
1909	First tests with 40 percent nicotine sulfate made in Colorado.
1910	Passage of Federal Insecticide Act.
1911	First publication of the use, outside the Orient, of derris as an insecticide, in British patents.
1912	Zinc arsenite first recommended as insecticide.
	p-Dichlorobenzene applied in the United States as a moth fumigant on clothes.
1917	Nicotine sulfate first used in a dry carrier for dusting.

(*continued*)

TABLE 1-7 (*continued*)

1921	Airplane first used for spreading insecticide dust for catalpa sphinx at Troy, Ohio.
1922	Calcium cyanide begins commercial use.
	First aerial application of an insecticide to cotton, Tallulah, La.
1923	Geraniol discovered to be attractive to the Japanese beetle.
1924	Cubé (derris) first tested as insecticide in the United States.
	First tests of cryolite against Mexican bean beetle.
1925	Selenium compounds tested as insecticides.
1927	Tolerance established for arsenic on apples by U.S. FDA.
	Ethylene dichloride discovered to have fumigant value.
1928	Pyrethrum culture introduced into Kenya.
	Ethylene oxide patented as insect fumigant.
1929	Alkyl phthalates patented as insect repellents.
	n-Butyl carbitol thiocyanate produced commercially as a synthetic contact insecticide.
	Cryolite introduced as an insecticide.
1930	First fixed nicotine compound, nicotine tannate, used as a stomach poison.
1931	Anabasine isolated from plants and synthesized in the laboratory.
	Thiram, first organic sulfur fungicide, discovered.
1932	Methyl bromide first used in France as fumigant.
	Ethylene and acetylene discovered to promote flowering in pineapples; first plant growth regulators.
1934	Nicotine-bentonite combination, first dependable nicotine dust, developed.
1936	Pentachlorophenol introduced as wood preservative against fungi and termites.
1938	TEPP, first organophosphate insecticide, discovered by Gerhardt Schrader.
	Passage of pesticide amendment to Pure Food Law (1906), preventing contamination of food.
	Bacillus thuringiensis first used as microbial insecticide.
	DNOC, first dinitrophenol herbicide, introduced to United States from France.
1939	Rutgers 612, first good insect repellent, introduced.
	DDT discovered to be insecticidal by Paul Müller in Switzerland.
1940	Sesame oil patented as synergist for pyrethrin insecticides.
1941	Hexachlorocyclohexane (BHC) discovered in France to be insecticidal.
	Introduction of aerosol insecticides propelled by liquified gases.
1942	First batch of DDT shipped to United States for experimental use.
	Introduction of 2,4-D, the first of the hormone (or phenoxy) herbicides.
1943	First dithiocarbamate fungicide, zineb, introduced commercially.
1944	Introduction of 2,4,5-T for brush and tree control and warfarin for rodent control.
1945	Early synthetic herbicide, ammonium sulfamate, introduced for brush control.
	Chlordane, the first of the persistent, chlorinated cyclodiene insecticides introduced.
	The first carbamate herbicide, propham, becomes available.
1946	Organophosphate insecticides, TEPP and parathion, developed by the Germans, made available to U.S. producers.
	First resistance in houseflies to DDT observed in Sweden.
1947	Toxaphene insecticide introduced; to become the most heavily used insecticide in U.S. agricultural history.
	The Federal Insecticide, Fungicide, and Rodenticide Act (FIFRA) becomes law.
1948	Appearance of aldrin and dieldrin, the best of the persistent soil insecticides.
1949	Captan, first of the dicarboximide fungicides, appears.
	First synthesis of a synthetic pyrethroid, allethrin.
1950	Malathion introduced, probably the safest organophosphate insecticide.
	Maneb fungicide introduced.
1951	Introduction of first carbamate insecticides: isolan, dimetan, pyramat, pyrolan.
1952	Fungicidal properties of captan first described.
1953	Insecticidal properties of diazinon described in Germany.
	Guthion insecticide introduced.
1954	Passage of Miller Amendment to Food, Drug, and Cosmetic Act (1906); set tolerances for all pesticides on raw food and feed products.
	Ronnel, first animal systemic organophosphate insecticide, introduced.
1955	Deet, the first truly satisfactory insect repellent, becomes available.
1956	Introduction of first successful carbamate insecticide, carbaryl.

1957	Gibberellic acid, plant growth stimulant, made available to horticulturists.
1958	Atrazine, first of the triazine herbicides, and paraquat, first of the bipyridylium herbicides, introduced.
1959	Cranberries embargoed by U.S. FDA for excessive residues of the herbicide aminotriazole.
	FIFRA (1947) amended to include all economic poisons, e.g., desiccants, nematicides.
1960	Treflan® herbicide becomes available.
	Bacillus thuringiensis first registered on lettuce and cole crops.
1961	Introduction of chlorophacinone rodenticide and mancozeb fungicide.
1962	Publication of *Silent Spring* by Dr. Rachel Carson.
1963	Appearance of Shell No-Pest Strip®, slow-release household fumigant.
1964	Fungicidal properties of thiabendazole described, also as anthelmintic in human and veterinary medicine.
1965	Killmaster®, first slow-release household formulation, introduced to United States from Europe.
	Appearance of Temik®, first soil applied insecticide-nematicide.
1966	Appearance of carboxin, the first systemic fungicide, methomyl insecticide, and chlordimeform acaricide-ovicide.
1967	Introduction of the second group of systemic fungicides with benomyl.
1968	Discovery of tetramethrin, resmethrin, and bioresmethrin, synthetic pyrethroids with greater activity than natural pyrethrins.
1969	Arizona places a moratorium on the use of DDT in agriculture.
	USDA adopts policy on pesticides to avoid use of persistent materials when effective, nonresidual methods of control are available.
	Publication of the Mrak Report which laid groundwork for concerted environmental protection, resulting in Environmental Protection Agency in 1970.
1970	Formation of Environmental Protection Agency (EPA), which becomes responsible for registration of pesticides (instead of USDA).
	All registrations suspended for alkylmercury compounds as seed treatments.
	Authority to establish tolerances for pesticides in foods and feeds transferred from FDA to EPA.
1971	Glyphosate herbicide first introduced.
1972	Passage of Federal Environmental Pesticide Control Act (FEPCA or FIFRA amended)
	Introduction of first microencapsulated insecticide, Penncap M®, methyl parathion.
1973	Development of first photo-stable synthetic pyrethroid, permethrin.
	Cancelation of virtually all uses of DDT by the EPA.
1974	First standards set for worker reentry into pesticide treated fields by EPA, e.g., reentry intervals of 24 or 48 hours dependent on dermal toxicity of pesticide.
1975	Cancelation of all uses of aldrin and dieldrin, except as termiticides.
	Registration of first virus for budworm-bollworm control on cotton.
	First insect growth regulator (methoprene) registered with EPA.
1976	Introduction of Ficam®, first household-grade wettable powder.
	Appearance of Insectape®, insecticide-impregnated adhesive tapes for household use.
	Rebuttable Presumption Against Registration (RPAR) issued for strychnine, endrin, Kepone®, 1080, and BHC.
	Passage of Toxic Substances Control Act (TSCA) on October 11.
	Most pesticidal uses of mercury compounds canceled by EPA.
1977	Introduction of microencapsulated pyrethrins plus synergist for household use.
	Use of dibromochloropropane (DBCP) suspended, and all registered uses for Mirex® canceled by EPA.
1978	EPA concludes the first full-scale RPAR of a pesticide, chlorobenzilate.
	Certification training completed for applicators, private and commercial, to use restricted-use pesticides.
	Additional amendments to FIFRA designed to improve pesticide registration process.
	First list of restricted-use pesticides issued by EPA.
	First registration of a pheromone (gossyplure for pink bollworm) for use on cotton.
1979	Suspension by EPA of most uses of 2,4,5-T and silvex.
1980	Through new legislation Congress assumes responsibility for EPA oversight.
1981	Passage of the Comprehensive Environmental Response Compensation and Liability Act (known as "Superfund") for cleanup of toxic substance spills.

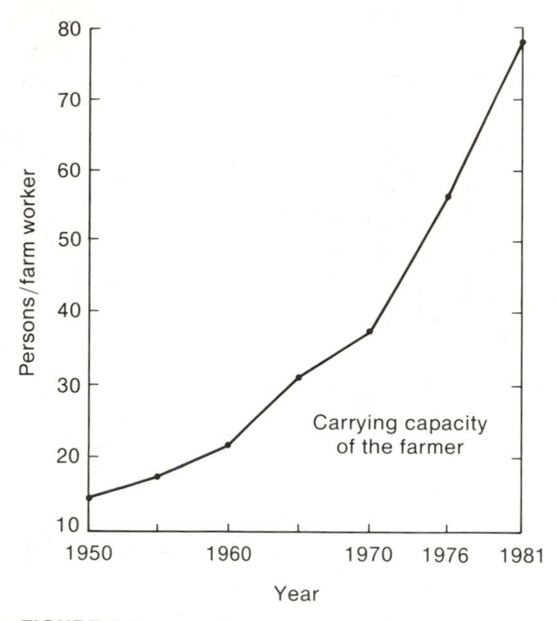

FIGURE 1-4
Persons supplied farm products per farmer, 1950–1981. (*Source:* U.S. Department of Agriculture, 1981.)

Because of our inability to recognize all the ecological relationships between nontarget organisms and the targets, we have made a few notable mistakes in the use of pesticides. Fortunately, however, these mistakes do not appear to have permanent or irrevocable consequences on the nontargets of the environment.

And the benefits of pesticides are many. These chemical tools are used as intentional additions to the environment to improve environmental quality for ourselves, our animals, and our plants. In agriculture pesticides are essential tools. Like the tractor, mechanical harvester, electric milker, and fertilizers, pesticides are part of modern agricultural technology. Farmers use pesticides to increase productivity, a benefit that favors the grower and, ultimately, the consumer of food and fiber products—the public. Pesticides have contributed significantly to the increased productivity of American farmers; in 1776 each farmer produced enough food and fiber for 3 people; in 1950, enough for 14 people; in 1981 enough for 78 people (Figure 1-4). It is estimated that insects, weeds, plant diseases, and nematodes account for losses of up to $20 billion annually in the United States alone. The use of pesticidal chemicals in agriculture enables farmers to save approximately an overall one-third of their crops. Because of the economic implications of such losses and savings pesticides have assumed major importance.

Food and Hunger

It is common knowledge that our current world food supply is inadequate. As much as 56 percent of the world's population is undernourished. And the situation is worse in undeveloped countries, where an estimated 79 percent of the inhabitants are undernourished.

The earth's population was estimated at 3.6 billion in 1970, 4.4 billion in 1980, and is expected to reach 5.4 billion in 1990 and 6.4 billion by the year 2000 (Figure 1-5). These numbers are not intended to evoke gloom about our ability to support such a population but to suggest that there will be great pressure to increase agricultural production, since this extra population must be fed and clothed.

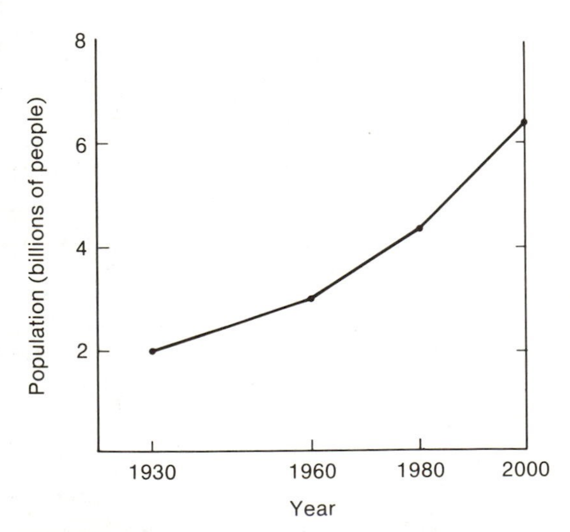

FIGURE 1-5
Growth of world population. (*Source:* FAO, United Nations. 1974. The Population Debate, Dimensions and Perspectives. Papers of the World Population Conference, Bucharest. Vol. 1, p. 5.)

Each Asian farm worker produces an average of 19,964 kg of food crops each year, a Russian farm worker 14,973 kg, a European farm worker 15,880 kg; while each American farm worker produces 170,145 kg. While achieving this impressive production, the United States grows 48 percent of the world's corn and 63 percent of the world's soybeans.

In the United States only 2 percent of the population is involved in agriculture, while only 14 percent of disposable income is spent for food (Table 1-8). In comparison, 17 percent of the Soviet Union's population is involved in agriculture and 34 percent of disposable income is spent for food.

The food production success of U.S. agriculture can be attributed to several factors, the more important of which are improved crop varieties, increased size and capacity of farm equipment, liberal use of fertilizers, and extensive use of pesticides (Table 1-9).

TABLE 1-8
Percentage of population involved in agriculture and percentage of disposable income spent for food.

Country	*Percentage of population involved in agriculture*[a]	*Percentage of disposable income spent for food*[b]
United States	2.3	14
Australia	6.0	17
United Kingdom	2.1	19
France	9.1	20
Japan	11.8	23
Soviet Union	17.3	34
Brazil	38.9	41
South Korea	39.9	44
India	64.	59
China	60.6	60

[a] Percentage of economically active population directly involved in agriculture, forestry, hunting, and fishing.
[b] Includes expenditures for food consumed away from home; figures for Australia and Japan include alcoholic beverages and tobacco.
Source: Agriculture Council of America (1981).

TABLE 1-9
Increased production per hectare of field crops grown in Arizona from 1950 to 1980.

Crop	*Yield per hectare (metric tons)*		*Increase (percent)*
	1950	*1980*	
Alfalfa hay	6.27	17.93	186
Wheat	1.61	5.38	236
Barley	2.15	4.25	98
Grain sorghum	2.37	4.69	98
Corn, for grain	.69	6.27	818

Source: Arizona Crop and Livestock Reporting Service, 1981.

Need for Pesticides

When millions of humans are killed or disabled annually from insect-borne diseases and world losses from insects, diseases, weeds, and rats are estimated at $90 billion annually, it becomes obvious that control of various harmful organisms is vital for the future of agriculture, industry, and human health. Pesticides thus become indispensable in feeding, clothing, and protecting the world's population, which will approach 6.4 billion by the year 2000.

CHAPTER **2**

I am a little world made cunningly
of elements, and an angelic sprite.
John Donne

The Chemistry and Vocabulary of Pesticides

To better understand pesticides, you need to learn to identify the different chemical elements with which pesticides are made. If you have had a course or two in chemistry, so much the better; if not, this chapter provides a brief introduction to the subject.

Everything about us, including the earth, is composed of chemical *elements.* These include such familiar elements as oxygen, nitrogen, iron, sulfur, and hydrogen. The smallest part or unit of an element is the *atom.* An element is pure in that all of its atoms are alike. It is not possible to chemically change an element.

A *compound* is a combination of two or more different elements. Water, salt, and DDT are examples of compounds. When different atoms combine or join, they form a *molecule,* the smallest part or unit of a compound. And because a compound is a combination of different elements, chemical changes can be made in it by changing the combinations of elements. Elements have a certain way of combining with other elements, and only certain elements will combine. Conversely, certain elements will not combine.

Combinations of two or more elements bound together by chemical bonds are chemical compounds. The most familiar compound is water, a mixture of hydrogen and oxygen, bound together by chemical bonds, which everyone recognizes as H_2O. Proper chemical terminology notes that H_2O is composed of two atoms of hydrogen bonded to one atom of oxygen to form one molecule of water.

To aid in the writing of chemistry, a chemical shorthand has been developed by chemists. Each element was given a *symbol,* which makes both the writing and the setting of type for printed matter simpler and easier to read. These symbols are frequently the first letter alone or the first letter plus another letter in the element's name.

Table 2-1 shows some of the elements or atomic names and the chemical symbols used in writing and illustrating molecules. The table contains only those elements used in the making or synthesis of pesticides. Indeed, only 21 of the more than 105 chemical elements are used to produce pesticides. This list can be further re-

TABLE 2-1
Chemical elements from which pesticides are made.

Element	*Symbol*	*Symbol derivation*[a]
Arsenic	As	
Boron	B	
Bromine	Br	
Cadmium	Cd	
Carbon	C	
Chlorine	Cl	
Copper	Cu	Cuprum
Fluorine	F	
Hydrogen	H	
Iron	Fe	Ferrum
Lead	Pb	Plumbum
Magnesium	Mg	
Manganese	Mn	
Mercury	Hg	Hydrargyrum
Nitrogen	N	
Oxygen	O	
Phosphorus	P	
Sodium	Na	Natrium
Sulfur	S	
Tin	Sn	Stannum
Zinc	Zn	

[a] Some atomic symbols are based on the Latin names.

duced by considering those elements used most frequently in pesticides: carbon, hydrogen, oxygen, nitrogen, phosphorus, chlorine, and sulfur. Some, however, may include the metallic and semimetallic elements, iron, copper, mercury, zinc, arsenic, and others. And because only about one-half the elements appearing in the table make up the bulk of the most frequently used pesticides, it becomes a simple matter to learn their symbols.

For practical generalizations, almost all pesticides are *organic* compounds, that is, they contain carbon in their molecules. Only a few contain no carbon and are thus identified as *inorganic* compounds.

CHEMICAL FORMULAS

A *chemical formula* is the printed description of one molecule of a chemical compound. There are several types of chemical formulas that must be distinguished from each other, since all are used in this book.

The *molecular formula* uses the symbols of the elements to indicate the number and kind of atoms in a molecule of the compound, for example, H_2O for water and C_6H_6 for benzene. The *structural formula* is written out, using symbols to indicate the way in which the atoms are located relative to one another in the molecule. The structural formulas for water and benzene can be presented as shown in the margin.

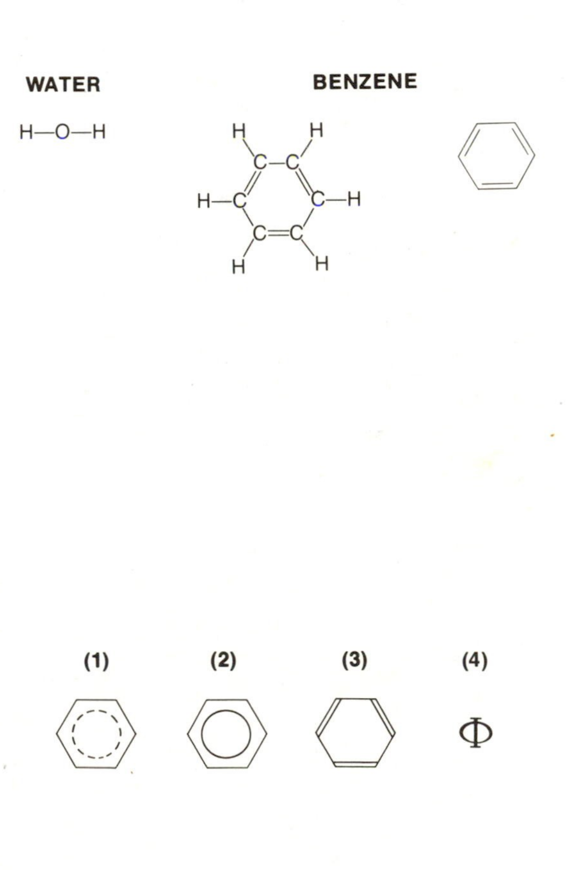

This six-carbon ring, with its six hydrogens as presented, is benzene and, for ease of representation, is usually indicated as a hexagon with double bonds. When other groups have replaced one or more of the hydrogens, the ring is referred to as the *phenyl radical* rather than as the *benzene radical*. For example, turn to the section on DDT (p. 36). The old chemical name of DDT contains the words *di-phenyl*, which refers to the two benzene rings with other groups attached.

The *benzene* or *phenyl ring* is found quite frequently in pesticide structures, and, because it is an awkward structure for printers to set from type, simpler ways of presentation have been devised and are commonly used in pesticide literature. Two of these have already been shown. The four remaining designs are: (1) the hexagon containing a broken circle or oval; (2) similar but with an unbroken circle or oval; (3) the hexagon with the three double bonds extending to its sides; and, finally, (4) the Greek letter *phi*, Φ.

The structural formula is a two-dimensional representation of a three-dimensional object. The actual shapes of organic molecules are not shown by structural formulas, but one can become familiar with these structures without knowing their shape and spatial design. The structural formulas are shown for most of the pesticides discussed in this book.

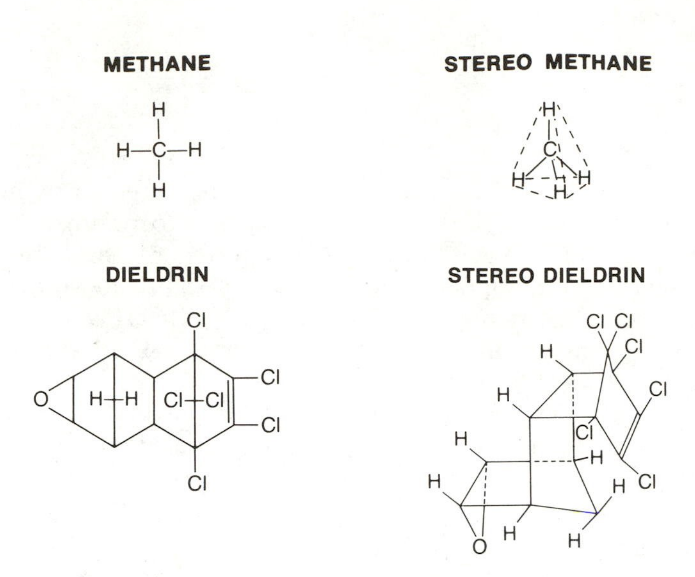

Stereo or three-dimensional formulas are printed so that the reader can visualize the depth and spatial conformation of certain molecules. The structural and stereo formulas for methane and dieldrin are illustrated.

GENERIC TERMS, TRADE NAMES, AND CHEMICAL NAMES

A recent nationwide survey of more than 8000 households revealed that many users of pesticides did not know the chemical name of the pesticide they used, only the generic term ("bug spray") or a brand name (Savage, Keefe, and Wheeler, 1979).

The term *pesticide* is an all-inclusive but nondescript word meaning "killer of pests." The various generic words ending in *-cide* (from the Latin *-cida,* "killer") are classes of pesticides, such as insecticides and herbicides. Table 2-2 lists the various pesticides and other classes of chemical compounds not commonly considered pesticides. These others, however, are included among the pesticides as defined by federal and state laws.

The term *biocide* is frequently used by laypersons to connote the adverse effects of pesticides and is often used as a synonym for pesticide. But biocide is not a scientific word and has no exact meaning. It is usually used to describe a chemical substance that kills any living plant or animal, and its tone is sensational and emotional rather than scientific. In contrast, the terms in Table 2-2 are precise and offer useful distinctions between various types of chemical substances.

Pesticides are legally classed as *economic poisons* in most state and federal laws and are defined as "any substance used for controlling, preventing, destroying, repelling, or mitigating any pest." Should you ever pursue the subject of pesticides from a legal viewpoint, they would be discussed as economic poisons. Thus, pesticides include groups of chemicals that do not actually kill pests (bottom part of Table 2-2). However, because they fit practically as well as legally under this umbrella term, they are included.

TABLE 2-2
A list of pesticide classes, their use, and derivation.

Pesticide class	*Function*	*Root-word derivation*[a]
Acaricide	Kills mites	Gr. *akari*, "mite or tick"
Algicide	Kills algae	L. *alga*, "seaweed"
Avicide	Kills or repels birds	L. *avis*, "bird"
Bactericide	Kills bacteria	L. *bacterium*; Gr. *baktron*, "a staff"
Fungicide	Kills fungi	L. *fungus*; Gr. *spongos*, "mushroom"
Herbicide	Kills weeds	L. *herba*, "an annual plant"
Insecticide	Kills insects	L. *insectum*, "cut or divided into segments"
Larvicide	Kills larvae (usually mosquito)	L. *lar*, "mask or evil spirit"
Miticide	Kills mites	Synonymous with *Acaricide*
Molluscicide	Kills snails and slugs (may include oysters, clams, mussels)	L. *molluscus*, "soft- or thin-shelled"
Nematicide	Kills nematodes	L. *nematoda*; Gr. *nema*, "thread"
Ovicide	Destroys eggs	L. *ovum*, "egg"
Pediculicide	Kills lice (head, body, crab)	L. *pedis*, "louse"
Piscicide	Kills fish	L. *piscis*, "a fish"
Predicide	Kills predators (coyotes, usually)	L. *praeda*, "prey"
Rodenticide	Kills rodents	L. *rodere*, "to gnaw"
Silvicide	Kills trees and brush	L. *silva*, "forest"
Slimicide	Kills slimes	Anglo-Saxon *slim*
Termiticide	Kills termites	L. *termes*, "wood-boring worm"
	Chemicals classed as pesticides not bearing the -cide *suffix*	
Attractants	Attract insects	
Chemosterilants	Sterilize insects or pest vertebrates (birds, rodents)	
Defoliants	Remove leaves	
Desiccants	Speed drying of plants	
Disinfectants	Destroy or inactivate harmful microorganisms	
Growth regulators	Stimulate or retard growth of plants or insects	
Pheromones	Attract insects or vertebrates	
Repellents	Repel insects, mites and ticks, or pest vertebrates (dogs, rabbits, deer, birds)	

[a] *Gr.* indicates Greek origin; *L.* indicates Latin origin.

Toxicity is another term frequently used in reference to pesticides and other chemicals. Toxicity is given as the LD_{50}, expressed as milligrams (mg) of toxicant per kilogram (kg) of body weight, the dose that kills 50 percent of the test animals, usually the laboratory rat. A more detailed explanation of LD_{50} and the types of toxicity determined for pesticides are presented in Chapter 21, "The Toxicity and Hazards of Pesticides."

In this book, each pesticide is identified in five ways. For example, let us look at diazinon, a commonly known household insecticide.

At the top is the name, diazinon (1), the *common name* for the compound. Common names are selected officially by the appropriate professional scientific society and approved by the American National Standards Institute (formerly United States of America Standards Institute) and the International Organization for Standardization. Common names of insecticides are selected by the Entomological Society of America; herbicides by the Weed Science Society of America; and fungicides by the American Phytopathologi-

(1)
DIAZINON

(2)
(Spectracide®)

(3)

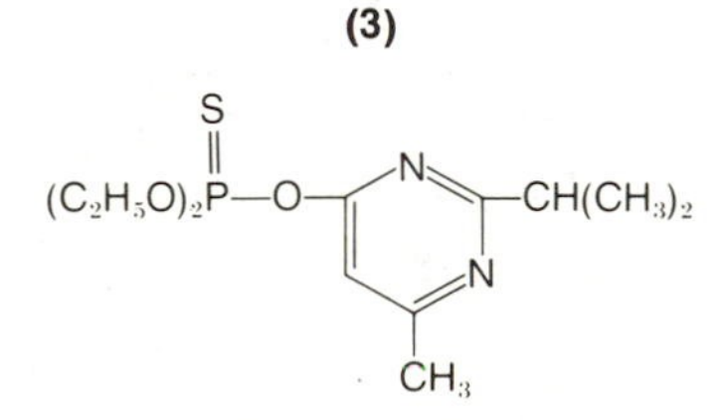

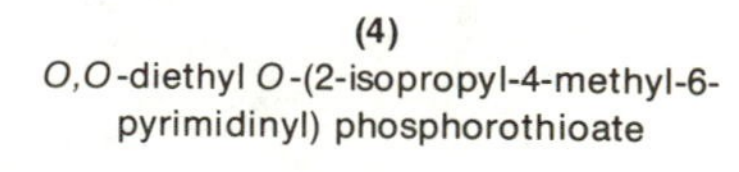
(4)
O,O-diethyl *O*-(2-isopropyl-4-methyl-6-pyrimidinyl) phosphorothioate

(5)
$C_{12}H_{21}N_2O_3PS$

TABLE 2-3
Abbreviations used in connection with the naming of pesticides.

ACS	American Chemical Society
ANSI	American National Standards Institute
AOAC	Association of Official Analytical Chemists
APS	American Phytopathological Society
BCPC	British Crop Protection Council
BIOS	British Intelligence Objectives Subcommittee
BP	British Patent
BPC	British Pharmacopoeia Commission
BSI	British Standards Institution
CIPAC	Collaborative International Pesticides Analytical Committee
ESA	Entomological Society of America
FAO	Food and Agricultural Organization (of the United Nations)
ISO	International Standardization Organization
IUPAC	International Union of Pure and Applied Chemistry
JMAF	Japanese Ministry for Agriculture and Forestry
USPharm	United States Pharmacopoeia
WHO	World Health Organization (of the United Nations)
WSSA	Weed Science Society of America

cal Society. The *proprietary name* (2), *trade name*, or *brand name* for the pesticide is given by the manufacturer or by the formulator. It is not uncommon to find several brand or trademark names given to a particular pesticide in various formulations by their formulators. To illustrate, diazinon is known as Dianon®, Diazide®, Diazol®, Knox Out®FM, Neocidal®, and Sarolex®, among others, in the United States and abroad.

Common names are assigned to avoid the confusion resulting from the use of several trade names, as just illustrated. The *structural formula* (3), as mentioned earlier, is the printed picture of the pesticide molecule. The long chemical name (4) beneath the structural formula is just that, the *chemical name*. It is usually presented according to the principles of nomenclature used in *Chemical Abstracts*, a scientific abstracting journal that is generally accepted as the world's standard for chemical names. And finally, we sometimes give the *molecular* or *empirical formula* (5) that indicates the various numbers of atoms for comparative purposes. The molecular formula is illustrated only when the structural formula is not known, which is rare.

Because pesticides are used in all countries and must be identified by some readily identifiable name, several organizations of scientific significance in the United States and abroad are involved in their naming, especially their common names. Should you pursue in more detail the background or biological effects of some pesticide, you may very well come across the abbreviations that designate these organizations (Table 2-3).

Some authors use the Wiswesser Line Notation (WLN), an extended standardization of the line-formula notations that first appeared with the beginning of structural chemistry in 1861. WLN employs the nine single-letter symbols established by Berzelius in 1813 for the nonmetals: B (boron), C (carbon), F (fluorine), H (hydrogen), I (iodine), N (nitrogen), O (oxygen), P (phosphorus), and S (sulfur). These are supplemented by five new single-letter symbols for related and very frequently cited nonmetallic or free-radical groups:

WLN Symbol	E	G	M	Q	Z
Meaning	Br	Cl	NH	OH	NH_2

This system of representing the formulas of pesticides is somewhat involved and we cannot cover it in detail here. For additional details, see Smith and Baker (1976).

CHAPTER 3

The Formulations of Pesticides

Soap emulsion:
kerosene, 2 gallons
whale-oil soap (or one quart soft soap)
1 to 2 pounds
water, one gallon

Milk emulsion:
kerosene, 2 gallons
sour milk, 1 gallon

California wash:
unslaked lime, 40 pounds
sulphur, 20 pounds
salt, 15 pounds

Recommendations of the Arizona Agricultural Experiment Station (1899)

After a pesticide is manufactured in its relatively pure form—the *technical grade material,* whether herbicide, insecticide, fungicide, or other classification—the next step is formulation. It is processed into a usable form for direct application or for dilution followed by application. The formulation is the final physical condition in which the pesticide is sold for use. The technical grade material may be formulated by its basic manufacturer or sold to a formulator. The formulated pesticide may be sold under the formulator's brand name, or it may be custom-formulated for another firm. As mentioned earlier, pesticides are sold in more than 35,000 formulations in the United States, and over 500 of these are used in American households.

Formulation is the processing of a pesticidal compound by any method that will improve its properties of storage, handling, application, effectiveness, or safety. The term formulation is usually reserved for commercial preparation prior to actual use and does not include the final dilution in application equipment.

The real test for a pesticide is acceptance by the user. And, to be accepted for use by the homeowner, the grower, or commercial applicator, a pesticide must be effective, safe, easy to apply, and relatively economic although householders commonly pay 10 to 30 times the price that growers pay for a given weight of a particular pesticide. Price depends to a great extent on the formulation; for instance, the most expensive form of insecticide is the pressurized aerosol.

Pesticides, then, are formulated into many usable forms for satisfactory storage, for effective application, for safety to the applicator and the environment, for ease of application with readily available equipment, and for economy. These goals are not always simply accomplished, due to the chemical and physical characteristics of the technical grade pesticide. For example, some materials in their "raw" or technical condition are liquids, others solids; some are stable to air and sunlight, whereas others are not; some are volatile, others not; some are water soluble, some oil soluble, and others may be insoluble in either water or oil. These characteristics pose problems to the formulator, since the final formulated product must meet the standards of acceptability by the user.

More than 98 percent of all pesticides used in the United States in 1982 are manufactured in the formulations appearing in the simplified classification presented in Table 3-1. Familiarity with the more important formulations is essential to the well-informed user. Let us examine the major formulations used in agriculture, in the home and garden, as well as those employed in structural and commercial pest control.

SPRAYS

Emulsifiable Concentrates

Formulation trends shift with time and need. Traditionally, pesticides have been applied as water sprays, water suspensions, oil sprays, dusts, and granules. Spray formulations are prepared for insecticides, herbicides, miticides, fungicides, algicides, growth regulators, defoliants, and desiccants. Consequently, more than 75 percent of all pesticides are applied as sprays. The bulk of these are currently applied as water emulsions made from emulsifiable concentrates, sometimes abbreviated as EC.

Emulsifiable concentrates, synonymous with emulsible concentrates, are concentrated oil solutions of the technical grade material with enough emulsifier added to make the concentrate mix readily with water for spraying. The emulsifier is a detergentlike material that makes possible the suspension of microscopically small oil droplets in water to form an emulsion.

When an emulsifiable concentrate is added to water, the emulsifier causes the oil to disperse immediately and uniformly throughout the water, if agitated, giving it an opaque or milky appearance. This oil-in-water suspension is a normal emulsion. There are a few rare formulations of invert emulsions, which are water-in-oil suspensions, and are opaque in the concentrated form, resembling salad dressing or face cream. Invert emulsions are employed almost exclusively as herbicide formulations. The thickened sprays result in very little drift and can be applied in sensitive situations.

Emulsifiable concentrates, if properly formulated, should remain suspended without further agitation for several days after dilution with water. A pesticide concentrate that has been held over from last year can be easily tested for its emulsifiable quality by adding 1 ounce to 1 quart of water and allowing the mixture to stand after shaking. The material should remain uniformly suspended for at least 24 hours with no precipitate. If a precipitate does form, the same condition may occur in the spray tank, resulting in clogged nozzles and uneven application. In the home, this can be remedied by adding 2 tablespoons of liquid dishwashing detergent to each pint of concentrate and mixing thoroughly. In an agricultural or other circumstance, where several gallons of costly pesticide are involved,

TABLE 3-1
Common formulations of pesticides.

1. Sprays (insecticides, herbicides, fungicides)
   a. Emulsifiable concentrates (also emulsible concentrates)
   b. Water-miscible liquids
   c. Wettable powders
   d. Water-soluble powders, e.g., prepackaged, tank drop-ins
   e. Oil solutions, e.g., barn and corral ready-to-use sprays
   f. Soluble pellets for water-hose attachments
   g. Flowable or sprayable suspensions
   h. Flowable microencapsulated suspensions
   i. Ultralow-volume concentrates (agricultural and forestry use only)
   j. Fogging concentrates, e.g., public health mosquito and fly abatement foggers
2. Dusts (insecticides, fungicides)
   a. Undiluted toxic agent
   b. Toxic agent with active diluent, e.g., sulfur, diatomaceous earth
   c. Toxic agents with inert diluent, e.g., home garden insecticide-fungicide combination in pyrophyllite carrier
   d. Aerosol "dust," e.g., silica aerogel or boric acid in aerosol form
3. Aerosols (insecticides, disinfectants, or "germicides")
   a. Pushbutton
   b. Total release
4. Granulars (insecticides, herbicides, algicides)
   a. Inert carrier impregnated with pesticide
   b. Soluble granules, e.g., swimming pool chlorine
5. Fumigants (insecticides, nematicides)
   a. Stored products and space treatment, e.g., liquids, gases, moth crystals
   b. Soil treatment liquids that vaporize
6. Impregnates (insecticides, fungicides, herbicides)
   a. Polymeric materials containing a volatile insecticide, e.g., No-Pest Strips®, pet collars
   b. Polymeric materials containing nonvolatile insecticides, e.g., pet collars, adhesive tapes, pet tags, livestock eartags
   c. Shelf papers, strips, cords containing a volatile or contact insecticide
   d. Mothproofing agents for woolens
   e. Wood preservatives
   f. Wax bars (herbicides)
   g. Insecticide soaps for pets
7. Fertilizer combinations with herbicides, insecticides, or fungicides
8. Baits
   a. Insecticides
   b. Molluscicides
   c. Rodenticides
   d. Avicides
9. Slow-release insecticides
   a. Microencapsulated materials for agriculture, mosquito abatement, and household
   b. Paint-on lacquers for pest control operators and homeowners
   c. Adhesive tapes for pest control operators and homeowners
   d. Resin strip containing volatile, organophosphate fumigant, e.g., No-Pest Strips®
10. Insect repellents
    a. Aerosols
    b. Rub-ons (liquids, lotions, paper wipes, and "sticks")
    c. Vapor-producing candles, torch fuels, smoldering wicks
11. Insect attractants
    a. Food, e.g., Japanese bettle traps
    b. Sex lures, e.g. pheromones for agricultural and forest pests
12. Animal systemics (insecticides, parasiticides)
    a. Oral (premeasured capsules or liquids)
    b. Dermal (pour-on or sprays)
    c. Feed-additive

additional emulsifier can be obtained from the formulator. This should be added to the concentrate at the rate of 0.2 to 0.5 pound for each gallon of outdated material. The bulk of pesticides available to the homeowner are formulated as emulsifiable concentrates and generally have a shelf life of about 3 years.

Every gallon of emulsifiable concentrate contains from 4 to 7 pounds of petroleum solvent, usually one of the more expensive aromatic solvents such as xylene. Within the past three years petroleum solvents have increased in price 300 percent or more, increasing significantly the costs of formulation and consequently the cost of the formulated product to the buyer. These increases are not likely to stabilize within the foreseeable future, and formulators are searching diligently for other, more economical formulations for their products.

The most recent innovation in emulsifiable concentrates is the transparent emulsion concentrates (TEC). These are the mixture of the pesticide and the emulsifier with little or no hydrocarbon solvent. In these formulations the emulsifier or surfactant serves as the solvent, replacing the aromatic solvents, yet yielding a product that can be readily diluted with water.

Water-Miscible Liquids

Water-miscible liquids are mixable in water. The technical grade material may be initially water miscible, or it may be alcohol miscible and formulated with an alcohol to become water miscible. These formulations resemble the emulsifiable concentrates in viscosity and color, but do not become milky when diluted with water. Few home and garden pesticides are sold as water miscibles, since few of the pesticides that are safe for home use have these physical characteristics. Water-miscible liquids are labeled as water-soluble concentrate (WSC), liquid (L), soluble concentrate (SC), or solution (S).

Wettable Powders

Wettable powders, abbreviated WP, are essentially concentrated dusts containing a wetting agent to facilitate the mixing of the powder with water before spraying. The technical material is added to the inert diluent, in this case a finely ground talc or clay, in addition to a wetting agent or surfactant and mixed thoroughly in a ball mill. Without the wetting agent, the powder would float when added to water, and the two would be almost impossible to mix. Because wettable powders usually contain from 50 to 75 percent clay or talc, they sink rather quickly to the bottom of spray tanks unless the spray mix is agitated constantly. Many of the insecticides sold for garden use are in the form of wettable powders because there is very little chance that this formulation will burn foliage, even at high concentrations. In contrast, the original carrier in emulsifiable concentrates is usually an aromatic solvent, which in relatively moderate concentrations can cause foliage burning at temperatures above 32.5°C.

Water-Soluble Powders

In water-soluble powders (SP) the technical grade material is a finely ground water-soluble solid and may contain a small amount of wetting agent to assist its solution in water. It is simply added to the spray tank, where it dissolves immediately. Unlike the wettable powders and flowables, these formulations do not require constant agitation; they are true solutions and form no precipitate. Because of their sometimes dusty quality, soluble powders may be packaged in convenient, water-soluble bags to be dropped into the spray tank.

Oil Solutions

In their commonest form, oil solutions are the ready-to-use household and garden insecticide sprays sold in an array of bottles, cans, and plastic containers, all usually equipped with a handy spray atomizer. Not to be confused with aerosols, these sprays are intended to be used directly on pests or places they frequent. Oil solutions may be used as roadside weed sprays, for marshes and standing pools to control mosquito larvae, in fogging machines for mosquito and fly abatement programs, or for household insect sprays purchased in supermarkets. Commercially they may be sold as oil concentrates of the pesticide to be diluted with kerosene or diesel fuel before application or in the dilute, ready-to-use form. In either case, the compound is dissolved in oil and is applied as an oil spray; it contains no dust diluent, emulsifier, or wetting agent.

Soluble Pellets

Despite their seeming convenience and ease of handling with a water hose, soluble pellets are not very effective. They are sold in kits, including the water-hose attachment, fertilizer, fungicide, insecticide, and even a car-wax pellet. The actual amount of active ingredients is very small, and even distribution with a watering hose is difficult.

Flowable or Sprayable Suspensions

There are several types of flowable (F) or sprayable (S) suspensions. Physically they all are thick and creamy, and they vary in appearance from tan to white. The original flowable was an ingenious solution to a formulation problem. Some pesticides are soluble in neither oil nor water, but *are* soluble in one of the exotic solvents, which makes the formulation quite expensive and may price it out of the marketing competition. To handle the problem, the technical material is wet-milled with a clay diluent and water, leaving the pesticide-diluent mixture finely ground but wet. This "pudding" mixes well with water and can be sprayed but has the same tank-settling characteristic as the wettable powders.

An example of a second kind of innovative flowable is the blending of the finely ground insecticide carbaryl with molasses. This formulation (Sevimol®) reduces to some extent the pesticide's drift off target during aerial application, increases its adherence to foliage and thus reduces removal by rain, and increases the mortality of moths that are attracted to feed on the molasses.

A third flowable is made by mixing an emulsifiable concentrate, containing a very high percentage of a water-stable toxicant and a thickener, with two to four volumes of water. The result is a thick, concentrated normal emulsion, which is then diluted for use with the appropriate volume of water just before using. This formulation responds to the need for an emulsifiable concentrate made with the least solvent possible to avoid foliage burn (phytotoxicity) in citrus groves. These formulations are occasionally identified as spray concentrates.

A fourth innovation is the flowable microencapsulated formulation. The insecticide is incorporated by a special process in small, permeable spheres of a polymer or plastic, 15 to 50 μm (1 μm = 10^{-6} m) in diameter. These spheres are then mixed with wetting agents, thickeners, and water to give the desired concentration of insecticide in the flowable, usually 2 pounds per gallon. (These are discussed in more detail in the section on slow-release insecticides.)

In 1981, Sevin® XLR, a fifth modification of flowables, appeared. It is an improved water- or rain-resistant formulation containing latex as a binding and sticking agent, which extends residual activity to 7–10 days. Additionally, it poses a reduced hazard to honeybees because it adheres to the foliage rather than to foraging bees. Particle size of XLR is reported to range from 5 to 10 μm in diameter, compared to an average particle size of 20 μm for the 80 percent wettable powder.

The most recent flowable is a newly patented flowable aqueous pesticide. It is made by wet-milling a water-soluble pesticide, wetting agents, thickeners, anticaking and antifoaming agents, and a freeze-point depressant in water, resulting in a thick, water-dispersible concentrate.

Ultralow-Volume Concentrates

Ultralow-volume concentrates (ULV) are available only for commercial use in the control of public health, agricultural, and forest pests. They are usually the technical product in its original liquid form or, if solid, the original product dissolved in a minimum of solvent. They are usually applied without further dilution, by special aerial or ground spray equipment that limits the volume from .6 liter (L) to a maximum of 4.7 L per hectare (ha), as an extremely fine spray. The ULV formulations are used where good results can be obtained while economizing through the elimination of the normally high spray volumes, varying from 11 to 38 L/ha. This technique has proved extremely useful where insect control is desired over vast areas.

A recent innovation in ULV formulations is the aerial application of a pyrethroid insecticide at .11 to .23 kg active ingredient (AI) in

2.4 L of semirefined cottonseed or soybean oil per hectare. The only crop for which this is currently registered by the EPA is cotton.

Fogging Concentrates

Fogging concentrates are the formulations sold strictly for public health use in the control of nuisance or disease vectors, such as flies and mosquitoes, and to pest control operators. Fogging machines generate droplets whose diameters are usually less than 10 μm but greater than 1 μm. They are of two types. The thermal fogging device utilizes a flash heating of the oil solvent to produce a visible vapor or smoke. The ambient fogger atomizes a tiny jet of liquid in a venturi tube through which passes an ultrahigh-velocity air stream. The materials used in fogging machines depend on the type of fogger. Thermal devices use oil only, whereas ambient generators use water, emulsions, or oils. In dwellings and food establishments the oil most commonly used is a deodorized kerosene or deobase.

DUSTS

Historically, dusts have been the simplest formulations of pesticides to manufacture and the easiest to apply. Examples of the undiluted toxic agent are sulfur dust, used on ornamentals, and one of the older household roach dusts, sodium fluoride. An example of the toxic agent with active diluent is any of the garden insecticides having sulfur dust as its carrier or diluent. A toxic agent with an inert diluent is the most common type of dust formulation in use today, both in the home garden and in agriculture. Insecticide-fungicide combinations are applied in this manner, the carrier being an inert clay, such as pyrophyllite. In this instance, particles small enough to pass through a 60-mesh screen are considered dusts. *Mesh* is a unit that refers to the number of grids per inch through which they will pass. The last type, the aerosol dust, is a finely ground silica or boric acid in a liquefied gas propellant that can be directed into crevices of homes and commercial structures for insect control.

Despite their ease in handling, formulation, and application, dusts are the least effective and, ultimately, the least economical of the pesticide formulations. The reason is that dusts have a very poor rate of deposit on foliage, unless it is wet from dew or rain. In agriculture, for instance, an aerial application of a standard dust formulation of pesticide will result in 10 percent to 40 percent of the material reaching the crop. The remainder drifts upward and downwind. Psychologically, dusts are annoying to the nongrower who sees great clouds of dust resulting from an aerial application, in contrast to the grower who believes he or she is receiving a thorough application for the very same reason. The same statement may be relevant to the avid user of dusts in his garden and the abstaining neighbor! Under similar circumstances, an aerial or garden hand sprayer application of a water emulsion spray will deposit 60 to 99 percent of the pesticide on target.

AEROSOLS

We have been raised in the aerosol culture: bug bombs, hair sprays, underarm deodorants, home deodorizers, oven cleaners, Mace, shaving cream, lubricants, mouth fresheners, furniture polish, car wax, electrostatic inhibitors, adhesives, starch, fabric finishers, pruning sealer, spot removers, engine cleaner, motor starting fluids, water repellents, window sprays, repellents, paints, garbage can and shower-tub disinfectants, and supremely, the anti-itch remedies—the foot and groin sprays.

In 1981, 2.20 billion aerosols were packaged in the United States. Of these, 173 million, or 8 percent, were insecticides. The use of insecticide aerosols has increased steadily over the years, an average of 4 percent annually from 1971 to 1980. The production of all kinds of aerosols peaked at 2.90 billion units in 1973, the year the fluorocarbon-ozone controversy began. With purchases dropping steadily, producers scrambled to replace fluorocarbon propellants with less ecologically suspect substitutes, such as carbon dioxide. They also mounted a publicity campaign to convince consumers of the safety and utility of pressurized products. Ozone is by now a reasonably dead issue, so far as aerosols are concerned.

To produce an aerosol, the active ingredients must be soluble in the volatile, petroleum solvent in its pressurized condition. The pressure is provided by a propellant gas. When the petroleum solvent is atomized, it evaporates rapidly, leaving the microsized droplets of toxicant suspended in air.

Developed during World War II for the GIs, the pushbutton variety of insecticide is used as a space spray to knock down flying insects. Aerosols are effective only against resident flying and crawling insects and provide little or no residual effect. More recently the total-release aerosol has been designed to discharge its entire contents in a single application. Once the nozzle is depressed, it locks into place, permitting the container's total emission while the occupants leave and remain away for a few hours. These products are available for homeowners as well as for commercial pest control operators.

Caution: Aerosols commonly produce droplets well below 10 μm in diameter. Such droplets are respirable, that is, they are absorbed by alveolar tissue in the lungs rather than impinging on the bronchioles, as do larger droplets. Consequently, aerosols of all varieties should be handled with discretion, and the user should inhale as little as possible.

GRANULAR PESTICIDES

Granular (G) pesticides overcome the disadvantages of dusts in their handling characteristics. The granules are small pellets formed from various inert clays and sprayed with a solution of the toxicant to give the desired content. After the solvent has evaporated, the granules are packaged for use. Granular materials range in size from 20 to 80 mesh. Only insecticides and a few herbicides are formulated as granules. They range from 2 to 25 percent active ingredient and are used almost exclusively in agriculture, although systemic insecticides as granules can be purchased for lawn and ornamentals. Granular materials may be applied at virtually any time of day, since they can be applied aerially in winds up to 20 mph without problems of drift, an impossible task with sprays or dusts. They also lend themselves to soil application in the drill at planting time to protect the roots from insects or to introduce a systemic to the roots for transport to aboveground parts in lawns and ornamentals.

Microgranules, mesh size ranging from 40 to 80 mesh, have been used in the United States on an experimental basis, somewhat unsuccessfully on cotton and field crops. They are used routinely in the Orient, particularly in Japan.

FUMIGANTS

Fumigants are a rather loosely defined group of formulations. The plastic insecticide-impregnated strips and pet collars of the same materials are really a slow-release formulation that permits the insecticide to work its way slowly to the surface and volatilize. Moth crystals and moth balls—paradichlorobenzene and naphthalene, respectively—are crystalline solids that evaporate slowly at room temperatures, exerting both a repellent as well as an insecticidal effect. (They can also be used in small quantity to keep cats and dogs off or away from their favorite parking places.) Soil fumigants are used in horticultural nurseries, greenhouses, and on high-value cropland to control nematodes, insect larvae, and adults and sometimes to control diseases and weed seeds. Depending on the fumigant, the treated soils may require covering with plastic sheets for several days to retain the volatile chemical, allowing it to exert its maximum effect.

IMPREGNATING MATERIALS

Impregnating materials include treatment of woolens for mothproofing and timbers against wood-destroying organisms. For several years, woolens and occasionally leather garments have been mothproofed in the final stage of dry cleaning (using chlorinated solvents). The last solvent rinse contains an ultralow concentration of the chlorinated, biodegradable insecticide ethylan, which has long residual qualities against moths and leather-eating beetle larvae. Railroad ties, telephone and light poles, fence posts, and other wooden objects that have close contact with or are actually buried in the ground soon begin to deteriorate as a result of attacks from fungal decay microorganisms and insects, particularly termites, unless treated with fungicides and insecticides. Such treatment extends the wooden object's functional life to some 40 or 60 years. The insecticides of choice for wood exposed to potential termite damage are dieldrin and chlordane.

Impregnated Shelf Papers

Impregnated shelf papers, strips, and cords containing insecticides have practically disappeared from the market. Thoroughly effective against stored products insect pests, they usually contain one of the chlorinated insecticides to give long residual activity. Because these insecticides, along with most others, cannot be used where food and food utensils are stored, according to regulations established by the Environmental Protection Agency, the use of impregnated materials has declined substantially in the past decade.

Impregnated Wax Bars

Impregnated wax bars contain a herbicide that is selective against broad-leaf plants. When dragged over grass lawns in a uniform pattern, enough rubs off on weeds to eliminate them, leaving the grass unaffected. This type of application is very selective, represents a spot application that is not disruptive to the environment, and is the type that should be strongly encouraged.

FERTILIZER COMBINATIONS

Fertilizer combinations are formulations fairly familiar to the urbanite who has purchased a lawn or turf fertilizer that contains a herbicide for crabgrass control, insecticides for grubs and sod webworms, or a fungicide for numerous lawn diseases. Fertilizer-insecticide mixtures have been made available to growers, particularly in the corn belt, by special order with the fertilizer distributor. The fertilizer and insecticide can then be applied to the soil during planting in a single, economical operation.

BAITS

Baits can be purchased or formulated at home. Those that are purchased contain low levels of the toxicant incorporated into materials that are relished by the target pests. Spot application—the placing of the bait in selected places accessible only to the target species—permits the use of very small quantities of oftentimes highly toxic materials in a totally safe manner, with no environmental disruption.

SLOW-RELEASE INSECTICIDES

Slow-release insecticides (Table 3-2) are relatively new and only a few are available to the homeowner. The first significant breakthrough was the appearance of the Shell No-Pest Strips® in 1963. Dichlorvos, a volatile organophosphate insecticide, was incorporated into panels of polychlorovinyl resin, which permitted the insecticide to volatilize at a slower rate, killing flying and some crawl-

TABLE 3-2
Slow-release insecticide formulations.

Product, active ingredient, and manufacturer	*Physical form*	*Use*	*Method of application*
No-Pest Strips® dichlorvos (20%) Texize, Division, Morton-Norwich	Polychlorovinyl resin strip	Fumigant for flying insects	Strips hung near ceiling
Killmaster II® chlorpyrifos (2%) Velsicol Chemical Co.	Ready-to-use solvent containing lacquers and plastics	Most crawling household pests	Professional application by brush
Positive Control® Pest Aid "180"®, chlorpyrifos (1%) Positive Formulators, Inc.	Ready-to-use solvent containing lacquers and plastics	Most crawling household pests	Homeowner application by brush or coarse spray
Penncap M® methyl parathion (22%) Pennwalt Corp.	Cross-linked nylon polymer, microencapsulated concentrate (30–50 μm diam.)[a]	Many agricultural pests	Professional application by air or ground equipment
Penncap E® ethyl parathion (22%) Pennwalt Corp.	Cross-linked nylon polymer, microencapsulated concentrate (30–50 μm diam.)	Several agricultural pests	Professional application by air or ground equipment
Knox Out 2FM® diazinon (23%) Pennwalt Corp.	Cross-linked nylon polymer, microencapsulated concentrate (30–50 μm diam.)	Many agricultural pests and most crawling household pests	Aerial or ground application for agriculture, and spray for household (professional use)
Sectrol #90 Concentrate® pyrethrins + synergists (1.1% + 5.9%) 3M, Industrial Tape Div.	Polyurea microencapsulated concentrate (15–20 μm diam.)	Most crawling and flying household insects	Professionally applied as spray or as ULV aerosol
Insectape® 10% diazinon, chlorpyrifos, or propoxur Herculite Products, Inc.	Ready-to-use, adhesive, laminated, multilayered, polymeric strips	Most crawling household pests	Tape applied to surfaces by hand (homeowner use—except propoxur by professionals)
Roach-Tape® propoxur (4%) Herculite Products, Inc.	Ready-to-use, adhesive, laminated, multilayered, polymeric strips	Most crawling household pests	Tape applied to surfaces by hand (homeowner use)
Altosid Briquet® 7.9% methoprene Zoecon Corp.	Ready-to-use briquets (68 g) of slowly disintegrating medium.	Mosquito growth regulator to prevent adult mosquito emergence.	Briquets placed in mosquito breeding bodies of water at beginning of mosquito season

[a] For purposes of comparison, an average human hair is approximately 50 μm in diameter.

ing insects in the vicinity. The most recent slow-release formulation appeared in 1979, a microencapsulated concentrate of diazinon (Knox Out 2FM®) for use in agriculture and for household pests. The principle of this form of slow-release involves the incorporation of the insecticide in a permeable covering, microcapsules or tiny spheres, with diameters ranging from 15 to 50 μm (Figure 3-1), that permits its release at a reduced but effective rate. The insecticide escapes through the sphere wall over an extended period, thus preserving its effectiveness much longer, usually two to four times longer, than if formulated as an emulsifiable concentrate.

A paint-on slow-release formulation containing chlorpyrifos, Killmaster II®, is a recent innovation in structural pest control. The insecticide is dissolved in a volatile petroleum solvent containing a unique combination of dissolved plastics and lacquers in small quantities. Following its paint-on application as a spot treatment in homes, restaurants, and food-handling establishments, the solvent quickly evaporates, leaving the insecticide incorporated in a thin

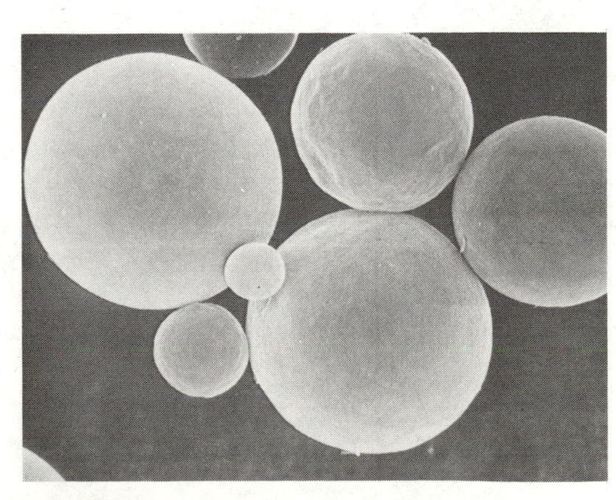

FIGURE 3-1
KNOX OUT 2FM microcapsules magnified 600 times. (Courtesy Pennwalt Corporation.)

transparent film. Over time the insecticide "blooms" or escapes to the surface at a constant rate, presenting at all times a freshly exposed surface to crawling insects.

Insecticidal adhesive tapes work as contact insecticides against crawling insects. The insecticide is laminated into the multilayered polymeric strips. The adhesive back is exposed by removing a protective film, and the tape is attached beneath counters, under shelves, and in other protected places. Adhesive and paint-on slow-release formulations are now available to homeowners under various brand names (Table 3-2).

CHEMICALS USED IN THE CONTROL OF INVERTEBRATES— ANIMALS WITHOUT BACKBONES

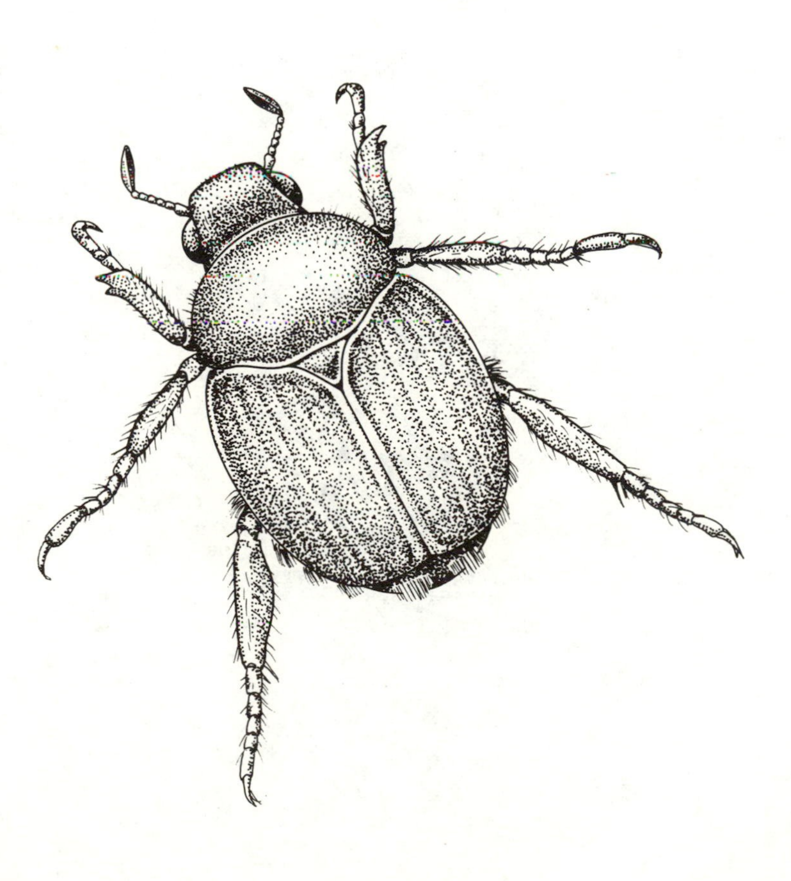

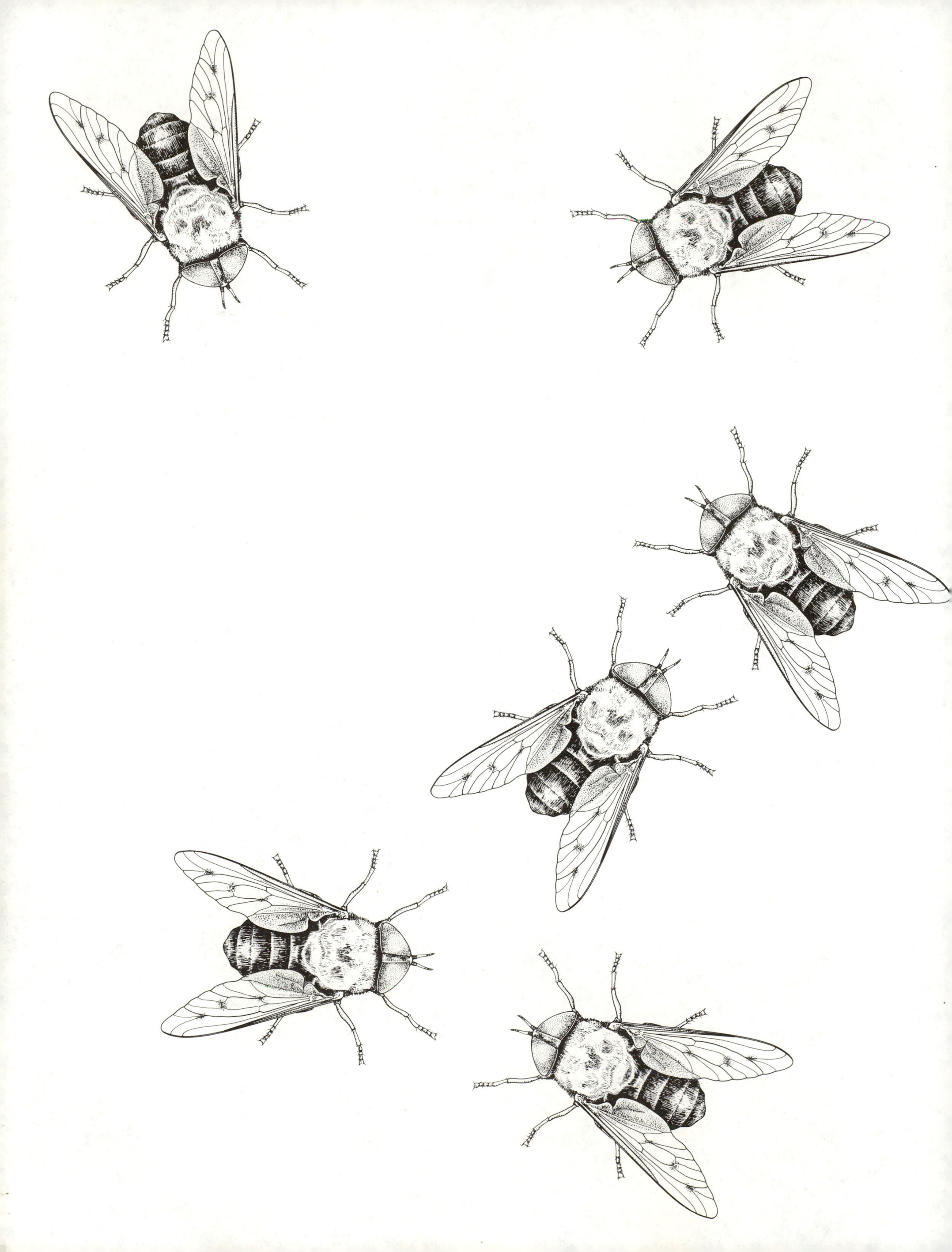

CHAPTER 4

Insecticides

> And the locusts came . . . and settled on the whole country of Egypt . . . and they ate all the plants in the land and all the fruit of the trees . . . not a green thing remained, neither tree nor plant of the field, through all the land.
>
> *Exodus* 10:14–15

According to recent discoveries humanoids have existed on earth more than 3 million years, while insects are known to have existed for 250 million years. Yet human beings have learned to live and compete with the insect world. We can guess that the first materials used by our primitive ancestors that could be classed as insecticides (repellents) in the crudest definition of the word were mud and dust spread over their skin to repel biting and tickling insects, a practice that resembles the habits of water buffalo, pigs, and elephants.

History doesn't tell us very much about chemicals used against insects. As noted in Table 1-7, the earliest records of insecticides pertain to the burning of "brimstone" (sulfur) as a fumigant. Pliny the Elder (AD 23–79) recorded most of the earlier insecticide uses in his *Natural History*. Included among these were the use of gall from a green lizard to protect apples from worms and rot. In the interim, a variety of materials have been used with doubtful results: extracts of pepper and tobacco, hot water, soapy water, whitewash, vinegar, turpentine, fish oil, brine, lye, and many others.

Even as recently as 1940, our insecticide supply was limited to several arsenicals, petroleum oils, nicotine, pyrethrum, rotenone, sulfur, hydrogen cyanide gas, and cryolite. World War II opened the Chemical Era with the introduction of a totally new concept of insect control chemicals—synthetic organic insecticides, the first of which was DDT.

ORGANOCHLORINES

The organochlorines are insecticides that contain carbon (thus the name *organo-*), chlorine, and hydrogen. They are also referred to by other names: *chlorinated hydrocarbons, chlorinated organics, chlorinated insecticides, chlorinated synthetics.*

DDT and Related Insecticides

DDT

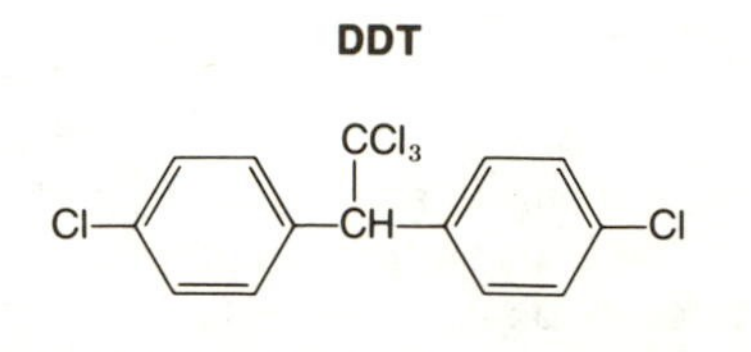

1,1,1-trichloro-2,2-bis(*p*-chlorophenyl)ethane

The EPA canceled all uses of DDT, effective January 1, 1973. In retrospect, DDT can now be considered the pesticide of greatest historical significance, as it affected human health, agriculture, and the environment. The story of its rise to stardom, carrying with it the Nobel Prize, and decline to infamy is rather sensational and should be briefly narrated for the uninitiated.

Easy to say, easy to remember, DDT is probably the best known and most notorious chemical of this century. It is the most fascinating, and remains to be acknowledged as the most useful insecticide developed. Surprisingly, DDT is more than 100 years old. It was first synthesized in 1873, by a German graduate student who had no idea of its tremendous insecticidal value, and after synthesis it was put on the shelf and forgotten. In 1939 a Swiss entomologist, Dr. Paul Müller, rediscovered DDT while searching for a long-lasting insecticide against the clothes moth. DDT proved to be extremely effective against flies and mosquitoes, ultimately bringing to Dr. Müller the Nobel Prize in medicine in 1948 for his lifesaving discovery. We should keep in mind that its most beneficial use was in public health, for malaria control, and in many nations it still is so used.

More than 1.8 billion kg of DDT have been used throughout the world for insect control since 1940, and 80 percent of that amount was used in agriculture. Production reached its maximum in the United States in 1961, when 73 million kg were manufactured. The greatest agricultural benefits from DDT have been in the control of the Colorado potato beetle and several other potato insects, the codling moth on apples, corn earworm, cotton bollworm, tobacco budworm, pink bollworm on cotton, and the worm complex on vegetables. It has been most useful against the gypsy moth and the spruce budworm in forests. From the standpoint of human medicine, DDT has been most successful against mosquitoes that transmit malaria and yellow fever, against body lice that can carry typhus, and against fleas that are vectors of plague.

One of the most amazing features of DDT was its low cost. Most of that sold to the World Health Organization went for less than 22 cents a pound. Without question, it was the most economical insecticide ever sold. A federal ban on the use of DDT, declared by the EPA on January 1, 1973, named DDT an environmental hazard due to its long residual life and to its accumulation, along with the metabolite DDE, in food chains, where it proved to be detrimental to certain forms of wildlife. It is no longer available to the grower or homeowner.

TDE (DDD)

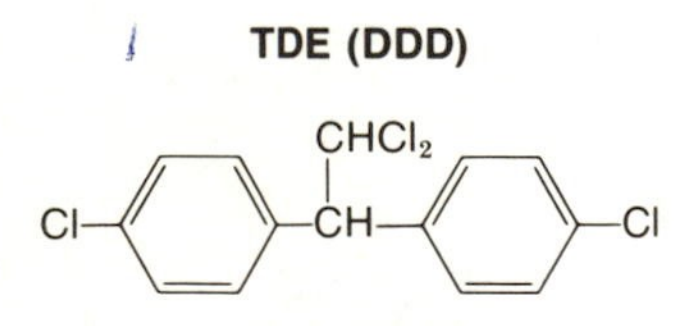

1,1-dichloro-2,2-bis(*p*-chlorophenyl)ethane

DDT belongs to the chemical class of diphenyl aliphatics, which consist of an aliphatic, or straight carbon chain, with two (di-) phenyl rings attached, as in the illustrations. DDT was first known chemically as dichloro diphenyl trichloroethane, hence DDT.

METHOXYCHLOR

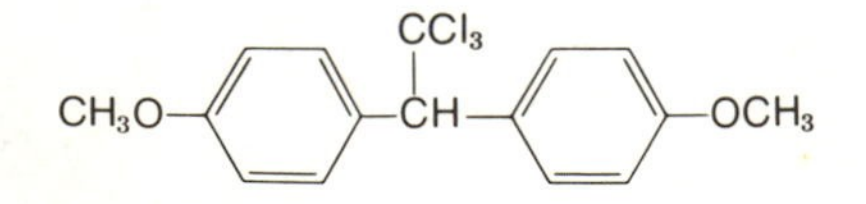

1,1,1-trichloro-2,2-bis(*p*-methoxyphenyl)ethane

The chemical structure for DDT is presented here for the reader, more as a matter of historical curiosity than any other. Five relatives of DDT should be mentioned because they all had an early role in pest control: TDE (or DDD), methoxychlor, ethylan, dicofol, and chlorobenzilate. The latter two are not really insecticides, but rather are acaricides (miticides). More information is given on these and other pesticides in Appendix A.

How does DDT kill? The mode of action has never been clearly worked out for DDT. In some complex manner it destroys the delicate balance of sodium and potassium within the neuron, thereby preventing it from conducting impulses normally.

Let us consider a few salient points concerning DDT in order to understand some of the well-documented evils attributed to it. The first point is DDT's chemical stability. DDT and TDE are *persistent*, that is, their chemical stability gives the products long lives in soil and aquatic environments and in animal and plant tissues. They are not readily broken down by microorganisms, enzymes, heat, or ultraviolet light. The remaining DDT relatives are considered nonpersistent.

Second, we note that DDT's solubility in water is only about six parts per billion parts (ppb) of water. DDT has been reported in the chemical literature to be probably the most water-insoluble compound ever synthesized. However, it is quite soluble in fatty tissue, and, as a consequence of its resistance to metabolism, it is readily stored in fatty tissue of any animal ingesting DDT alone or DDT dissolved in the food it eats, even when it is part of another animal.

Since DDT is not readily metabolized and thus not excreted, and it is freely stored in body fat, it accumulates in every animal that preys on other animals. It also accumulates in animals that eat plant tissue bearing even traces of DDT. For example, the dairy cow excretes (or secretes) a large share of the ingested DDT in its milk fat. Humans drink milk and eat the fatted calf, and thereby ingest DDT. The same story is repeated in food chains ending in the osprey, falcon, golden eagle, seagull, pelican, and so on.

The principle of these food chain oddities is this: Any chemical that possesses the characteristics of stability and fat solubility will follow the same biological magnification (*biomagnification*) as DDT. The polychlorinated biphenyls (PCBs), a group of chemicals that have no insecticidal properties, are stable and fat soluble and have climbed the food chain just as DDT has. Other insecticides incriminated to some extent in biomagnification, belonging to the organochlorine group, are TDE, DDE (a major metabolite of DDT), dieldrin, aldrin, several isomers of HCH, endrin, heptachlor, and mirex.

The rise and fall of DDT contains a lesson, and this is perhaps as good a place as any to moralize. Because of DDT's great success in World War II against body lice in Naples, during the typhus outbreak, and in the Pacific, against mosquitoes known to vector malaria, after the war it was rapidly adopted for agricultural use with inadequate basic knowledge. And, because of its effectiveness against a host of agricultural insect pests and its ridiculously low cost, it was overused, and abused, and then—when it caused problems—was banned in rather a panic.

The lesson we can learn from this is: There is an absolute need for an informed, cautious, and—according to available knowledge—correct way of employing a specific chemical for pest control. We need basic research performed in autonomous institutions not subjected to competition in the economic marketplace. And, because this re-

ETHYLAN (Perthane®)

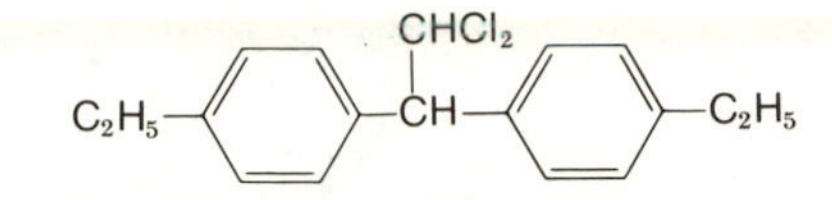

1,1-dichloro-2,2-bis(*p*-ethylphenyl)ethane

DICOFOL (Kelthane®)

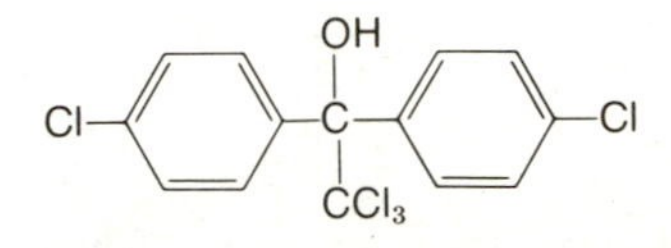

4,4′-dichloro-*a*(trichloromethyl)benzhydrol

CHLOROBENZILATE

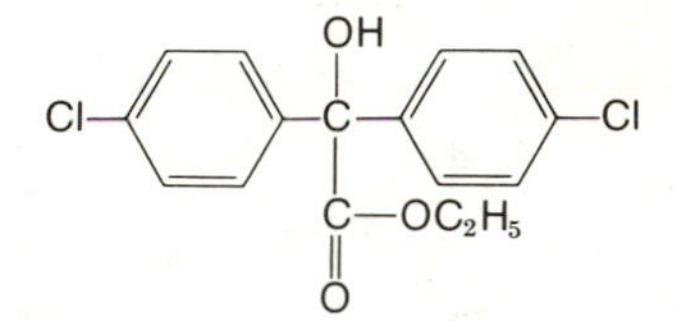

ethyl 4,4′-dichlorobenzilate

search is basic, as opposed to applied, it will of necessity be slow, expensive, long-term, and not immediately applicable. But our social policy must support such research, to enable us to cope with—if not to avoid—those chemicals that prove to pose dangers to our environment and health.

Hexachlorocyclohexane (HCH)

Hexachlorocyclohexane (HCH)—previously erroneously called benzenehexachloride (BHC)—was first discovered in 1825. But, like DDT, it was not known to have insecticidal properties until 1940, when French and British entomologists found the material to be active against all insects tested.

It is made by chlorinating benzene, which results in a product made up of several isomers, that is, molecules containing the same kinds and numbers of atoms but differing in the internal arrangement of those atoms. HCH, for instance, has five isomers, named, after the Greek letters, *alpha, beta, gamma, delta,* and *epsilon*. After much laboratory work in isolating and identifying these isomers, the chemists found to their great surprise that only the gamma isomer had insecticidal properties. In a normal mixture of HCH, the gamma isomer makes up only about 12 percent of the total, leaving the other four isomers as inert material or insecticidally inactive ingredients.

HCH

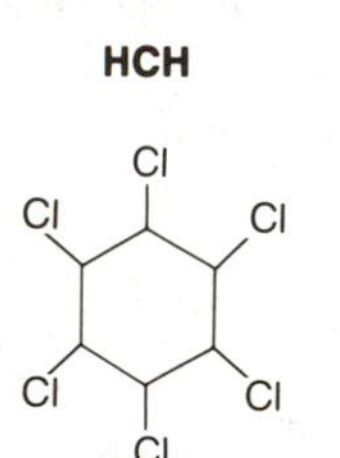

1,2,3,4,5,6-hexachlorocyclohexane

Since the gamma isomer is the only active ingredient, methods were developed to manufacture lindane, a product containing 99 percent gamma isomer, which is effective against most insects, but also quite expensive, making it impractical for crop use.

Technical grade HCH has one highly undesirable characteristic, a prominent musty odor and flavor. The odor is from the inert isomers, which are more persistent than the odorless gamma isomer in animal and plant tissues as well as in soil. As a result, root and tuber crops planted in soils previously treated with HCH retain its odor and are usually unsalable. The same problem is reported with leafy vegetables, poultry, eggs, and milk that directly or indirectly come in contact with HCH residues.

The effects of HCH on insects and mammals superficially resemble those of DDT. Lindane is a neurotoxicant whose effects are normally seen within hours and result in increased activity, tremors, and convulsions leading to prostration.

Lindane is odorless and has a high degree of volatility. It quickly became popular as a household fumigant sold as pellets to be attached to light bulbs or to small, decorative electric wall vaporizers. These were later found to be hazardous to humans and house pets and were removed from the market.

Cyclodienes

The cyclodienes, also known as the *diene-organochlorine insecticides,* were developed after World War II and are therefore of more recent origin than DDT (1939) and HCH (1940). The eight compounds listed here were first described in the scientific literature or patented in the year indicated: chlordane, 1945; aldrin and dieldrin,

1948; heptachlor, 1949; endrin, 1951; mirex, 1954; endosulfan, 1956; and chlordecone (Kepone®), 1958. Other cyclodienes developed in the United States and Germany are of minor importance in a general survey. These include isodrin, alodan, bromodan, and telodrin.

Generally, the cyclodienes are persistent insecticides and are stable in soil and relatively stable to the ultraviolet action of sunlight. Consequently, they have been used in greatest quantity as soil insecticides (especially chlordane, heptachlor, aldrin, and dieldrin), for the control of termites and soil-borne insects whose immature stages (larvae) feed on the roots of plants. Because of their persistence, the use of cyclodienes on crops was restricted; undesirable residues remained beyond the time for harvest. To understand the effectiveness of cyclodienes as termite control agents, consider that structures treated with chlordane, aldrin, and dieldrin in the year of their development are still protected from damage more than thirty years later. These insecticides are the most effective, long-lasting, economical, and safest termite control agents known. However, several other soil insects became resistant to these materials in agriculture, resulting in a rapid decline in their use. Most agricultural uses of the cyclodienes were canceled by the EPA between 1975 and 1980.

CHLORDANE

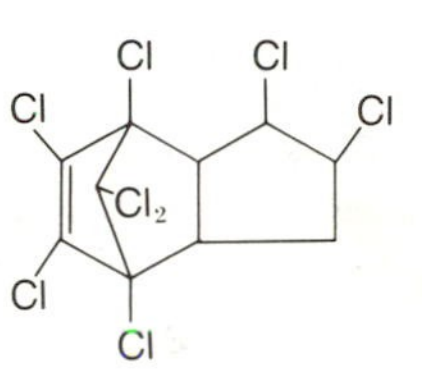

1,2,4,5,6,7,8,8-octachloro-3*a*,4,7,7*a*-tetrahydro-4,7-methanoindane

HEPTACHLOR

1,4,5,6,7,8,8-heptachloro-3*a*,4,7,7*a*-tetrahydro-4,7-methanoindene

ALDRIN

1,2,3,4,10,10-hexachloro-1,4,4*a*,5,8,8*a*-hexahydro-1,4-*endo-exo*-5,8-dimethanonaphthalene

ENDOSULFAN (Thiodan®)

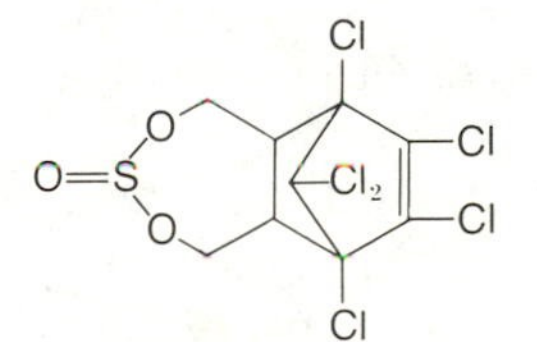

6,7,8,9,10,10-hexachloro-1,5,5*a*,6,9,9*a*-hexahydro-6,9-methano-2,4,3-benzodioxathiepin 3-oxide

MIREX

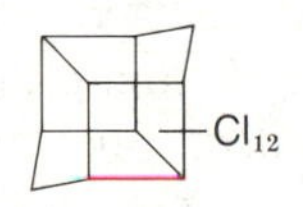

dodecachlorooctahydro-1,3,4-metheno-1*H*-cyclobuta[*cd*] pentalene

CHLORDECONE (Kepone®)

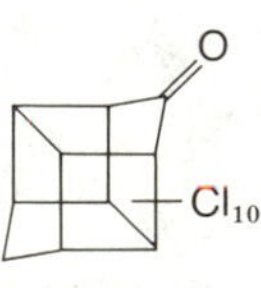

decachlorooctahydro-1,3,4-metheno-2*H*-cyclobuta[*cd*] pentalen-2-one

DIELDRIN

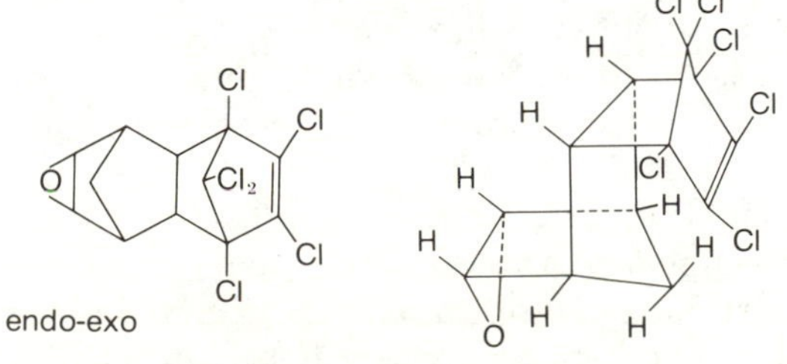

1,2,3,4,10,10-hexachloro-6,7-epoxy-1,4,4*a*,5,6,7,8,8*a*-octahydro-1,4-*endo-endo*-5,8-dimethanonaphthalene

ENDRIN

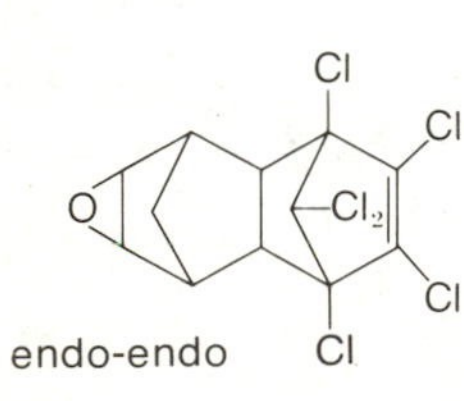

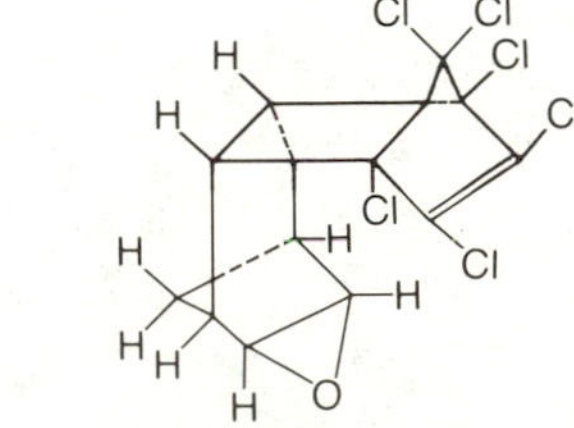

1,2,3,4,10,10-hexachloro-6,7-epoxy-1,4,4*a*,5,6,7,8,8*a*-octahydro-1,4-*endo-endo*-5,8-dimethanonaphthalene

The most valuable, and produced in the greatest quantity, were chlordane and dieldrin. Structures of the common cyclodienes are presented to illustrate their similarity and complexity.

The nomenclature and chemistry of the cyclodienes are rather complicated. The cyclodienes have three-dimensional structures and thus possess stereoisomers; that is, forms that have the same kinds and numbers of atoms, but their atoms differ in their spatial location and structure. For instance, endrin is a stereoisomer of dieldrin.

The cyclodienes have about equal toxicity or toxic effects on insects, mammals, and birds. They are, however, much more toxic to fish, perhaps because when the compound is introduced into water the fish continually respire and ingest any toxic compound contained in their aquatic environment.

The modes of action of the cyclodienes are not clearly understood. It is known that they are neurotoxicants that have effects similar to those of DDT and HCH. They appear to affect all animals in generally the same way, first with nervous activity followed by tremors, convulsions, and prostration. The cyclodienes undoubtedly disturb the delicate balance of sodium and potassium within the neuron but in a way differing from that of DDT and HCH.

Polychloroterpene Insecticides

There are only two polychloroterpene materials, toxaphene, discovered in 1947, and strobane, introduced in 1951. Neither have ever been considered urban insecticides. Toxaphene is manufactured by the chlorination of camphene, a pine tree derivative.

Toxaphene had by far the greatest use of any single insecticide in agriculture. It was used on cotton, first in combination with DDT, for alone it has a low order of toxicity to insects. In 1965, after several cotton insects became resistant to DDT, toxaphene was formulated in combination with methyl parathion, an organophosphate insecticide discussed later in this chapter. As late as 1976 some 11.8 million kg of toxaphene was used on cotton, or 41 percent of all insecticides used on cotton that year.

TOXAPHENE

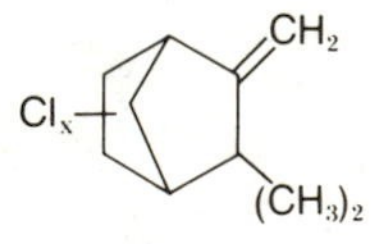

chlorinated camphene containing 67 to 69 percent chlorine

Toxaphene is an amazing mixture of more than 177 polychlorinated derivatives, which are 10-carbon compounds including Cl_6, Cl_7, Cl_8, Cl_9, and Cl_{10} constituents. Most are probably isomeric Cl_{7-}, Cl_{8-}, and Cl_{9-} bornanes. No single component makes up more than a small percentage of the technical mixture. The toxaphene components of greater concern are those most toxic to mammals and fish. One of these is toxicant A, shown in its three-dimensional form (Saleh and Casida, 1978). Toxicant A makes up only 3 percent of technical toxaphene, but it is 18 times more toxic to mice, 6 times more toxic to houseflies, and 36 times more toxic to goldfish than technical toxaphene.

TOXAPHENE TOXICANT A

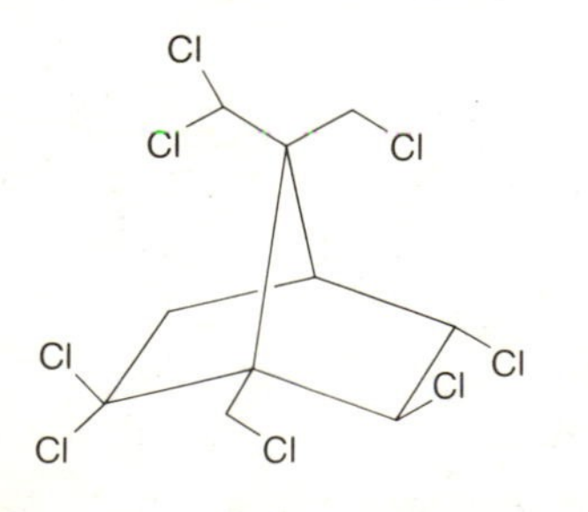

2,2,5-endo,6-exo,8,9,9,10-octachlorobornane
$C_{10}H_{10}Cl_8$

These materials are persistent in the soil, though not as persistent as the cyclodienes, and disappear in three to four weeks from the surfaces of most plant tissues. This disappearance is attributed more to volatility than to actual metabolism or photolysis (disintegration from the effects of ultraviolet light in sunlight). They are fairly easily metabolized by mammals and birds, and are not stored in body fat to any great extent, as are DDT, HCH, or the cyclodienes. Despite low toxicity to insects, mammals, and birds, fish are highly susceptible to toxaphene poisoning, in the same order of magnitude as to the cyclodienes.

The modes of action for toxaphene and strobane are similar to the cyclodiene insecticides, acting on the neurons and causing an imbalance in sodium and potassium ions.

ORGANOPHOSPHATES

The chemically unstable organophosphate (OP) insecticides have virtually replaced the persistent organochlorine compounds, especially with regard to use around the home and garden. *Organophosphate* is usually used as a generic term to include all the insecticides containing phosphorus.

The OPs have several other commonly used names, any of which is correct: organic phosphates, phosphorus insecticides, nerve gas relatives, phosphates, phosphate insecticides, and phosphorus esters or phosphoric acid esters. They are all derived from phosphoric acid and are generally the most toxic of all pesticides to vertebrate animals. Because of their chemical structures and mode of action, they are related to the "nerve gases." Their insecticidal action was observed in Germany during World War II in the study of materials closely related to the nerve gases sarin, soman, and tabun. Initially, the discovery was made in search of substitutes for nicotine, which was in critically short supply in Germany.

WORLD WAR II GERMAN NERVE GASES

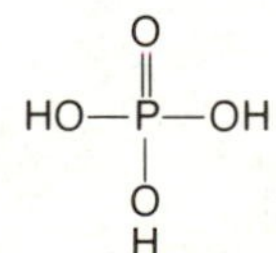

SARIN

(CH₃)₂CHO(O)P(F)CH₃

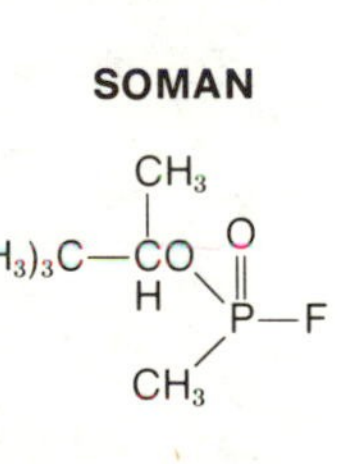

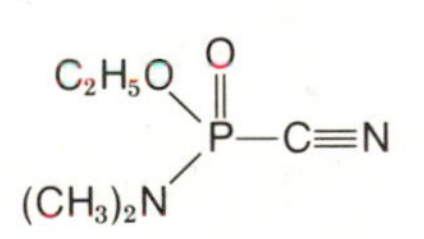

The OPs have two distinctive features. First, they are generally much more toxic to vertebrates than are the organochlorine insecticides, and, second, they are chemically unstable or nonpersistent. It is this latter quality that brings them into agricultural use as substitutes for the persistent organochlorines, particularly DDT.

The OPs exert their toxic action by inhibiting certain important enzymes of the nervous system, cholinesterases (ChE). This inhibition results in the accumulation of acetylcholine(ACh), which interferes with the neuromuscular junction, producing rapid twitching of voluntary muscles and finally paralysis.

OPs that have attached to their phosphorus atoms combinations of different alcohols and different phosphorus acids are termed *esters*. Esters of phosphorus have varying combinations of oxygen, carbon, sulfur, and nitrogen, and so have different identities. The nuclei of the six subclasses of OPs are shown to help explain some of the seemingly odd chemical names given to these insecticides.

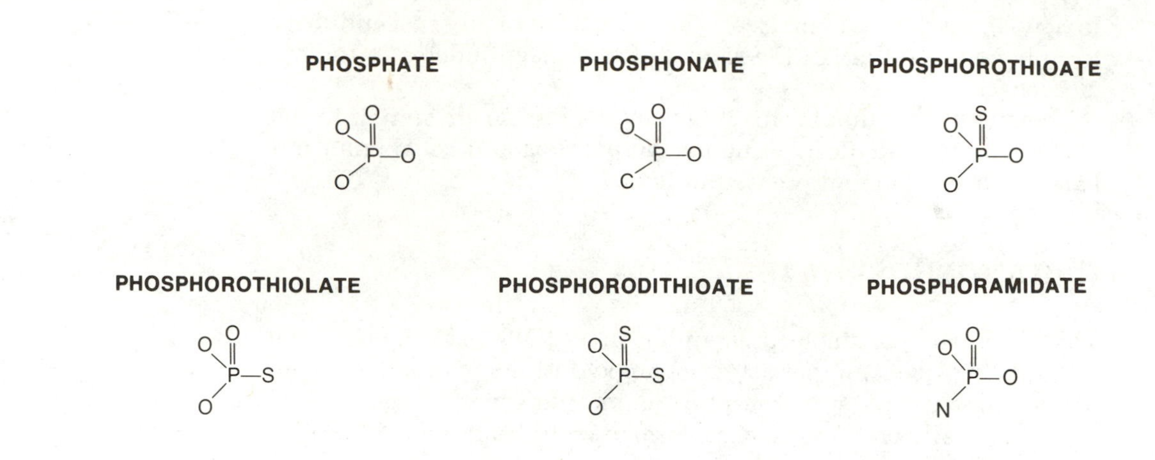

The OPs are divided into three groups—the aliphatic, phenyl, and heterocyclic derivatives.

Aliphatic Derivatives

The term *aliphatic* literally means "carbon chain," and the linear arrangement of carbon atoms differentiates them from ring or cyclic structures. All the aliphatic OPs are simple phosphoric acid derivatives bearing short carbon chains.

The first OP introduced into agriculture was TEPP in 1946. It is the only useful pyrophosphate and is probably the most toxic. It was never available for home use. Because TEPP is very unstable in water, it hydrolyzes (breaks down) quickly after spraying on crops and disappears within 12 to 24 hours.

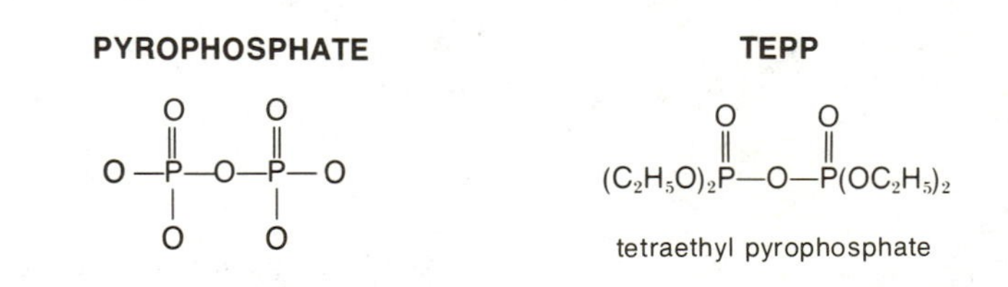

tetraethyl pyrophosphate

MALATHION

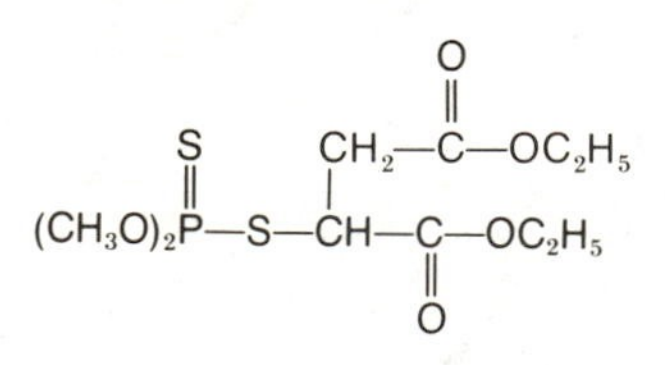

O,O-dimethyl-*S*-1,2-di(carboethoxy) ethyl phosphorodithioate

Malathion is the oldest and most heavily used aliphatic OP. Introduced in 1950, it was quickly adopted by agriculture for use on most vegetables, fruits, and forage crops for control of an extensive range of insect pests. It was soon recognized as suitable for home use for it was safe to use around pets, fast-acting, and controlled practically every kind of garden and household insect including aphids and cockroaches. It is so safe that it is prescribed by physicians for use on humans for the control of head, body, and crab lice. It commonly appears in flea powders for dogs, cats, and other domestic animals, and is used in dips for the control of mange mites.

In 1981, and again in 1982, malathion became the insecticide of choice in the control of the Mediterranean fruit fly, which had invaded the rich fruit-growing areas of California. Malathion was mixed with a protein bait made of molasses and yeast and sprayed from ground equipment and by helicopter over the infested and surrounding areas. Both male and female fruit flies are attracted to the bait and die a few hours after feeding on the tasty morsels.

The bait formulation applied by helicopter is the most selective and inoffensive of all forms and methods of malathion use. With this technique malathion is applied at the astonishingly low rate of 171 g/ha mixed with 684 g of the bait. When applied by ground equipment it is used at the rate of 2 kg of active ingredient in 378 L of spray per hectare as a foliar or vegetation spray. This malathion-bait mixture was used successfully in the eradication of the Medfly from Florida in 1956–1957, and again in 1962–1963, from Texas in 1966, and from Los Angeles in 1975–1976. It was again placed in service in the brief 1981 Florida outbreak of the Medfly, because of its exceptionally low acute toxicity to humans and other warm-blooded animals.

Trichlorfon is a chlorinated OP, which has been useful for crop pest control and fly control around barns and other farm buildings.

Monocrotophos is an aliphatic OP containing nitrogen. It is a plant-systemic insecticide, but it has had limited use in agriculture because of its high mammalian toxicity. It is not available to the home gardener.

Systemic insecticides are those that are taken into the roots of plants and translocated to the above-ground parts, where they are toxic to any sucking insects feeding on the plant juices. Normally caterpillars and other plant tissue-feeding insects are not controlled, because they do not ingest enough of the systemic-containing juices to be affected.

Contained among the aliphatic derivatives are several plant systemics, dimethoate, dicrotophos, oxydemetonmethyl, and disulfoton, all of which can be used safely by the homeowner.

TRICHLORFON (Dylox®)

$(CH_3O)_2P(=O)—CH(OH)CCl_3$

dimethyl (2,2,2-trichloro-1-hydroxyethyl) phosphonate

MONOCROTOPHOS (Azodrin®)

$(CH_3O)_2P(=O)—O—C(CH_3)=CHC(=O)—NH—CH_3$

3-hydroxy-*N*-methyl-*cis*-crotonamide dimethyl phosphate

plant systemic insecticide

DIMETHOATE (Cygon®)

$(CH_3O)_2P(=S)—S—CH_2C(=O)—NH—CH_3$

O,O-dimethyl *S*-[2-(methylamino)-2-oxoethyl] phosphorodithioate

OXYDEMETONMETHYL (Meta Systox R®)

$(CH_3O)_2P(=O)—S—CH_2CH_2—S(=O)—C_2H_5$

S-[2-(ethylsulfinyl)ethyl]*O,O*-dimethyl phosphorothioate

DICROTOPHOS (Bidrin®)

$(CH_3O)_2P(=O)—O—C(CH_3)=CHC(=O)—N(CH_3)_2$

O,O-dimethyl *O*-1-methylvinyl-*N,N*-dimethyl carbamoyl phosphate

DISULFOTON (Di-Syston®)

$(C_2H_5O)_2P(=S)—S—CH_2CH_2—S—C_2H_5$

O,O-diethyl *S*-2-[(ethylthio)ethyl] phosphorodithioate

Dichlorvos is an aliphatic OP with a very high vapor pressure, giving it strong fumigant qualities. It has been incorporated into polychlorovinyl resin pet collars and pest strips, from which it is

DICHLORVOS (Vapona®)

$$(CH_3O)_2P(=O)-O-CH=CCl_2$$

O,O-dimethyl-*O*-2,2-dichloro-vinyl phosphate

MEVINPHOS (Phosdrin®)

$$(CH_3O)_2P(=O)-O-C(CH_3)=CHC(=O)-OCH_3$$

methyl (E)-3-hydroxycrotonate dimethyl phosphate

released slowly. It lasts several months and is useful for insect control in the home and other closed areas (see p. 30).

Mevinphos is a highly toxic OP used in commercial vegetable production because of its very short insecticidal life. It can be applied up to one day before harvest for insect control, yet it leaves no residues on the crop to be eaten by the consumer.

Two of the recent arrivals in the aliphatic organophosphate structures are methamidophos and acephate. Both have proved highly useful in agriculture, especially for vegetable insect control.

In summary, the aliphatic organophosphate insecticides are the simplest in structure of the organophosphate molecules. They have a wide range of toxicities, and several possess a relatively high water solubility, giving them plant-systemic qualities several of which are useful around the home.

METHAMIDOPHOS (Monitor®)

$$(CH_3O)(CH_3S)P(=O)-NH_2$$

O,S-dimethyl phosphoramidothioate

ACEPHATE (Orthene®)

$$(CH_3O)(CH_3S)P(=O)-NH-C(=O)-CH_3$$

O,S-dimethyl acetylphosphoramidothioate

Phenyl Derivatives

In Chapter 2 we mentioned that when the benzene ring is attached to other groups it is referred to as *phenyl*. The phenyl OPs contain a benzene ring with one of the ring hydrogens displaced by attachment to the phosphorus moiety and others frequently displaced by Cl, NO_2, CH_3, CN, or S. The phenyl OPs are generally more stable than the aliphatic OPs; consequently their residues are longer lasting.

Parathion is the most familiar of the phenyl OPs, being, in 1947, the second phosphate insecticide introduced into agriculture. The first, TEPP, was introduced in 1946. As a result of its age and utility, parathion's total usage is greater than that of many of the less useful materials combined. Ethyl parathion was the first phenyl derivative used commercially and, because of its hazard, has not been available for home use.

Methyl parathion became available in 1949 and proved to be more useful than (ethyl) parathion because of its lower toxicity to humans and domestic animals and broader range of insect control. Its shorter residual life also makes it more desirable in certain instances. This material is also not used by the layperson.

Systemic insecticides are also found in the phenyl OPs. They are, however, usually animal systemics used for the control of the cattle grub; ronnel and crufomate are examples.

Stirofos is a home-safe OP much like malathion in its overall usefulness against home and livestock pests.

Two of the more recently registered phenyl derivatives are profenophos and sulprofos. Both materials have a broad spectrum of insecticidal activity and are currently labeled for use only on field crops.

ETHYL PARATHION

$$(C_2H_5O)_2P(=S)-O-C_6H_4-NO_2$$

O,O-diethyl *O*-*p*-nitrophenyl phosphorothioate

METHYL PARATHION

$$(CH_3O)_2P(=S)-O-C_6H_4-NO_2$$

O,O-dimethyl *O*-*p*-nitrophenyl phosphorothioate

RONNEL (Korlan®)

O,O-dimethyl *O*-2,4,5-trichlorophenyl phosphorothioate

CRUFOMATE (Ruelene®)

4-tert-butyl-2-chlorophenyl methyl methylphosphoramidate

STIROFOS (Gardona®)

O,O-dimethyl *O*-2-chloro-1-(2,4,5-trichlorophenyl) vinyl phosphate

PROFENOFOS (Curacron®)

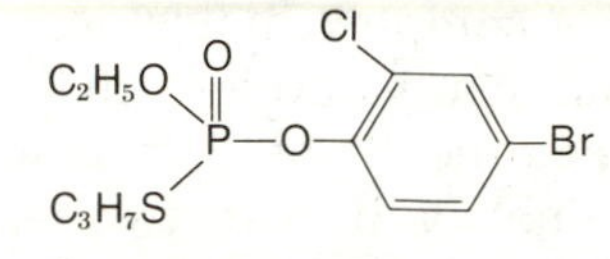

O-(4-bromo-2-chlorophenyl)-*O*-ethyl-S-propyl phosphorothioate

SULPROFOS (Bolstar®)

O-ethyl *S*-propyl *O*(4-methylthio) phenyl phosphorodithioate

ISOFENPHOS (Amaze®)

1-methylethyl 2-([ethoxy (1-methylethyl) amino phosphinothioyl] oxy)benzoate

Isofenphos is used as a soil insecticide in field crops and vegetables, against corn rootworm and onion maggot, and also for white grubs, chinch bugs, and sod webworms in turf.

Heterocyclic Derivatives

The term *heterocyclic* means that the ring structures are composed of different or unlike atoms. In a heterocyclic compound, for example, one or more of the carbon atoms is displaced by oxygen, nitrogen or sulfur, and the ring may have three, five, or six atoms.

DIAZINON

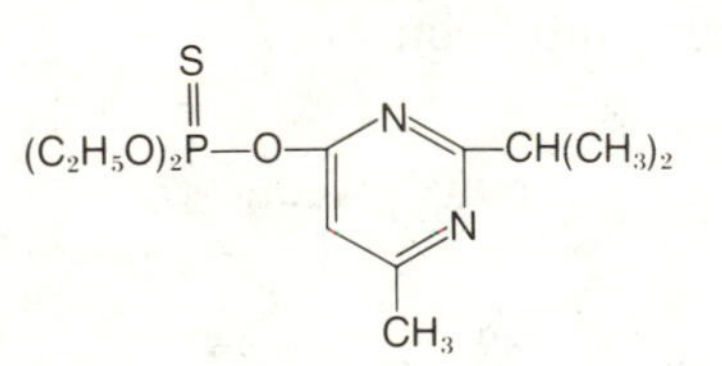

O,O-diethyl *O*-(2-isopropyl-4-methyl-6-pyrimidinyl) phosphorothioate

AZINPHOSMETHYL (Guthion®)

O,O-dimethyl *S*(4-oxo-1,2,3-benzotriazin-3(4*H*)-ylmethyl) phosphorodithioate

CHLORPYRIFOS (Dursban®, Lorsban®)

O,O-diethyl *O*-(3,5,6-trichloro-2-pyridyl) phosphorothioate

The first insecticide made available in this group was probably diazinon, in 1952. Note that the six-membered ring contains two nitrogen atoms, very likely the source of its proprietary name, since one of the constituents used in its manufacture is pyrimidine, a diazine.

Diazinon is a relatively safe OP that has an amazingly good track record around the home. It has been effective for practically every conceivable use: insects in the home, lawn, garden, ornamentals, around pets, and for fly control in stables and pet quarters.

Azinphosmethyl is the second oldest member of this group (1954) and is used in U.S. agriculture. It serves both as an insecticide and acaricide in cotton production and is not available to the layperson.

Chlorpyrifos has become the most frequently used insecticide by

METHIDATHION (Supracide®)

O,O-dimethyl phosphorodithioate *S*-ester with 4-[mercaptomethyl]-2-methoxy-Δ^2-1,3,4-thiadiazolin-5-one

PHOSMET (Imidan®)

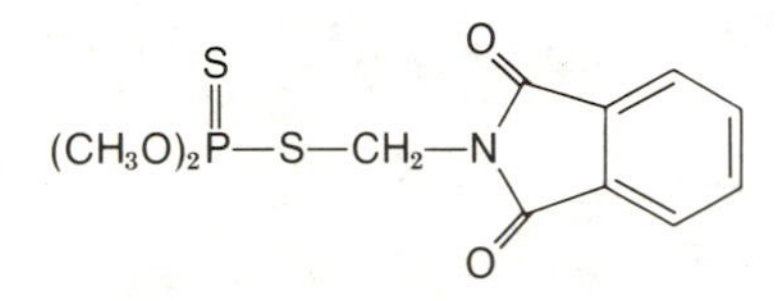

N-(mercaptomethyl)-phthalimide *S*-(*O,O*-dimethylphosphorodithioate)

DIALIFOR (Torak®)

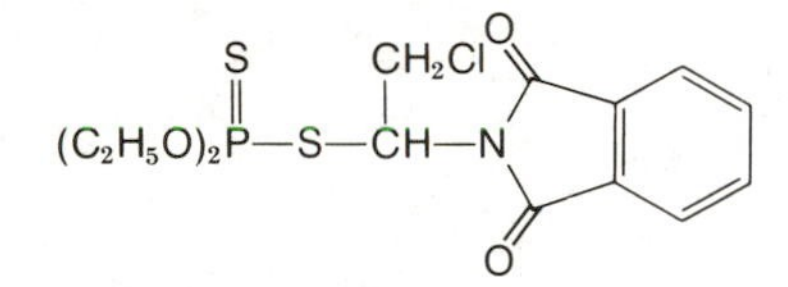

O,O-diethyl phosphorodithioate *S*-ester with *N*-(2-chloro-1-mercaptoethyl) phthalimide

pest control operators in homes and restaurants for controlling cockroaches and other household insects.

Methidathion is not particularly new, but it has in the last few years acquired registrations for forage and field crops, true fruits, and nut crops for an exceptionally wide variety of insect and mite pests.

Phosmet has a set of registration credentials similar to methidathion, including the infamous boll weevil and the plum curculio, two closely related weevil pests.

Dialifor was first introduced in the mid-1960s as were methidathion and phosmet. Its uses are somewhat more limited, however, to apples, grapes, pecans, and citrus.

The heterocyclic organophosphates are complex molecules and generally have longer-lasting residues than many of the aliphatic or phenyl derivatives. Also, because of the complexity of their molecular structures, their breakdown products (metabolites) are frequently many, making their residues sometimes difficult to measure in the laboratory. Consequently, their use by growers on food crops is somewhat less than either of the other two groups of phosphorus-containing insecticides.

ORGANOSULFURS

The organosulfurs, as the name suggests, have sulfur as their central atom. They resemble the DDT structure in that most have two phenyl rings.

Dusting sulfur alone is a good acaricide (miticide), particularly in hot weather. The organosulfurs, however, are far superior, requiring much less material to achieve control because sulfur in combination with phenyl rings is particularly toxic to mites. Of greater interest, however, is that the organosulfurs have very low toxicity to insects. As a result, they are used for selective mite control.

TETRADIFON (Tedion®)

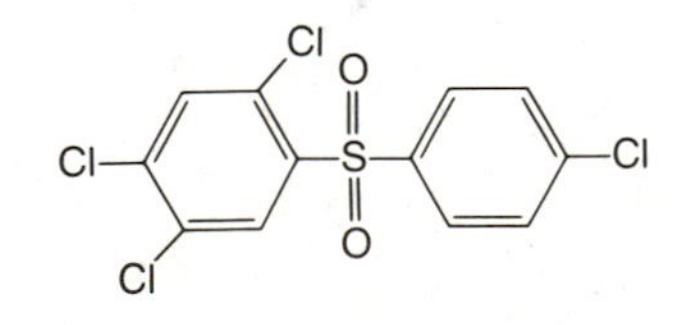

p-chlorophenyl 2,4,5-trichlorophenyl sulfone

FENSON

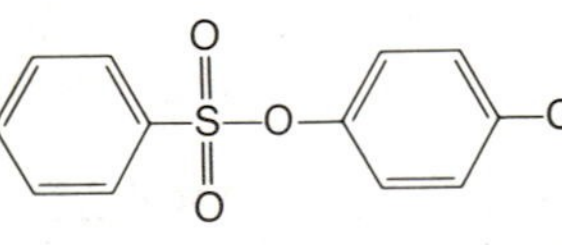

p-chlorophenyl benzenesulfonate

OVEX (Ovotran®)

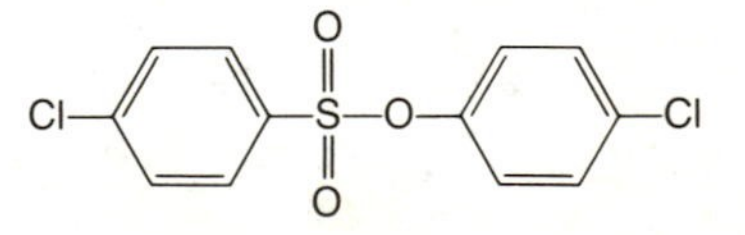

p-chlorophenyl *p*-chlorobenzenesulfonate

Aramite®

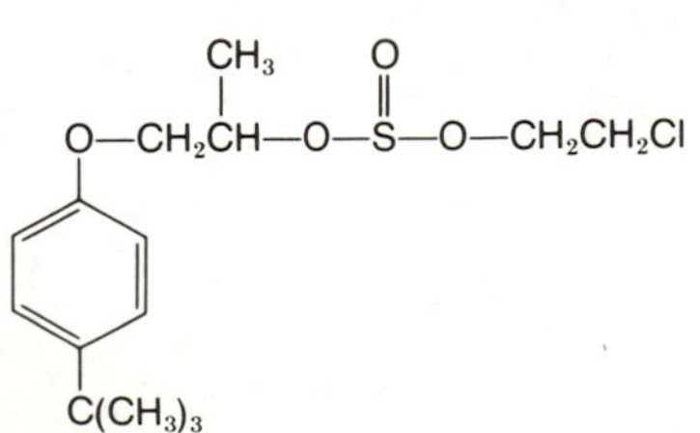

2-(*p-tert*-butylphenoxy)-1-methylethyl 2-chloroethyl sulfite

TETRASUL (Animert V-101®)

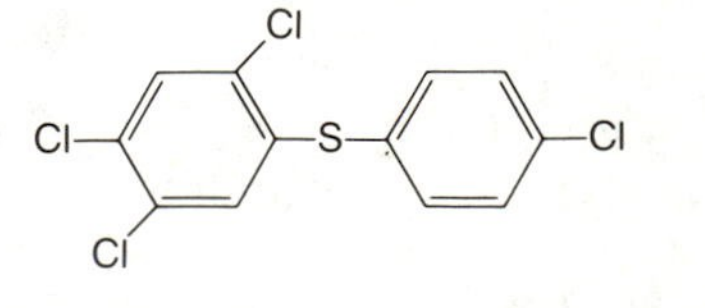

p-chlorophenyl 2,4,5-trichlorophenyl sulfide

PROPARGITE (Omite®)

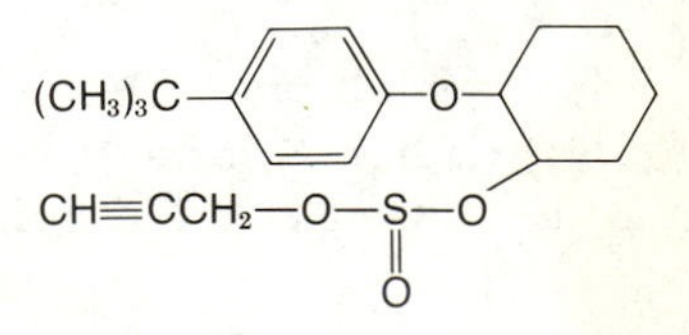

2-(*p-tert*-butylphenoxy)cyclohexyl 2-propynyl sulfite

This group has one other valuable property: They are usually ovicidal as well as being toxic to the young and adult mites.

Tetradifon is one of the older acaricides and typically bears the sulfur and twin phenyl rings, as do most of the organosulfurs.

No doubt the oldest of this group is Aramite®, introduced in 1951. Notice that Aramite has only one phenyl ring and is, therefore, an exception to the general rule that organosulfurs have two phenyl rings.

CARBAMATES

Since the organophosphate insecticides are derivatives of phosphoric acid, the carbamates must be derivatives of carbamic acid $HO-\overset{O}{\overset{\|}{C}}-NH_2$. And like the organophosphates, the mode of action of the carbamates is that of inhibiting the vital enzyme cholinesterase (ChE).

In 1951 the carbamate insecticides were introduced by the Geigy Chemical Company in Switzerland. They fell by the wayside because the first ones were not very effective, while being quite costly. The early carbamates are shown in the margin at the right.

At that time it was not known that the *N,N*-dimethyl carbamates, as shown in these structures, were generally less toxic to insects than the *N*-methyl carbamates, which were developed later and which make up the bulk of the currently used materials.

Carbaryl, the first successful carbamate, was introduced in 1956. More of it has been used worldwide than all the remaining carbamates combined. Two distinct qualities have made it the most popular material: very low mammalian oral and dermal toxicity and a rather broad spectrum of insect control. This has led to its wide use as a lawn and garden insecticide. Notice that carbaryl is an *N*-methyl carbamate.

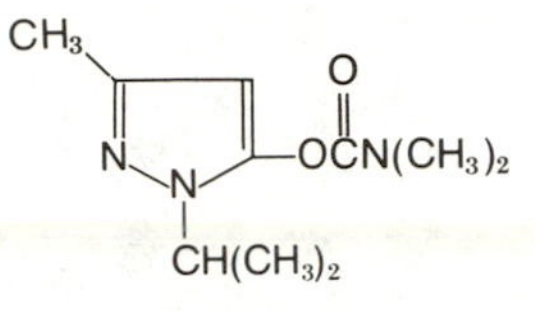

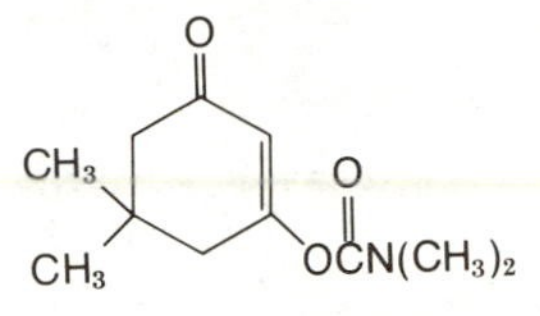

PYRAMAT

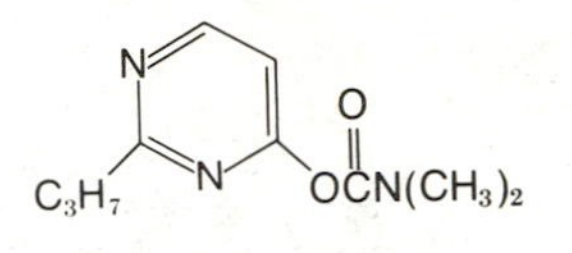

PYROLAN

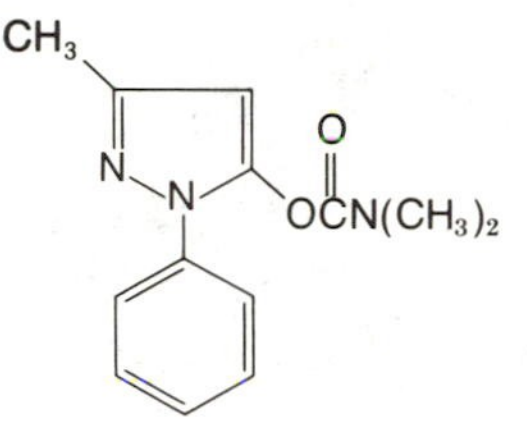

CARBARYL (Sevin®) **METHOMYL (Lannate®, Nudrin®)**

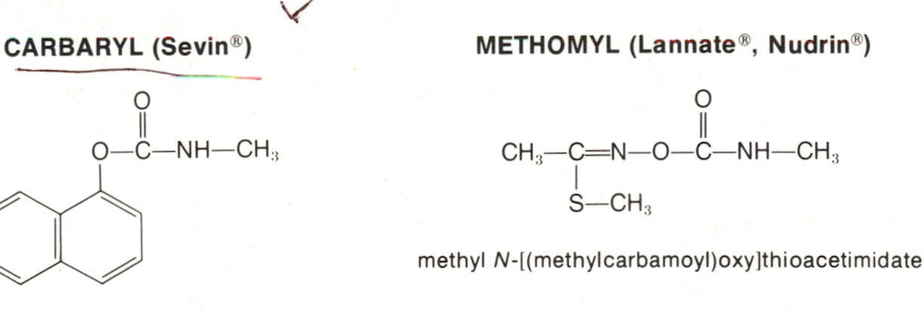

1-naphthyl methylcarbamate

methyl *N*-[(methylcarbamoyl)oxy]thioacetimidate

CARBOFURAN (Furadan®)

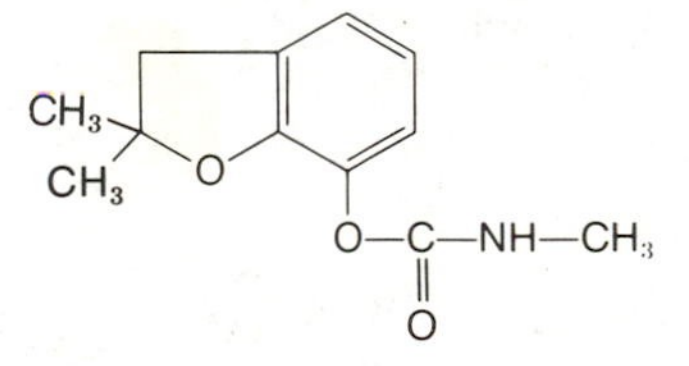

2,3-dihydro-2,2-dimethyl-7-benzofuranyl methylcarbamate

ALDICARB (Temik®)

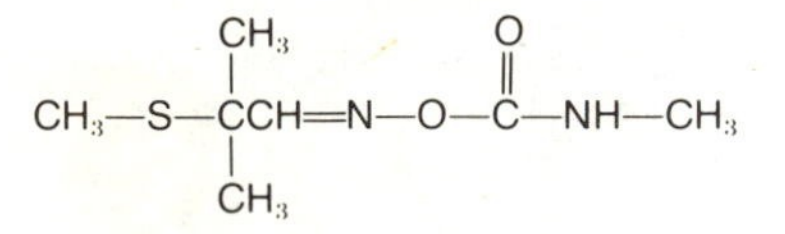

2-methyl-2-(methylthio) propionaldehyde *O*-(methylcarbamoyl) oxime

OXAMYL (Vydate®)

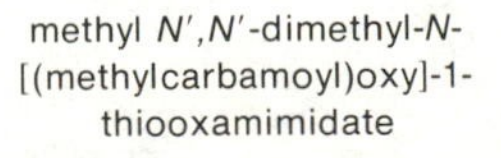

methyl *N′,N′*-dimethyl-*N*-[(methylcarbamoyl)oxy]-1-thiooxamimidate

Several of the carbamates are plant systemics, indicating that they have a high water solubility, which allows them to be taken into the roots or leaves. They are also not readily metabolized by the plants. Methomyl, oxamyl, aldicarb, and carbofuran have distinct systemic characteristics, making them useful also as nematicides. Of these, only aldicarb and carbofuran are used as soil insecticides and nematicides. Under rare circumstances, aldicarb has been detected in shallow groundwater following certain uses.

Methomyl has proved especially effective for worm control on vegetables. Bufencarb is used in agriculture exclusively as a soil insecticide, becoming a replacement for the long-residual organochlorine insecticides, aldrin, dieldrin, and heptachlor. Methiocarb, aminocarb, and promecarb are effective against foliage- and fruit-eating insects. Methiocarb and aminocarb are both excellent molluscicides, used for slug and snail control in flower gardens and ornamentals. Methiocarb is also registered as a bird repellent for cherries and blueberries and as a seed dressing.

Propoxur is highly effective against cockroaches that have developed resistance to the organochlorines and organophosphates. It is used by most structural pest control operators for roaches and other household insects in restaurants, kitchens, and homes. For home use it is formulated in bottled sprays. Similarly, bendiocarb has found its greatest use in the United States as a residual household insecticide.

In summary, the carbamates are inhibitors of cholinesterase, are plant systemics in several instances, and are, for the most part, broad-spectrum in effectiveness, being used as insecticides, miticides, and molluscicides.

BUFENCARB (Bux®)

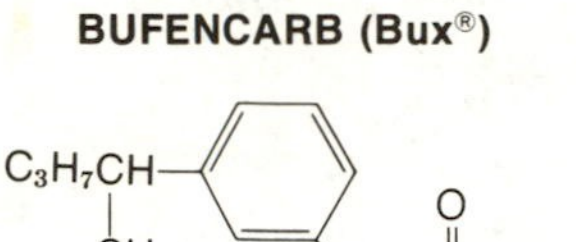

and

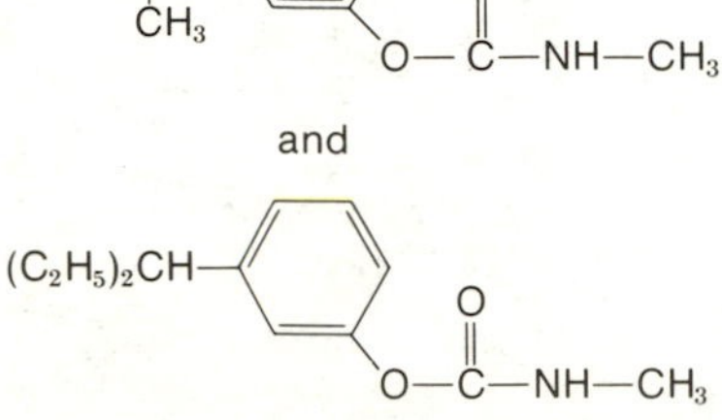

mixture of *m*-(ethylpropyl)phenyl methylcarbamate and *m*-(1-methylbutyl)phenyl methylcarbamate (ratio 1 : 3)

METHIOCARB (Mesurol®)

CH_3, $CH_3—S—$, $—O—C(=O)—NH—CH_3$, CH_3

4-(methylthio)3,5-xylyl methylcarbamate

AMINOCARB (Matacil®)

$(CH_3)_2N—$, $—O—C(=O)—NH—CH_3$, CH_3

4-(dimethylamino)-3-methylphenol methylcarbamate

PROMECARB (Carbamult®)

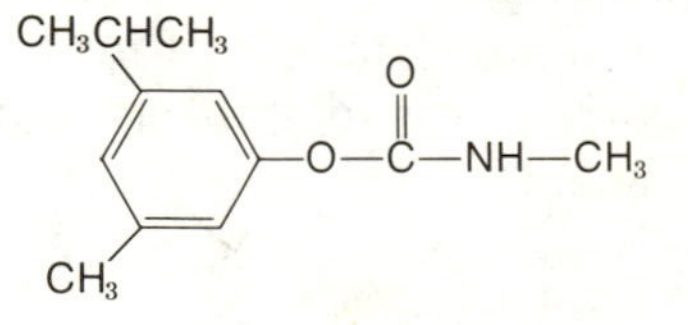

3-methyl-5-(1-methylethyl)phenyl methylcarbamate

PROPOXUR (Baygon®)

$—O—C(=O)—NH—CH_3$, $OCH(CH_3)_2$

o-isopropoxyphenyl methylcarbamate

BENDIOCARB (Ficam®)

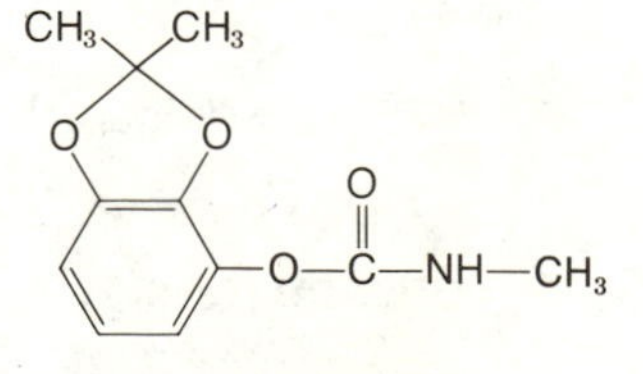

2,2-dimethyl-1,3-benzo dioxol-4-yl methylcarbamate

FORMAMIDINES

The formamidines comprise a new, small, promising group of insecticides. Three examples are chlordimeform, formetanate, and amitraz. They are effective against the eggs and very young caterpillars of several moths of agricultural importance and are also effective against most stages of mites and ticks. Thus, they are classed as ovicides, insecticides, and acaricides. Late in 1976, chlordimeform was removed from the market by its manufacturers, Ciba-Geigy Corporation, and Nor-Am Agricultural Products, Inc., because it proved to be carcinogenic to a cancer-prone strain of laboratory mice during high-level, lifetime feeding studies. In 1978 it was returned for use on cotton, but under very strict application restrictions.

Their present value lies in the control of organophosphate- and carbamate-resistant pests. Poisoning symptoms are distinctly different from other materials. It has been proposed that one possible mode of action is the inhibition of the enzyme monoamine oxidase. This results in the accumulation of compounds termed *biogenic amines*. Thus the formamidines introduce a new mode of action for the insecticides and acaricides. This fact alone makes them extremely useful, for we are slowly losing ground in the battle of insect resistance to the modes of action of the older insecticide groups.

THIOCYANATES

Thiocyanates have easily recognized structural formulas. Remembering that *theion* is the Greek word for "sulfur" and that the cyanides or cyanates end in *-CN*, we have molecules that bear *-SCN*, or *thiocyanate* endings. These insecticides have very distinct, creosote-like odors, are relatively safe to use around humans and animals, and give astonishingly quick knockdown of flying insects. Their mode of action is somewhat complex and can be said simply to interfere with cellular respiration and metabolism. These materials may be found in aerosols to be used around horses and other farm animals. However, with the appearance of the synthetic pyrethroids, the thiocyanates have declined in appeal and will probably soon disappear.

DINITROPHENOLS

The dinitrophenols are another group possessing easily recognized structural formulas:

CHLORDIMEFORM (Galecron®, Fundal®)

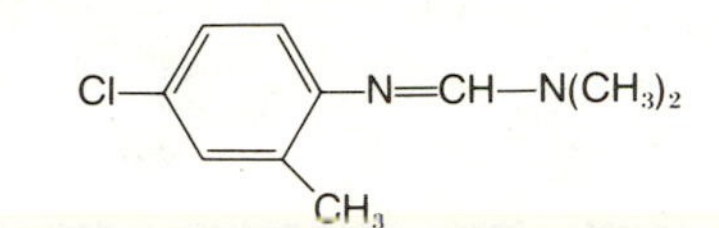

N'-(4-chloro-*o*-tolyl)-*N,N*-dimethylformamidine

FORMETANATE (Carzol®)

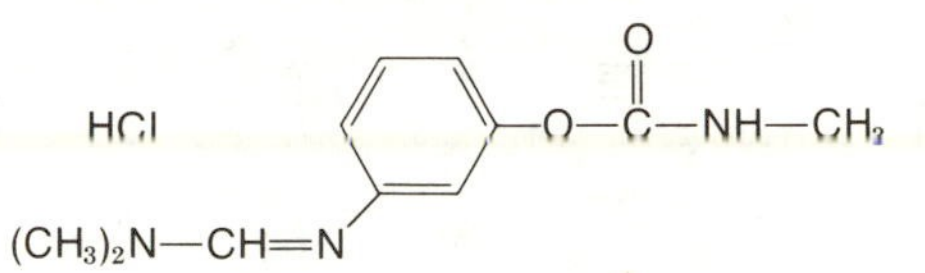

[3-dimethylamino-(methylene-iminophenyl)]-*N*-methylcarbamate hydrochloride

AMITRAZ (Baam®)

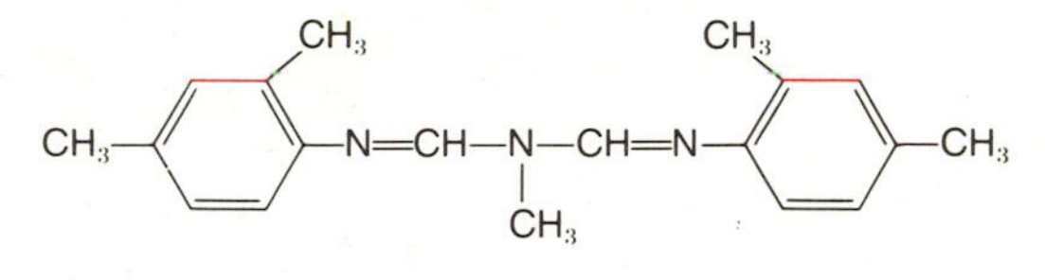

N'-(2,4-dimethylphenyl)-*N*-[[2,4-dimethylphenyl) imino]methyl]-*N*-methylmethanimidamide

LETHANE 384®

$CH_2CH_2—O—C_2H_4—SCN$
$|$
$O—C_4H_9$

2-(2-butoxyethoxy)ethyl thiocyanate

THANITE®

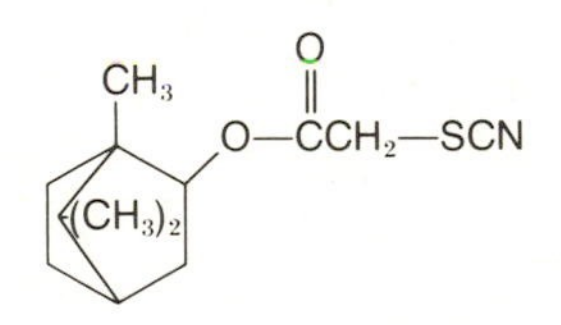

isobornyl thiocyanoacetate

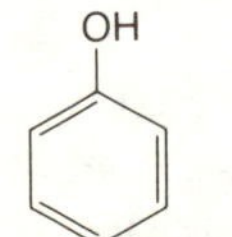

Di (two) nitro (NO_2) phenol

DINITROPHENOL

OH
NO_2 NO_2

The basic dinitrophenol molecule has a broad range of toxicities. Compounds derived from it are used as herbicides, insecticides, ovicides, and fungicides. They act by uncoupling oxidative phosphorylation or basically by preventing the utilization of nutritional energy. In the 1930s, certain dinitrophenols were given by uninformed physicians to their overweight patients to induce rapid weight loss. They were extremely effective, but quite toxic, and their use resulted in several widely publicized deaths.

The oldest of this group is DNOC (3,5-dinitro-o-cresol), introduced as an insecticide in 1892. DNOC has also been used as an ovicide, herbicide, fungicide, and blossom-thinning agent. Its use has declined today to herbicidal applications in which all plants are to be killed. Dinoseb is used as a dormant fruit spray for control of many insects and mites.

DINITROCRESOL (DNOC)

DINOSEB

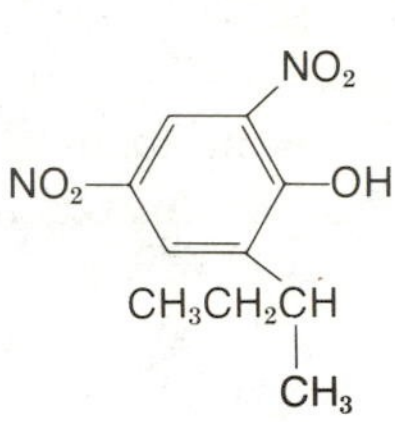

2-*sec*-butyl-4,6-dinitrophenol

BINAPACRYL (Morocide®)

2-*sec*-butyl-4,6-dinitrophenyl 3-methyl-2-butenoate

DINOCAP (Karathane®)

2-(1-methylheptyl)-4,6-dinitrophenyl crotonate

Binapacryl is used exclusively as an acaricide and was introduced in 1960. Dinocap was developed in 1949 as an acaricide and fungicide and is one of the rare materials made up of several related molecular structures, only one of which is shown. Dinocap is particularly effective against powdery mildew fungi. Owing to its safety to green plants, it has often replaced the phytotoxic sulfur that is so effective against powdery mildews.

In summary, the dinitrophenols have been used as pesticides in practically all classifications: ovicides, insecticides, acaricides, herbicides, fungicides, and blossom-thinning agents.

CYHEXATIN (Plictran®)

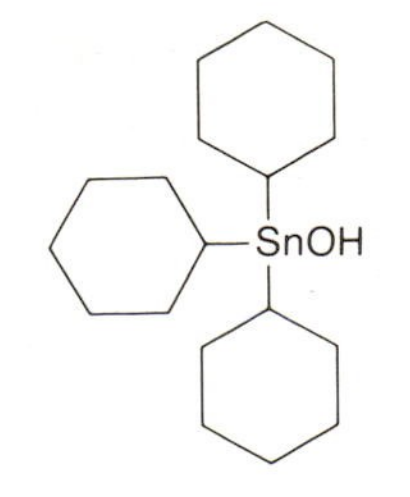

tricyclohexylhydroxytin

ORGANOTINS

The organotins are a relatively new group of acaricides, which double as fungicides, as you will see in Chapter 14. Of particular interest here is cyhexatin, one of the most selective acaricides presently known, introduced in 1967. Introduced somewhat later, fenbutatin-oxide has proved to be most effective against mites on deciduous fruits, citrus, greenhouse crops, and ornamentals. The mode of

FENBUTATIN-OXIDE (Vendex®)

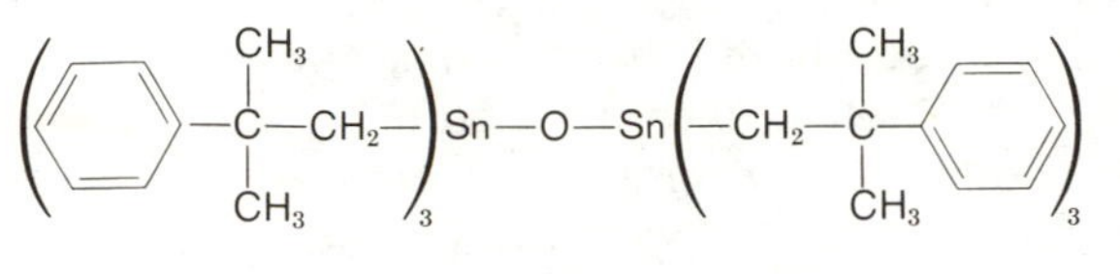

hexakis (2-methyl-2-phenylpropyl)distannoxane

action of this group is not completely known but is believed to be the inhibition of oxidative phosphorylation at the site of dinitrophenol uncoupling (the production of energy in the form of adenosine triphosphate, ATP). These trialkyl tins also inhibit photophosphorylation in chloroplasts (the chlorophyll-bearing subcellular units) and can thus serve as algicides.

BOTANICALS

Botanical insecticides are of great interest to many, because they are "natural" insecticides, toxicants derived from plants. Historically, the plant materials have been in use longer than any other group, with the possible exception of sulfur. Tobacco, pyrethrum, derris, hellebore, quassia, camphor, and turpentine were some of the more important plant products in use before the organized search for insecticides began.

Some of the most widely used insecticides come from plants. The flowers, leaves, and roots are finely ground and used in this form, or the toxic ingredients are extracted and used alone or in mixtures with other toxicants.

There are five natural or botanically derived insecticides that are of interest to gardeners in general, but especially to the organic gardener: pyrethrum, rotenone, sabadilla, ryania, and nicotine. All except nicotine are exempt from the requirement of a tolerance when applied to growing fruit and vegetables. That is, they can be eaten anytime after application, but these botanical insecticides must be used according to label directions.

Botanical insecticide use reached its maximum in the United States in 1966 and has declined steadily since. Pyrethrum is now the only botanical of significance in use, typically in rapid knockdown sprays in combination with synergists and one or more synthetic organic insecticides formulated for use in the home and garden.

Botanical insecticides are naturally occurring chemicals, synthesized by plants. They are really no safer than most of the currently available synthetic insecticides, at least as compared to those available to the layman. Botanicals are expensive to extract from plant tissues, but these chemicals would be unaffordable as insecticides if they were synthesized in the laboratory, which is also possible.

Nicotine

NICOTINE SULFATE

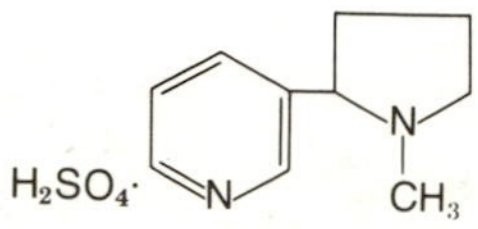

$H_2SO_4\cdot$ N, N, CH_3

3-(1-methyl-2-pyrrolidyl) pyridine sulfate

Smoking tobacco was introduced to England in 1585 by Sir Walter Raleigh. As early as 1690, water extracts of tobacco were reported as being used to kill sucking insects on garden plants. As early as about 1890, the active principle in tobacco extracts was known to be nicotine, and, from that time on, extracts were sold as commercial insecticides for home, farm, and orchard. Today organic gardeners may soak a cigar or two in water overnight and spray insect-infested plants with the extract, achieving some success. "Black Leaf 40," which has long been a favorite garden spray, is a concentrate containing 40 percent nicotine sulfate. Today, nicotine is commercially extracted from tobacco by steam distillation or solvent extraction.

Nicotine is an alkaloid; it is a heterocyclic compound containing nitrogen and having prominent physiological properties. Other well-known alkaloids, which are not insecticides, are caffeine (found in tea and coffee), quinine (from cinchona bark), morphine (from the opium poppy), cocaine (from coca leaves), ricinine (a poison in castor oil beans), strychnine (from *Strychnos nux vomica*), coniine (from spotted hemlock, the poison that killed Socrates), and, finally, LSD (a hallucinogenic derived from the ergot fungus attacking grain).

As its mode of action, nicotine mimics acetylcholine (ACh) at the neuromuscular (nerve-muscle) junction in mammals, and results in twitching, convulsions, and death, all in rapid order. In insects, the same action is observed, but only in the ganglia of their central nervous systems.

Nicotine sulfate, as it is commonly sold, is highly toxic to all warm-blooded animals, as well as insects; for example, the LD_{50} (dose that proves lethal for 50 percent of the test population) for rats is 50–60 mg/kg. This makes it the most hazardous of the botanical insecticides to home gardeners. It has been used with great success since before the turn of the century.

Dusts are not available for garden use because they are too toxic to humans. Nicotine sulfate is used primarily for piercing-sucking insects such as leafhoppers, aphids, scales, thrips, and whiteflies, but it can kill all insects and spider mites on which it is sprayed directly. Many caterpillar pests, however, are very resistant to nicotine. It is more effective during warm weather, but degrades quickly. It is registered for use on flowering plants, and ornamental shrubs and trees to control aphids, mealybugs, scales and thrips, lace bugs, leafminers, leafhoppers, rose slugs, and spider mites. Its use for most greenhouse pests is also acceptable.

Nicotine sulfate is registered for use on a variety of vegetables and fruit trees, but it cannot be used in most cases within seven days of harvest, as required by EPA. There are several ornamental plants that are sensitive to nicotine, such as roses. The label should identify those sensitive plants.

Nicotine is also registered for out-of-door use as a dog and cat repellent as well as for furniture in the home. Additionally, tobacco dust is acceptable as a dog and rabbit repellent out-of-doors.

Rotenone

Rotenoids, the rotenone-related materials, have been used as crop insecticides since 1848, when they were applied to plants to control leaf-eating caterpillars. However, they have been used for centuries (at least since 1649) in South America to paralyze fish, causing them to surface.

Rotenoids are produced in the roots of two genera of the legume (bean) family: *Derris*, grown in Malaya and the East Indies, and *Lonchocarpus* (also called *cubeb* or *cubé*), grown in South America.

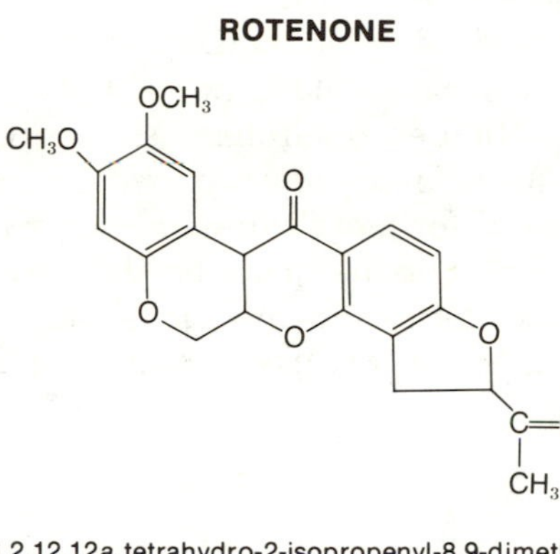

1,2,12,12*a*,tetrahydro-2-isopropenyl-8,9-dimethoxy-[1]benzopyrano-[3,4-*b*]furo[2,3-*b*][1]benzopyran-6(6*aH*)one

Rotenone has an oral LD_{50} of approximately 350 mg/kg (in rats) and has been used for generations as the ideal general garden insecticide. It is harmless to plants, highly toxic to fish and many insects, especially caterpillars, moderately toxic to warm-blooded animals, and leaves no harmful residues on vegetables. There is no waiting interval between application and harvest of a food crop.

Rotenone is both a contact and a stomach poison to insects and is sold as spray concentrates and ready-to-use dust. It kills insects slowly, but causes them to stop their feeding almost immediately. Like all the other botanical insecticides its life in the sun is short, one to three days. It is useful against caterpillars, aphids, beetles, true bugs, leafhoppers, thrips, spider mites, ants, rose slugs, whiteflies, sawflies, bagworms, armyworms, cutworms, leafrollers, midges, and a host of other pests.

Next to pyrethrum, rotenone is probably second in the number of approved uses, exceeding 1000.

Rotenone is the most useful piscicide available for reclaiming lakes for game fishing. It eliminates all fish, closing the lake to reintroduction of rough species. After treatment, the lake can be restocked with the desired species. Rotenone is a selective piscicide in that it kills all fish at dosages that are relatively nontoxic to fish food organisms. It also breaks down quickly, leaving no residues harmful to the fish used for restocking. The recommended rate is 0.5 part of rotenone to one million parts of water (ppm), or 5.1 kg per hectare-meter of water (1.36 pounds per acre-foot).

Sabadilla

Sabadilla is extracted from the seeds of a member of the lily family. Its oral LD_{50} is approximately 5000 mg/kg, making it the least toxic to warm-blooded animals of the five botanical insecticides discussed. It acts as both contact and stomach poison for insects.

Sabadilla contains two known active alkaloids, cevadine ($C_{32}H_{49}NO_9$) and veratridine ($C_{36}H_{51}NO_{11}$). Neither chemical structure has been established. It is irritating to human eyes and causes violent sneezing in some sensitive individuals. It deteriorates rapidly in sunlight and can be used safely on food crops with no waiting interval required by the EPA. Sabadilla is probably the most difficult of the five botanical insecticides to purchase, simply because there was hardly any demand for it for about 15 years.

Sabadilla is registered for most commonly grown vegetables and will control caterpillars, grasshoppers, beetles, leafhoppers, thrips, chinch bugs, stink bugs, harlequin and squash bugs, other true bugs, and potato psyllids. It is not very useful against aphids and will not control spider mites.

Ryania

Ryania is another botanical or plant-derived insecticide that is quite safe for humans and domestic animals—so safe that no waiting is required between the time of application to food crops and harvest, as there is for most other insecticides. Ryania is made from the ground roots of the ryania shrub grown in Trinidad and, like nicotine, belongs to the chemical class of alkaloids. It has an oral LD_{50} of approximately 750 mg/kg. It is a slow-acting insecticide, requiring as long as 24 hours to kill. Insects exposed to ryania usually stop their feeding almost immediately, making it particularly useful for caterpillars.

The active principle of ryania is the alkaloid ryanodine ($C_{25}H_{35}NO_9$), whose chemical structure has not been determined. Ryanodine affects insect muscles directly by preventing contraction, resembling the effects of strychnine in mammals.

The preferred uses for ryania are against fruit- and foliage-eating caterpillars on fruit trees, especially the codling moth on apple trees. However, it is useful against almost all plant-feeding insects, making it an ideal material for small orchards of deciduous fruits. It is not effective against spider mites.

Ryania is exempt from a waiting period between application and harvest, and is registered by the EPA for the control of a host of insect pests on a wide variety of plants, shrubs, and trees. Vegetable garden pests include aphids, cabbage loopers, Colorado potato beetles, corn borers, cucumber beetles, diamond back moths, flea beetles, leafhoppers, Mexican bean beetle, spittle bugs, and tomato hornworms.

Ryania is registered for deciduous fruit trees to control aphids (except the woolly aphid), codling moth, Japanese beetle, and cherry fruit fly. It has long been used to control citrus thrips on all citrus. Though not very effective, ryania can be used in the home to control ants, silverfish, cockroaches, spiders, and crickets. On ornamentals it is registered for aphids and lace bugs. It can be used on roses against aphids, Japanese beetles, thrips, and whiteflies; on brambles, for aphids, raspberry fruitworms, and sawflies; and on grapes, for aphids (except the woolly aphid) and the Japanese beetles.

Ryania is difficult to obtain since its importation into the United States has been discontinued by its major distributor.

Pyrethrum

Pyrethrum is extracted from the flowers of a chrysanthemum grown in Kenya, Africa, and Ecuador, South America. It has an oral LD_{50} of approximately 1500 mg/kg and is one of the oldest household insecticides available. The ground, dried flower heads were used back in the nineteenth century as the original louse powder to control body lice in the Napoleonic Wars. Pyrethrum acts on insects with phenomenal speed causing immediate paralysis, thus its popularity in fast knockdown household aerosol sprays. However, unless it is formulated with one of the synergists, most of the paralyzed insects recover to once again become pests. Pyrethrum is formulated as household sprays and aerosols and is available as spray concentrates and dusts for use on vegetables, fruit trees, ornamental shrubs, and flowering plants at any stage of growth. Vegetables and fruit sprayed or dusted with pyrethrum may be harvested or eaten immediately; there is no waiting interval required between application and harvest of the food crop.

Because of its general safety to humans and domestic animals and its effectiveness against practically every known crawling and flying insect pest, pyrethrum has more uses approved by the EPA than any other insecticide, numbering in the thousands.

Pyrethrum is a mixture of four compounds: pyrethrins I and II and cinerin I and II. Their structures can be assembled by attaching the R_1 and R_2 in their proper positions on the large ester structure to the left.

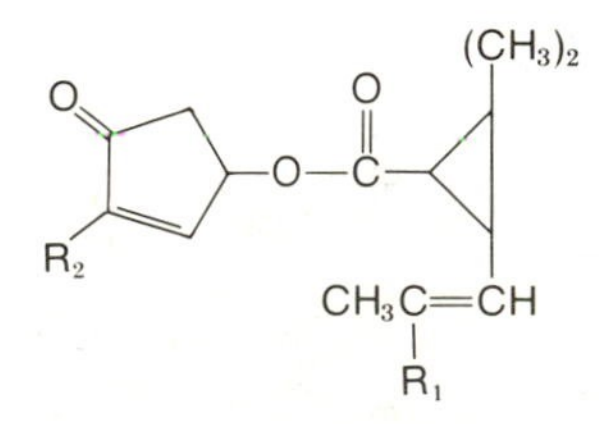

PYRETHRIN I

$R_1 = —CH_3$
$R_2 = —CH_2CH{=}CHCH{=}CH_2$

CINERIN I

$R_1 = —CH_3$
$R_2 = —CH_2CH{=}CHCH_3$

PYRETHRIN II

$R_1 = —C({=}O)—OCH_3$

$R_2 = —CH_2CH{=}CHCH{=}CH_2$

CINERIN II

$R_1 = —C({=}O)—OCH_3$

$R_2 = —CH_2CH{=}CHCH_3$

SYNTHETIC PYRETHROIDS

The natural insecticide pyrethrum has seldom been used for agricultural purposes because of its cost and instability in sunlight. Recently, however, several synthetic pyrethrin-like materials have become available and are referred to as synthetic pyrethroids. These materials are very stable in sunlight and are generally effective against most agricultural pests when used at the low rate of 0.11 to 0.23 kg/ha. Examples are permethrin (Ambush® or Pounce®) and fenvalerate (Pydrin®).

The synthetic pyrethroids, or more correctly *pyrethroids*, have a rather long and successful history. For ease of classification, they are placed in four categories, or generations.

The first generation contains but one pyrethroid, allethrin (Pynamin®). Commercially available in 1949, it marked the beginning of an era of complex syntheses, involving as many as 22 chemical reactions to produce the final insecticide. Allethrin is merely a synthetic duplicate of cinerin I (a component of pyrethrum), with a slightly more stable side chain, and it is more persistent than pyrethrum. Equally effective against houseflies and mosquitoes, but less so against cockroaches and other insects, it was readily synergized by the common pyrethrum synergists.

The second generation includes tetramethrin (Neo-Pynamin®), which appeared in 1965. It gives stronger knockdown of flying insects than allethrin and is readily synergized. Resmethrin (NRDC-104, SBP-1382, and FMC-17370) appeared in 1967, is approximately 20 times more effective than pyrethrum in housefly knockdown, and is not synergized to any appreciable extent with pyrethrum synergists. Bioresmethrin (NRDC-107, FMC-18739, and RU-11484), also described in 1967, is 50 times more effective than pyrethrum against normal (susceptible to insecticides) houseflies, and also not synergized with pyrethrum synergists. Both resmethrin and bioresmethrin are more stable than pyrethrum, but decompose fairly rapidly on exposure to air and sunlight, which explains why they were never developed for agricultural use. (Resmethrin has become the most used of the second generation pyrethroids for sprays and aerosols to control flying and crawling insects indoors.)

Bioallethrin® (*d-trans*-allethrin) was introduced in 1969. It is more potent than allethrin and readily synergized, but is not as effective as resmethrin. The last of this period was phenothrin

FENVALERATE (Pydrin®)

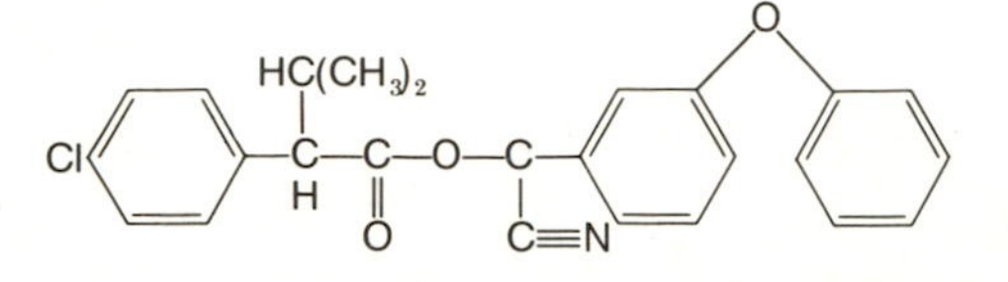

cyano (3-phenoxyphenyl) methyl 4-chloro-α-(1-methylethyl) benzeneacetate

PERMETHRIN (Ambush®, Pounce®)

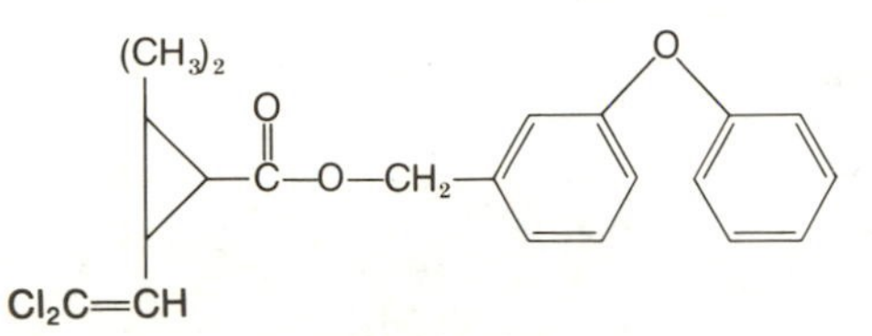

m-phenoxybenzyl (±)-*cis, trans*-3-(2,2-dichlorovinyl)-2,2-dimethylcyclopropanecarboxylate

ALLETHRIN (Pynamin®)

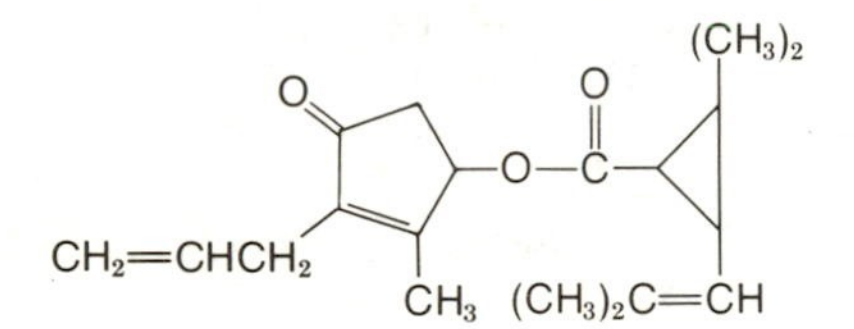

2-methyl-4-oxo-3-(2-propenyl)-2-cyclopenten-1-yl 2,2-dimethyl-3-(2-methyl-1-propenyl)cyclopropanecarboxylate

TETRAMETHRIN (Neo-Pynamin®)

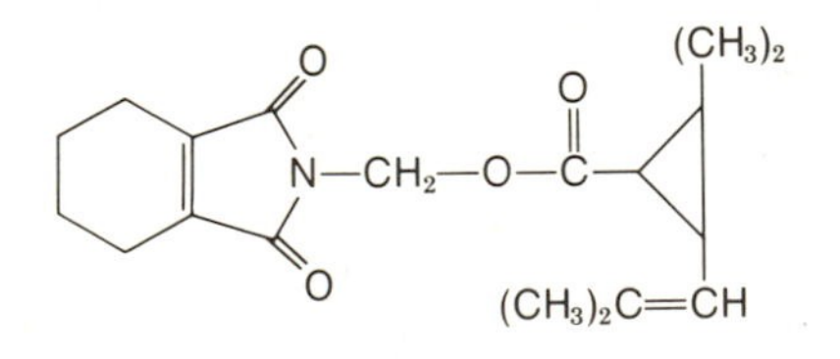

(1,3,4,5,6,7-hexahydro-1,3-dioxo-2*H*-isoindol-2-yl)methyl 2,2-dimethyl-3-(2-methyl-1-propenyl)cyclopropanecarboxylate

RESMETHRIN (Synthrin®)

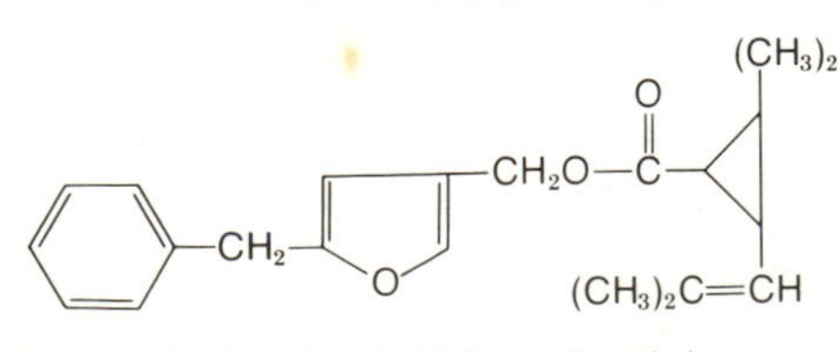

(5-phenylmethyl-3-furanyl)methyl 2,2-dimethyl-3-(2-methyl-1-propenyl) cyclopropanecarboxylate

BIORESMETHRIN

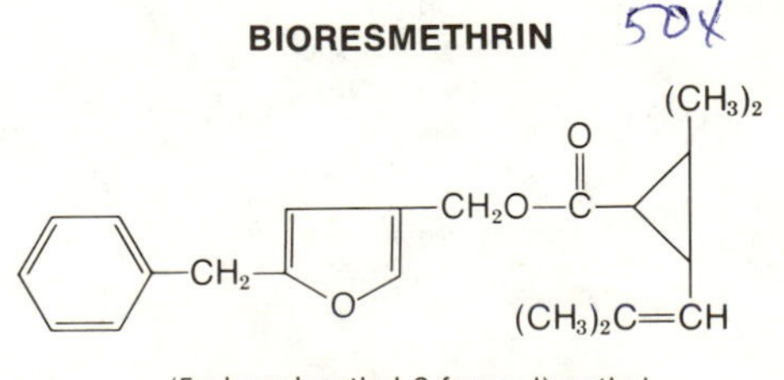

(5-phenylmethyl-3-furanyl)methyl (1R-*trans*)-2,2-dimethyl-3-(2-methyl-1-propenyl)cyclopropanecarboxylate

Stereoisomer of RESMETHRIN

d-trans-**ALLETHRIN (Bioallethrin®)** **PHENOTHRIN (Sumithrin®)**

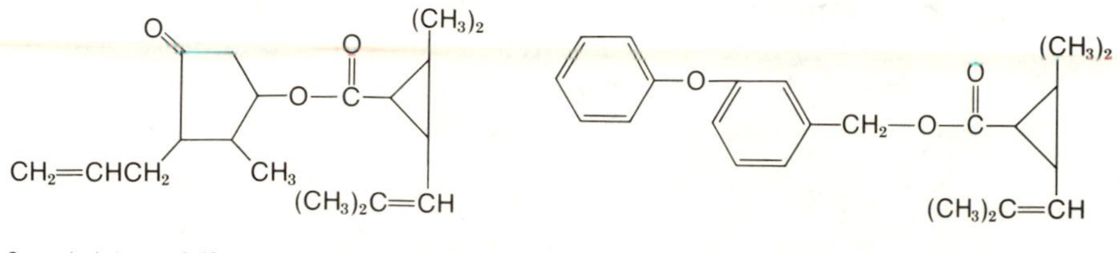

2-methyl-4-oxo-3-(2propenyl)-2-cyclopenten-1-yl 2,2-dimethyl-3-(2-methyl-1-propenyl) cyclopropanecarboxylate

(3-phenoxyphenyl)methyl 2,2-dimethyl-3-(2-methyl-1-propenyl)cyclopropane-carboxylate

(Sumithrin®), introduced in 1973. It, too, is intermediate in quality and slightly enhanced by synergists.

The third generation includes fenvalerate (Pydrin®) and permethrin (Ambush®, Pounce®, and Pramex®), which appeared in 1972 and 1973 respectively. These became the first agricultural pyrethroids because of their exceptional insecticidal activity (0.11 kg AI/ha) and their *photostability*. They are seemingly unaffected by ultraviolet in sunlight, lasting four to seven days on crop foliage as effective residues.

The contemporary fourth generation, still being developed and registered, is truly exciting, for their rates of application are again reduced to one-tenth of the previous generation, or to the order of 0.01 to 0.06 kg AI/ha. This is truly phenomenal compared to the rate of 1.1 to 2.3 kg AI/ha required of the organophosphate, carbamate, and organochlorine insecticides. Emerging in this truly revolutionary form of chemical insect control are the pyrethroids described and illustrated in Chapter 24: cypermethrin (Ammo®, Cymbush®, and Ripcord®), fenpropathrin, flucythrinate (Pay-Off®), fluvalinate (Mavrik®), and decamethrin (Decis®). All these insecticides are photostable, providing long residual effectiveness in the field, and are not significantly improved with the addition of pyrethrin synergists.

DECAMETHRIN (Decis®)

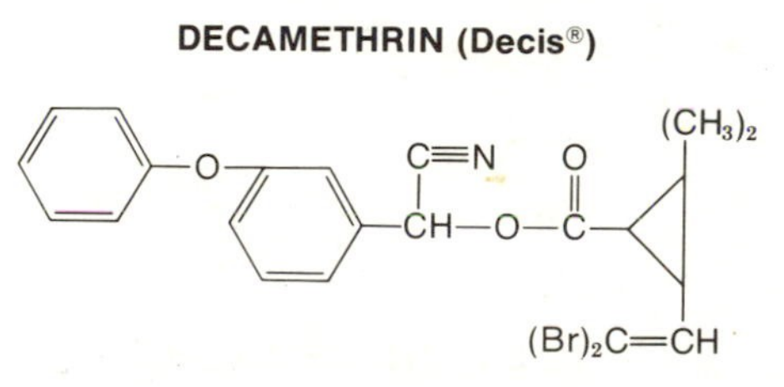

(1R(1 (*S**),3))-cyano(3-phenoxyphenyl) methyl 3-(2,2-dibromoethenyl)-2,2-dimethyl-cyclopropanecarboxylate

SYNERGISTS OR ACTIVATORS

Synergists are not in themselves considered toxic or insecticidal, but are materials used with insecticides to synergize or enhance the activity of the insecticides. Synergists are added to certain insecticides in the ratio of 8 : 1 or 10 : 1. The first synergist was introduced in 1940 to increase the effectiveness of pyrethrum. Since then many materials have been introduced, but only a few are still marketed because of cost and ineffectiveness. Synergists are found in practically all the "bug-bomb" aerosols to enhance the action of the fast knockdown insecticides pyrethrum, allethrin, and sometimes resmethrin against flying insects.

Although initially developed for use with pyrethrum, they have since been observed to synergize some, but not all, organophosphates, organochlorines, carbamates, as well as a few of the botanical, or plant-derived, insecticides. The mode of action of the synergists is to inhibit mixed-function oxidases, enzymes that metabolize foreign compounds, which in this instance would be the pyrethrum.

The most popular synergists belong to only two molecular groups, or moieties. The first is the methylenedioxyphenyl moiety. The R_1 and R_2 are simple or oxygenated carbon chains or other groups of varying combinations.

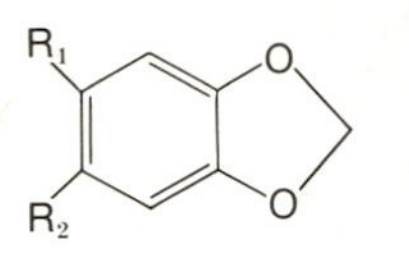

The second synergistic moiety does not have a single name but is characterized by either of the following structures.

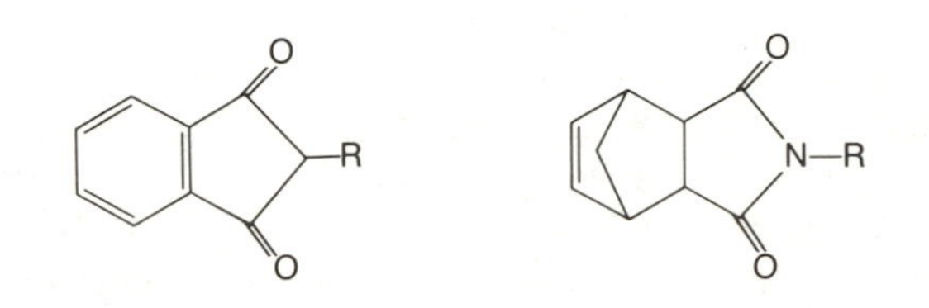

Notice that all three moieties involve a five-membered ring associated with two oxygens. Because their mode of action is the inhibition of insecticide-metabolizing enzymes, it is likely that this steric three-dimensional structure is generally the most effective in enzyme binding.

The synergists are usually used in sprays prepared for the home and garden, stored grain, and on livestock, particularly in dairy barns. Synergists are quite expensive, thus seldom if ever used on crops.

PIPERONYL BUTOXIDE

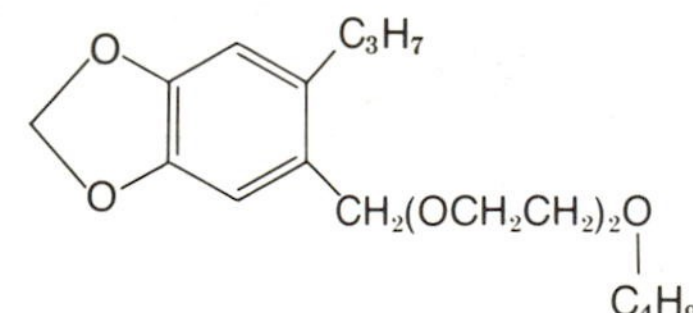

α-[2-(2-butoxyethoxy)ethoxy]-4,5-methylenedioxy-2-propyltoluene

SESAMEX

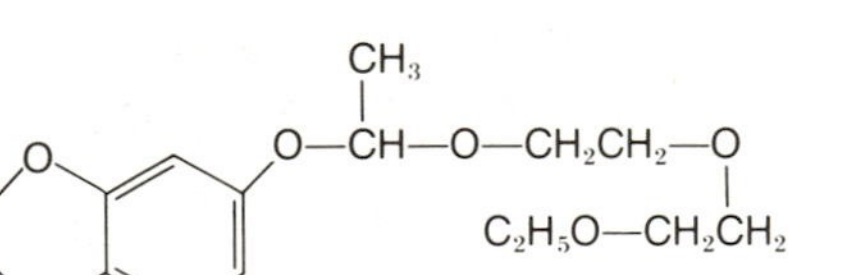

2-(2-ethoxyethoxy)ethyl-3,4-(methylenedioxy) phenyl acetal of acetaldehyde

SULFOXIDE

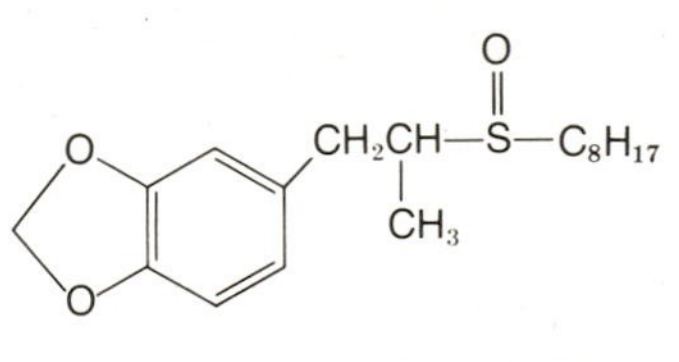

1,2-methylenedioxy-4-[2-(octylsulfinyl)propyl] benzene

First discovered in sesame oil, a material containing the methylenedioxyphenyl group was given the name *sesamin*. As mentioned earlier, many compounds having this moiety are synergistic, but only the structures of piperonyl butoxide, sesamex, and sulfoxide are shown.

MGK 264®

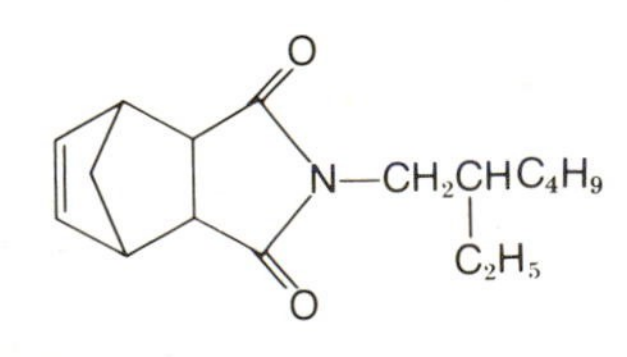

N-(2-ethylhexyl)-5-norbornene-2,3-dicarboximide

MGK 264 belongs to the "no-name" moiety and appeared in 1944. It has been used in great quantity, mostly in livestock and animal shelter sprays.

In summary, the synergists are used in many of the insecticide mixtures for home, garden, and barn. Their mode of action is their binding to oxidative enzymes that would otherwise degrade the insecticide.

INORGANICS

Inorganic insecticides are those that do not contain carbon. Usually they are white and crystalline, resembling the salts. They are stable chemicals, do not evaporate, and are frequently soluble in water. They are mentioned here for their historical significance.

Sulfur, as mentioned in Chapter 1, is very likely the oldest known *effective* insecticide. Sulfur and sulfur candles were burned by our great-grandparents for every conceivable purpose, from bedbug fumigation to the cleansing of a house just removed from medical quarantine of smallpox. Sulfur is a highly useful pesticide in integrated pest management programs where target pest specificity is important. Sulfur dusts are especially toxic to mites of every variety, such as chiggers and spider mites, thrips, newly-hatched scale insects, and as a stomach poison for some caterpillars. Sulfur dusts and sprays are also fungicidal, particularly against powdery mildews.

Several other inorganic materials have been used as insecticides. These include compounds of mercury, boron, thallium, arsenic, antimony, selenium, and fluoride. The only one of these used extensively today is arsenic, which is used in two forms, the arsenites (salts of arsenious acid) and the arsenates (salts of arsenic acid).

Paris green (green because of its copper content) was the first commonly used arsenical, a water-soluble arsenite. Next was lead arsenate. Finally, the third and last of these arsenicals was calcium arsenate, which was used for a time on vegetables in the 1930s and on cotton in the 1930s and 1940s.

Arsenicals are truly stomach poisons, exerting their toxic action following ingestion by the insects. Their action is attributed to the arsenite or arsenate ion, as explained in Chapter 17.

The arsenicals have a rather complex mode of action. First, they uncouple oxidative phosphorylation (by substitution of the arsenite ion for phosphorus), a major energy-producing step of the cell. Second, the arsenate ion inhibits certain enzymes that contain sulfhydryl (—SH) groups. And, finally, both the arsenite and arsenate ions coagulate protein by causing the shape or configuration of proteins to change.

Arsenical insecticides were very useful agricultural tools from 1930 until 1956, as we were making the transition from the simple to the complex synthetic molecules. They were, in fact, responsible for the initiation of large-scale insecticide applications eventually leading to the intensive use of fungicides and herbicides in modern agriculture.

Fluorine insecticides also included organic fluorine compounds, but these were of little importance and seldom used. The inorganic fluorides were sodium fluoride, used for cockroach and ant control around the home, and barium fluosilicate, sodium silicofluoride, and cryolite (NaF, $BaSiF_6$, and Na_3AlF_6, respectively). The last three were used for a time in plant protection. The fluoride ion inhibits many enzymes that contain iron, calcium, and magnesium. Several of these enzymes are involved in energy production in cells, as in the case of phosphatases and phosphorylases.

Boric acid (H_3BO_4), used as an insecticide against cockroaches and other crawling household pests in the 1930s and 1940s, has returned in the 1980s. Although manufacturers claim it is "safe, does not evaporate, and continues to kill for years," in fact, it is only moderately effective, acting as both a stomach poison and adsorber of insect cuticle wax.

The last group of inorganics are the silica gels or silica aerogels. These are light, white, fluffy silicates used for household insect control. The silica aerogels kill insects by adsorbing waxes from the insect cuticle, permitting the continuous loss of water from the insect body. The insects then gradually become desiccated and die from dehydration. These include Dri-Die®, Drianone®, and Drione®. The latter two are fortified with pyrethrum and synergists, which enhance their effectiveness.

FUMIGANTS

The fumigants are small, volatile, organic molecules that become gases at temperatures above 5°C. They are usually heavier than air and commonly contain one or more of the halogens (Cl, Br, or F). Most are highly penetrating, reaching through large masses of material. They are used to kill insects, insect eggs, and certain microorganisms in buildings, warehouses, grain elevators, soils, and greenhouses and in packaged products such as dried fruits, beans, grain, and breakfast cereals.

Fumigants, as a group, are narcotics. That is, their mode of action is more physical than chemical. The fumigants are liposoluble (fat soluble); they have common symptomology; their effects are reversible; and their activity is altered very little by structural changes in their molecules. Narcotics induce narcosis, sleep, or unconsciousness, which in effect is their action on insects.

Liposolubility appears to be an important factor in the action of fumigants, since these narcotics lodge in lipid-containing tissues, which are found in the nervous system.

Some of the more common fumigants' names and structures are presented in the list to the left.

Name	Formula
Methyl bromide	CH_3Br
Ethylene dibromide	$BrCH_2CH_2Br$
Ethylene dichloride	$ClCH_2CH_2Cl$
Hydrogen cyanide	HCN
Chloropicrin	Cl_3CNO_2
Sulfuryl fluoride (Vikane®)	SO_2F_2
Vapam	$CH_3NHC(=S)-S-Na$
Telone	$ClCH=CH-CH_2Cl$
D-D®	$CHCl=CHCH_2Cl$ plus $ClCH_2CH(Cl)CH_3$
Chlorothene	CH_3CCl_3
Nemagon® (DBCP)*	$BrCH_2CH(Br)CH_2Cl$
Ethylene oxide	H_2C-CH_2 bridged by O
Naphthalene (crystals)	
p-dichlorobenzene (PDB crystals)	Cl— —Cl

* Registered uses were canceled by the EPA in 1980.

MICROBIALS

Microbial insecticides obtain their name from microbes or microorganisms that are used to control certain insects. Like mammals, insects are susceptible to diseases caused by fungi, bacteria, and viruses. In several instances, these have been isolated, cultured, and mass-produced for use as pesticides.

The insect disease-causing microorganisms do not harm other animals or plants. The reverse of this is also true. This method of insect control is ideal in that the diseases are usually rather specific. Undoubtedly the future holds many such materials in the arsenal of insecticides, since several new insect pathogens are identified each year. However, at the present only a few are produced commercially and approved by the EPA for use on food and feed crops. There is still some concern regarding the very remote chance of human susceptibility to these diseases of invertebrate animals, hence the slow advances into this relatively new field and exceptional precautionary testing.

The microorganism *Bacillus thuringiensis* is a disease-causing bacterium whose spores are necessary for disease induction. These spores produce compounds that injure the gut of the insect larva in such a way that invasion of the body cavity follows. This organism produces four substances toxic to insects. The first, and most important, is a crystalline protein whose ingestion by caterpillars results in paralysis of their gut. The second is a toxin, a water-soluble nucleotide derivative, that passes unchanged through the gut of a cow to kill certain fly maggots breeding in the manure. The remaining two substances are enzymes that are lost in the commercial preparation of *Bacillus*.

Only one insect virus has been registered for agricultural use by the EPA. Known as the *Heliothis* nuclear polyhedrosis virus, it is specific for *Heliothis zea* (corn earworm, cotton bollworm) and *Heliothis virescens* (tobacco budworm), two of the most destructive pest caterpillars in agriculture. Recently registration has been extended not only for cotton but also for all crops attacked by *Heliothis* species, including beans, corn, lettuce, okra, peppers, sorghum, soybeans, tobacco, and tomatoes. Proprietary names for this naturally occurring viral pathogen are Elcar® and Biotrol VHZ®.

Several other insect viruses are in the development stage, all of which are aimed at caterpillars and are only experimentally available. Viruses are highly specific and have modes of action that may not be identical throughout. Generally, the viruses result in crystalline proteins that are eaten by the larva and begin their activity in the gut. The virus units then pass through the gut wall and into the blood. There the units multiply rapidly and take over complete genetic control of the cells, causing their death.

A recent and innovative development in the agricultural use of microbial insecticides is the addition of feeding or gustatory stimulants. The feeding stimulants apparently attract the caterpillars to treated foliage, which increases their consumption of the microbial. Two successfully marketed products are Coax® and Gustol®, both of which are wettable powders and used at 1.1 to 2.3 kg/ha.

Mycar® miticide is a new biological control, a mycoacaricide, developed by Abbott Laboratories. The microorganism is *Hirsutella thompsonii*, a parasitic fungus that infects and kills the citrus rust mite. Under optimum conditions *H. thompsonii* can infect spider mites and other nontarget mites. It is, however, consistently effective only against the citrus rust mite; thus it is a selective miticide. When sprayed on plants the spores grow into colonies that attach to the mite. In the presence of ample free moisture the spore germinates and infects the mite. The mite dies in about three days, and the fungus spreads, continuing to propagate itself. In this manner it offers potentially long residual effectiveness. Because the active agent is a fungus, all commercial fungicides, copper chemicals, and certain other metal salts such as zinc, lead, and manganese are detrimental to *H. thompsonii*. Because of its specificity *H. thompsonii* should be highly compatible with other suppression techniques used in the integrated pest management of citrus crops.

Nosema locustae is a new biological developed by Sandoz, Inc., for the control of grasshoppers. Marketed under the name NOLOC® and Trojan 10®, the microorganism is a protozoan. Depending on the method of application, climatic conditions, and grasshopper density, the protozoa kill up to 50 percent of the insects and sterilize up to 30 percent of the survivors. Maximum effect occurs over a two- to four-week period. The residual effect of a single treatment continues to control grasshoppers through subsequent generations by transmission through the eggs up to three or four years. This biological insecticide is most effective when applied as a bait and is available for rangeland use as well as for yard and garden.

INSECT GROWTH REGULATORS

The first-generation insecticides were the stomach poisons, such as the arsenicals, heavy metals, and fluorine compounds. The second generation includes the familiar contact insecticides: organochlorines, organophosphates, carbamates, and formamidines.

The third generation of insecticides are the biorationals. These chemicals are environmentally sound, closely resembling or identical to chemicals produced by insects and plants. Among these are the insect growth regulators (IGRs), a relatively new group of chemical compounds that alter growth and development in insects. Their effects have been observed on embryonic and larval and nymphal development, on metamorphosis, on reproduction in both males and females, on behavior, and on several forms of diapause. They include ecdysone (the molting hormone), juvenile hormone (JH), JH mimic, JH analog (JHA), and their broader synonyms, juvenoids and juvegens. More recently another growth effect, chitin inhibition, has been identified and is discussed later, with the EPA-registered IGRs.

alpha-ECDYSONE

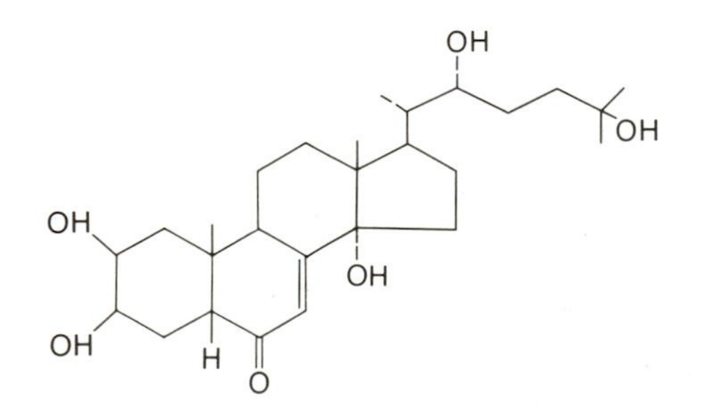

The IGRs are effective when applied in very minute quantities and apparently have no undesirable effects on humans and wildlife. They are, however, nonspecific, since they affect not only the target species, but most other arthropods as well. Consequently, when used with precision IGRs may play an important role in future insect control.

TYPICAL JUVENILE HORMONE

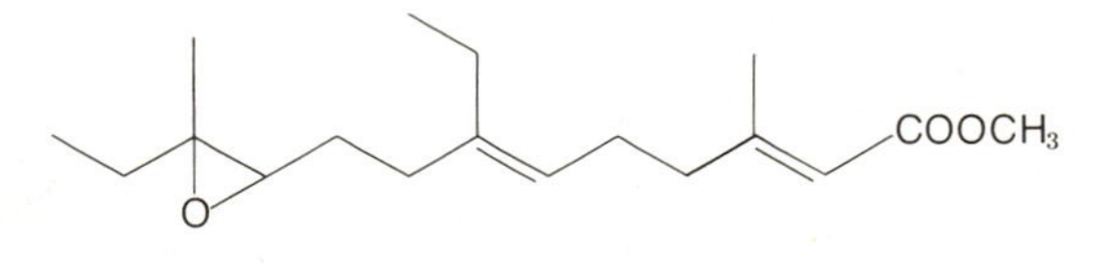

Several glands in insects are known to produce hormones, the principal functions of which are the control of reproductive processes, molting, and metamorphosis. Here we are interested only in the hormone ecdysone, which is responsible for molting, and JH, which inhibits or prevents metamorphosis.

When insects are treated with ecdysones, they usually die in all stages of growth, making ecdysones similar to second-generation insecticides. One attractive feature of ecdysones as potential tools is their widespread distribution in plants, and these may play as yet unrecognized roles in insect-plant relationships.

Keen interest has been directed toward JHs. These are not, in the usual sense, toxic to insects. Instead of killing directly, they interfere in the normal mechanisms of development and cause the insects to die before reaching the adult stage. One JH is the classical juvabione, found in the wood of balsam fir. Its effect was discovered quite by accident when paper towels made from this source were used to line insect-rearing containers, and the insects' development was suppressed.

Some of these plant-derived substances actually serve to inhibit the development of insects feeding thereon, thus protecting the host plant. These are referred to broadly as antijuvenile hormones, or more accurately, *antiallatotropins*. Recently, antiallatotropins, called *precocenes*, have been discovered. These are considered a potential fourth generation of insect control agents. Incidentally, the name *precocene* results from the *precocious* metamorphosis stimulated by compounds having the *chromene* nucleus. Although the mode of action of the precocenes is still unclear, it is known that they depress the level of juvenile hormone below that normally found in immature insects.

CHROMENE NUCLEUS

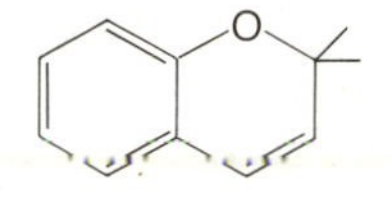

Dramatic results with JHs have been obtained in the laboratory, the most promising effects being on mosquito larvae, caterpillars, and hemipterans (bugs), although effects have been observed on practically all insect orders. Most insect species respond to treatment with JHs by producing extra larval, nymphal, or pupal forms that vary from giant, almost perfect, forms to intermediates of all sorts between the immatures and the adults. For the most part, the periods of greatest sensitivity for metamorphic inhibition are the last larval or nymphal stages and the pupa, in those having complete metamorphosis. One recognizable problem is the precision of timing applications to achieve maximum damaging effect on the upcoming life stage of a particular insect.

For practical purposes, IGRs could be used on crops to suppress damaging insect numbers. They would be applied with the purpose of preventing pupal development or adult emergence, thus keeping the insects in the immature stages, resulting eventually in their deaths.

To date, only three IGRs are registered by the EPA. The first, methoprene (Altosid®), manufactured by the Zoecon Corporation, was registered early in 1975 as a mosquito growth regulator, for use against second– through fourth–larval-stage floodwater mosquitoes at 0.11 to 0.14 kg/ha to prevent adult emergence. Larvae exposed to methoprene continue their development to the pupal stage, when they die. Methoprene has no effect when applied to pupae or adult mosquitoes. Additionally, methoprene has been formulated as Precor® for indoor control of dog and cat fleas. It acts by interrupting the flea's life cycle at the larval stage, preventing emergence of adult fleas for up to 75 days. Its use in combination with a conventional insecticide is usually necessary to control adult fleas not affected by the growth regulator.

METHOPRENE (Altosid®)

isopropyl (2*E*-4*E*)-11-methoxy-3, 7, 11-trimethyl-2, 4-dodecadienoate

Methoprene is also registered for use on tobacco to control cigarette beetles (Kabat®), on cattle feed to control horn flies, and on mushroom-growing media to control fungus gnats (Apex®). In addition, it has been approved by the World Health Organization for use in drinking water to control mosquitoes.

The second IGR is diflubenzuron (Dimilin®), manufactured by Thompson-Hayward Chemical Company. Currently registered for gypsy moth caterpillars in forests and for the boll weevil in cotton, it is awaiting registration for a wide variety of pests. Experimentally, it

DIFLUBENZURON (Dimilin®)

1-(4-chlorophenyl) 3-(2,6-difluorobenzoyl)urea

KINOPRENE (Enstar®)

2-propynyl (*E,E*)-3,7,11-trimethyl-2,4-dodecadienoate

DIMETHYL PHTHALATE

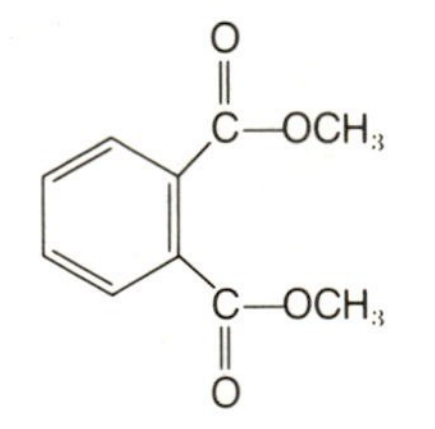

dimethyl 1,2-benzenedicarboxylate

Indalone®

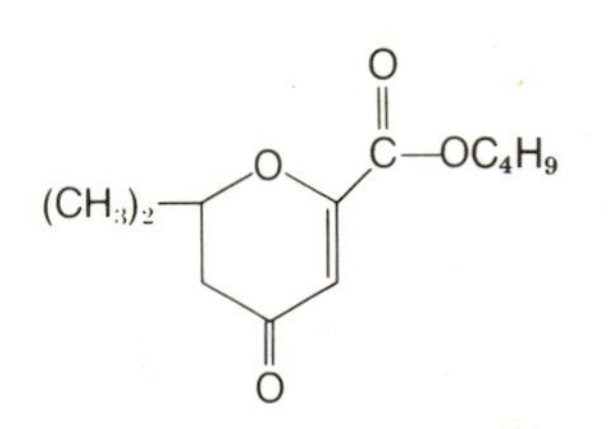

butyl 3,4-dihydro-2,2-dimethyl-4-oxo-2H-pyran-6-carboxylate

RUTGERS 612

OH
$C_3H_7CHCHCH_2OH$
C_2H_5

2-ethyl-1,3-hexanediol

has controlled insect pests in forests, woody ornamentals, fruit, vegetables, cotton, soybeans, citrus, larvae of flies, midges, gnats, and mosquitoes; on mosquitoes, a dose as small as 1.5 g/ha is effective.

Diflubenzuron is not truly a growth regulator of the juvenoid class but rather another insecticide with a different mode of action. However, it is tentatively classed with the IGRs. It acts on the larval stages of most insects by inhibiting or blocking the synthesis of chitin, a vital and almost indestructible part of the hard outer covering of insects, the exoskeleton.

The third, and most recent, IGR is kinoprene (Enstar®), also developed by the Zoecon Corporation. It is effective against aphids, whiteflies, mealybugs, and scales (both soft and armored) on ornamental plants and vegetable seed crops grown in greenhouses and shadehouses. It is specific for insects in the order Homoptera, and results in a gradual reduction rather than an immediate kill, by inhibiting development, reducing egg-laying, killing eggs already laid, and sterilizing mature whiteflies and aphids. Because it is specific for insects, like all other IGRs, it is virtually nontoxic to humans and other warm-blooded animals.

Can IGRs become successful pest control agents? Certainly they can in time. They will, however, have to meet the general criteria for other pest control agents; thus, they must be effective in reducing insect populations below economic damage levels, be competitive with second-generation insecticides in cost, and have no undesirable side effects. In summary, IGRs hold intriguing possibilities for future use in practical insect control. It should be kept in mind that IGRs are insect-controlling chemicals and thus fall within the same legal confines as other insecticides. Their great distinction, however, is that they are toxic to populations of insects rather than to individuals. The ultimate success of any pest control agent depends on its ability to control the fecundity (reproductive capacity) of the pest.

INSECT REPELLENTS

Historically, repellents have included smokes, plants hung in dwellings or rubbed on the skin as the fresh plant or brews, oils, pitches, tars, and various earths applied to the body. Camel urine sprinkled on the clothing has been of value, although questionable, in certain locales. Camphor crystals sprinkled among woolens have been used for decades to repel clothes moths. Before humans developed a more edified approach to insect olfaction and behavior, it was assumed that if a substance was repugnant to humans it would likewise be repellent to annoying insects.

Prior to World War II, there were only four principal repellents: (1) oil of citronella, discovered in 1901, used also as a hair-dressing fragrance by certain Eastern cultures, probably to control lice and other head ectoparasites; (2) dimethyl phthalate, discovered in 1929; (3) Indalone®, introduced in 1937, and (4) Rutgers 612, which became available in 1939.

With the onslaught of World War II and the introduction of American military personnel into new environments, particularly the tropics, it became necessary to find new repellents that would survive both time and dilution by perspiration. The ideal repellent would be nontoxic and nonirritating to humans, nonplasticizing, and long-lasting (12 hours) against mosquitoes, biting flies, ticks, fleas, and chiggers. Unfortunately, the ideal repellent has still not been found. Some repellents have unpleasant odors, require massive dosages, are oily or effective only for a short time, irritate the skin, or soften paint and plastics.

Insect repellents come in every conceivable formulation—undiluted, diluted in a cosmetic solvent with added fragrance, aerosols, creams, lotions, treated cloths to be rubbed on the skin, grease sticks, powders, suntan oils, and clothes-impregnating laundry emulsions. Regardless of the formulation, the periods of protection they offer vary with the chemical, the individual, the general environment, insect species, and avidity of the insect.

What happens to repellents once applied? Why aren't they effective longer than one or two hours? No single answer is satisfactory—but generally it is because they evaporate, are absorbed by the skin, are lost by abrasion of clothing or other surfaces, and are diluted by perspiration.

The following chemical structures are those of the most commonly found in today's repellents. Of these, diethyl toluamide (Delphene®, deet) is by far superior to all others against biting flies and mosquitoes.

DIBUTYL PHTHALATE

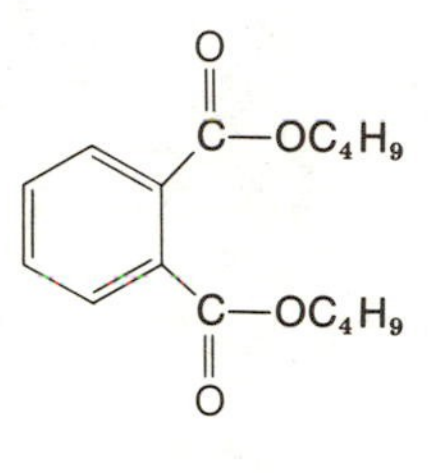

di-n-butyl phthalate

STA-WAY®

$CH_3\overset{O}{\overset{\|}{C}}OCH_2CH_2OCH_2CH_2OC_4H_9$

2-(2-butoxyethoxy)ethyl acetate

MGK® REPELLENT 874 (ants, cockroaches)

$CH_3(CH_2)_7SCH_2CH_2OH$

2-(octylthio)ethanol

CLOTHING IMPREGNANT (ticks, chiggers)

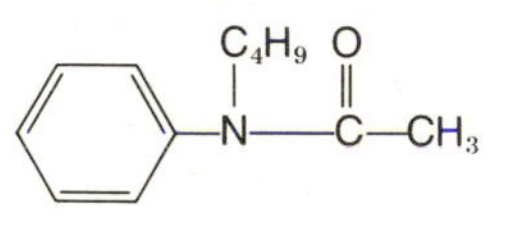

N-butyl acetanilide

MGK® REPELLENT 11

1,5α,6,9α9β-hexahydro-4α(4*H*)-dibenzofuran carboxaldehyde

MGK® REPELLENT 326

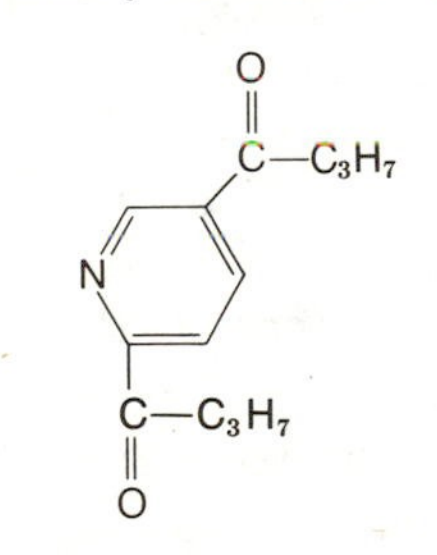

di-n-propyl 2,5-pyridinedicarboxylate

DIMETHYL CARBATE (Dimelone®)

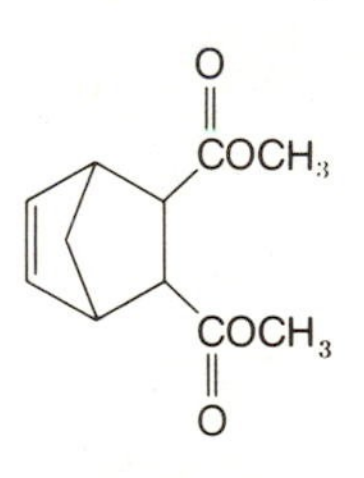

dimethyl cis-bicyclo(2,2,1)-5-heptane-2,3-dicarboxylate

DEET (Delphene®)

N,N-diethyl-*m*-toluamide

BENZYL BENZOATE (ticks, chiggers)

benzyl benzoate

TABATREX® (buildings only: ants, cockroaches, and flies)

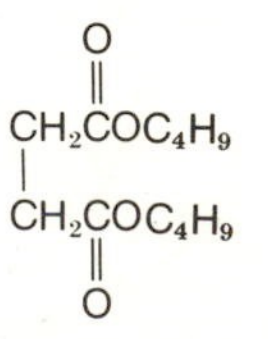

di-n-butylsuccinate

INSECTICIDE SELECTIVITY

Insecticides are our main tool for insect control, at least for the foreseeable future, yet we must supplement them with other forms of insect control. Selective insecticides must be used and insecticides must be used in a selective way, lest the effect of beneficial insects and natural enemies be minimized in any program requiring multiple insecticide applications.

Insecticides affect various insects differently. An insecticide that is lethal against one group may have little, if any, effect on another. Generally, however, most insecticides kill many kinds of insects; broad-spectrum insecticides are those that kill more kinds than others.

A selective pesticide is one that is toxic to some pests, but has little or no effect on other similar species. For example, a selective herbicide is one that kills certain undesirable species of weeds while harming the crop plants little or not at all. Certain fungicides are selective to the point that they control only powdery mildews and no other fungi. A selective insecticide kills selected insects, but spares many or most of the other insects, including beneficial species, either through different toxic action or the manner in which the insecticide is used. While it might be desirable to have a species-specific insecticide, the development of such a product would be uneconomical for the manufacturer, since not enough of the species-specific insecticide could be sold to recoup research, development, and marketing costs. Thus the chemical industry seeks rather to develop a general purpose, broad-spectrum chemical that will control not only elm leaf beetles, but boll weevils on cotton, caterpillars on vegetables, the gypsy moth in deciduous forests, green bugs in small grains, alfalfa weevils, the codling moth in apples, and scales on citrus. General-purpose insecticides offer the best possible incentives in the economic market: satisfied growers and contented stockholders.

A selective insecticide, in the narrowest use of the term, simply means that the active ingredient is inherently more toxic to one group of insects than to others; in other words, most groups of insects are physiologically more tolerant to the chemical than are the few readily killed by it. More commonly, *selective* is used to designate an insecticide that does not harm a beneficial species of insect while killing pest species.

This physiological selectivity results from differences between target and nontarget species in (1) penetration rate of the toxicant, (2) tissue binding and loss of the toxicant, (3) the speed of toxicant and metabolite excretion, (4) metabolic alterations or detoxification, (5) target sites of action or biochemical lesion, or (6) polyfactorial selectivity where more than one of the above processes is involved. (For a clear, detailed discussion of physiological selectivity, see Hollingworth, 1976.)

Physiologically nonselective insecticides may achieve use selectivity if they are applied in such a way that certain insect groups are more adversely affected than others. Use selectivity may result from the timing of the application, the formulation used, the dosage level, or numerous other techniques. From a practical standpoint this approach offers the most immediate hope of being included in precision integrated pest management programs. The selective use of an insecticide to kill a particular pest while permitting other insects to escape, particularly beneficials, achieves the same selective effect as the use of physiologically selective insecticides.

Use selectivity requires that the planner take into account the existence of both the pest and the desirable species, and utilize some difference in their habits, distribution, or biology to devise a discriminatory method of application. Although some insecticides are intrinsically more toxic physiologically to certain insects, and some are equally toxic to different species, they can be used selectively to affect pests more than beneficial insects.

Insecticide selectivity thus can be achieved not only by restricted toxicity but also by precision in timing, the calculation of effective rates of application, the use of materials with short residual lives, methods of application such as seed treatment, spot treatments restricted to areas in which pests vastly outnumber beneficials, or use of those formulations that tend to enhance the survival of beneficials. In general, selectivity is obtained by one of the following practices: (1) use of physiologically selective insecticides that are more toxic to pest species than to others; (2) timing of applications such that detrimental effects on the pest's natural enemies are minimized; and (3) reduction of dosages to levels that adequately control the target pest while sparing relatively large numbers of natural enemies of the target pest and other potential pests (Watson, 1975).

The selective use of insecticides is not simple. In most instances it requires an intimate knowledge of not only the target pest but also the secondary pests and beneficial species. Researchers are directing their efforts toward integrating the use of insecticides and beneficial insects. Effective insect control will require both toxicological research in the development of physiologically selective insecticides and applied ecological research in the use of available insecticides to achieve selective action.

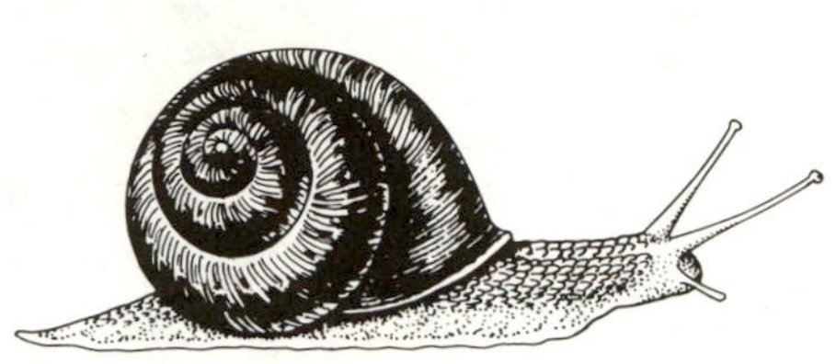

Molluscicides

Four and twenty tailors went to
kill a snail,
The best man among them durst not
touch her tail.
She put out her horns like a little
Kyloe cow,
Run, tailors, run, or she'll kill
you all e'en now.

Old English Nursery Rhyme

Molluscicides are compounds used to control snails, which are the intermediate hosts of parasites of medical importance to humans and which feed in gardens, greenhouses, and fields. Medically, snails are extremely important, especially freshwater snails, such as those that serve as intermediate hosts of organisms causing schistosomiasis and fascioliasis in humans, and lung and liver fluke intermediates of humans, dogs, cats, and domestic animals. Economically, the giant African snail, *Achatina fulica*, is considered to be the most important land snail pest known. This particular snail was introduced into Miami, Florida, in 1966, by an eight-year-old boy who returned from a vacation in Hawaii with three of these cute pests in his pockets.

METALDEHYDE

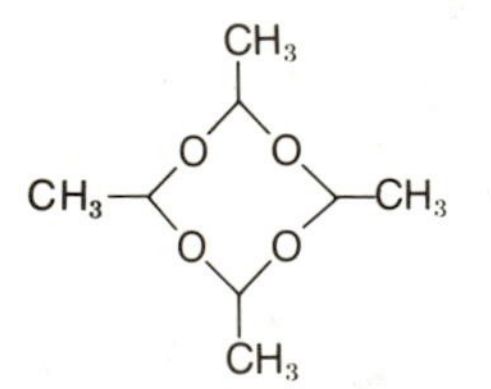

polymer of acetaldehyde, or metacetaldehyde

CLONITRALID (Bayluscide®)

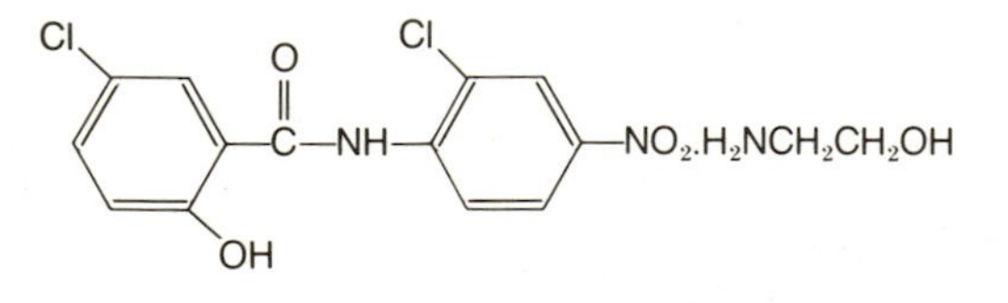

2′,5-dichloro-4′-nitrosalicylanilide, 2-aminoethanol salt

The molluscicide known as *metaldehyde* has been used commercially and around the home as bait for the control of slugs and snails since its discovery in 1936. Its continued use and success can be attributed both to its attractant and toxicant qualities. Other materials used in such formulations as baits, sprays, fumigants, and contact toxicants include metaldehyde plus calcium arsenate, sodium arsenite, ashes, copper sulfate, carbon disulfide, chlordane, coal tar, DDT, lindane, hydrogen cyanide, kerosene emulsion, methyl bromide gas, sodium chloride, and sodium dinitroorthocresylate.

Clonitralid (Bayluscide®) is one of the most promising molluscicides discovered. It is especially toxic to freshwater snails that serve as intermediate hosts of the organisms causing schistosomiasis and fascioliasis, two tragic diseases. It is also useful as a piscicide but has undesirable toxic side effects on other aquatic organisms, such as crayfish, frogs, and clams, with little effect on plankton and aquatic vegetation.

Triphenyltin acetate is also used as a fungicide and algicide in addition to its molluscicide properties. In snail control it may be applied as a spray to infested areas, or as a bait to be consumed after dark by these nocturnal pests.

Formetanate (Carzol®) showed great promise as a garden snail and slug bait, but for economic reasons its manufacturer never had it registered for this purpose. It is, however, registered as an acaricide-insecticide.

Trifenmorph (Frescon®) has been used very successfully to control aquatic snails that transmit bilharzia in humans and the aquatic and

FENTIN ACETATE (Brestan®) **FORMETANATE (Carzol®)** **TRIFENMORPH (Frescon®)** **PCP**

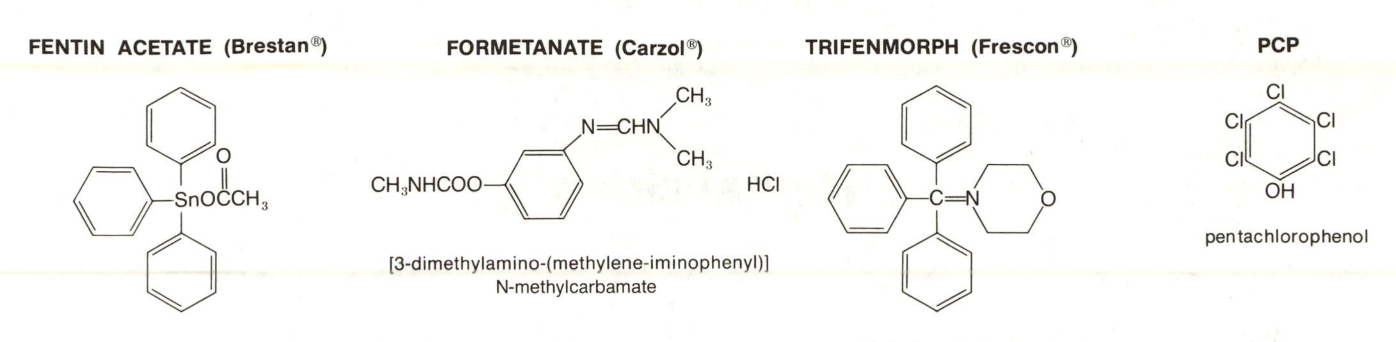

triphenyltin acetate

[3-dimethylamino-(methylene-iminophenyl)] N-methylcarbamate

N-tritylmorpholine

pentachlorophenol

semiaquatic snails that are the intermediate hosts of the organisms causing fascioliasis. However, its production has been discontinued by the manufacturer.

PCP, or pentachlorophenol, also a fungicide and a herbicide, is one of the universally toxic phenolic compounds. It is applied as a molluscicide in Egypt to control snails that carry larval human blood flukes causing schistosomiasis. PCP is available as a formulated product to be applied with petroleum solvent or as an emulsion.

Polystream is a mixture of chlorinated benzenes proved effective against oyster drill, a predatory snail on oysters. It is both molluscicidal and repellent and is used as a selective molluscicide on oyster grounds or beds in Connecticut and New York waters.

Methiocarb (Mesurol®) is a carbamate insecticide also registered for use against snails and slugs in and around home flower gardens and ornamentals. It is highly effective against these pests and has also demonstrated repellency to several bird species.

Carbaryl, the universal yard and garden carbamate insecticide, is also formulated as slug and snail baits. It offers good residual, moderately effective control, and is safe for use in the vegetable garden.

Mexacarbate (Zectran®), another carbamate insecticide, is also formulated as a bait for slug and snail control in and around home flower gardens and ornamentals. Though one of the better molluscicides, its production has been discontinued.

POLYSTREAM

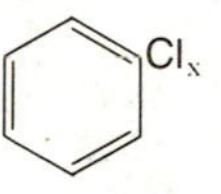

chlorinated benzenes

METHIOCARB (Mesurol®)

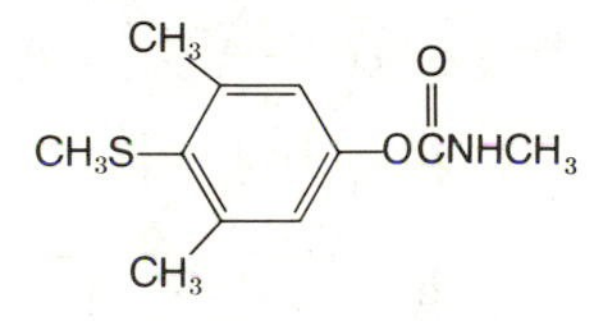

3,5-dimethyl-4-(methylthio)phenol methyl carbamate

CARBARYL

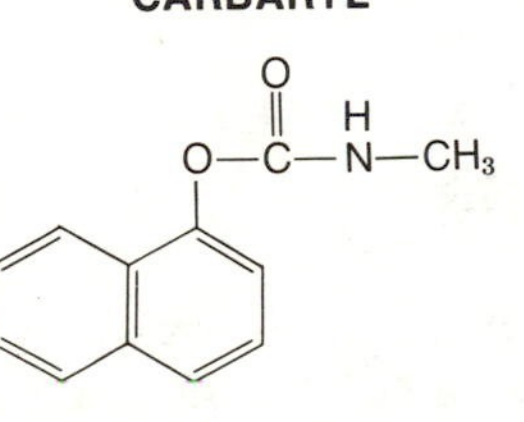

1-naphthyl *N*-methylcarbamate

MEXACARBATE (Zectran®)

4-dimethylamino-3,5-xylyl *N*-methylcarbamate

O Rose, thou are sick!
The invisible worm
That flied in the night,
In the howling storm,

Has found out thy bed
Of crimson joy,
And his dark secret love
Does thy life destroy.

William Blake

Nematicides

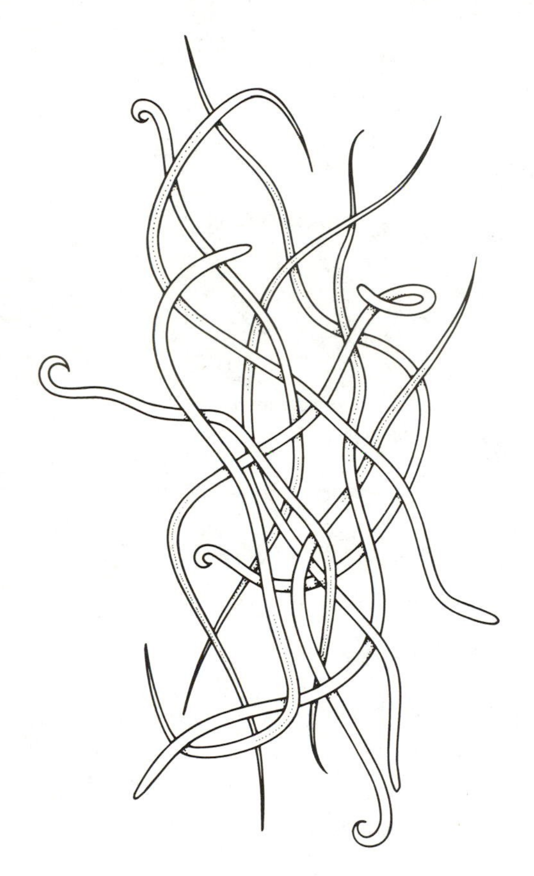

The microscopic roundworms that live in soil or water are known as *nematodes*. Many are free-living, whereas others are parasitic on plants or animals. Some species of nematodes inadvertently introduce pathogenic root-invading microorganisms into the plants while feeding. Nematodes may also predispose crop-plant varieties to other disease-causing agents, such as wilts and root rots. In other instances, the nematodes themselves cause the disease, disrupting the flow of water and nutrients in the xylem system, resulting in rootknot or deprivation of the above-ground parts, and ultimately causing stunting.

Nematodes are covered with an impermeable cuticle, which provides them with considerable protection. Chemicals with outstanding penetration characteristics are therefore required for their control.

Nematicides are seldom used by homeowners except in a greenhouse or cold frames. For the most part, nematicides are not and should not be used by the layperson, mainly because of their hazard. Those that are available commercially fall into four groups: (1) halogenated hydrocarbons; (2) isothiocyanates; (3) organophosphate insecticides; and (4) carbamate or oxime insecticides.

Most of today's nematicides are soil fumigants, volatile halogenated hydrocarbons, that is, hydrocarbons that have halogens (chlorine, bromine, and fluorine) replacing some of their hydrogen atoms. To be successful, they must have a high vapor pressure, to spread through the soil and to contact nematodes in the water films surrounding soil particles.

DICHLOROPROPENE-DICHLOROPROPANE (DD®)

$HC(Cl){=}CH{-}CH_2Cl$ and $H_2C(Cl){-}CH(Cl){-}CH_3$

1,3-dichloropropene and 1,2-dichloropropane

EDB

$Br{-}CH_2{-}CH_2{-}Br$

ethylenedibromide

HALOGENATED HYDROCARBONS

The nematicidal properties of DD and EDB were discovered in 1943 and 1945 and effectively launched the use of volatile nematicides on a field-scale basis. Previously only seedbeds, greenhouse beds, and potting soil had been treated, with materials such as chloropicrin, carbon disulfide, and formaldehyde. These were very expensive, in some instances explosive, and usually required a surface seal, because of their relatively high vapor pressures.

DD and EDB are both injected into the soil several days before planting, to kill nematodes, eggs, and soil insects. Beyond these generalizations, the two soil fumigants differ considerably in human hazard, soil retention, and uses.

DBCP was one of the easier-to-use nematicides because it was formulated as both an emulsifiable concentrate and as granules. The only truly effective nematicide that could be used near living plants, it was used for postplant treatment and was registered for use on numerous fruit and vegetable crops. However, in 1977 its manufacturers announced that DBCP was found to cause sterility in male workers involved in its manufacture. Subsequently it was removed from the market, and all its registrations, except for pineapple grown in Hawaii, were canceled by the EPA in 1979. An extensive search is now underway by nematologists to find a replacement, for DBCP was virtually indispensable in grape and citrus production.

Methyl bromide is a first-class, all-around fumigant, lethal to all plant and animal life, hence classified as a sterilant. The gas is a soil fumigant for control of weeds, weed seeds, nematodes, insects, and soil-borne diseases. It is also used for dry-wood termite control above ground, agricultural commodity fumigation, and rodent control. As a nematicide, methyl bromide must be used as a preplant application because of its phytotoxicity, followed by adequate aeration time. A two-week waiting period after fumigation is an acceptable rule of thumb.

Tetrachlorothiophene was used as a preplant treatment of tobacco beds for control of root knot, meadow stunt, and dagger species of nematodes. Its manufacture has been discontinued.

Most of these halogenated hydrocarbon nematicides have 1 to 2 percent chloropicrin added to serve as a warning agent. Their mode of action is that of a narcotic fumigant on nematodes. They are liposoluble and, as such, lodge in the primitive nervous systems of nematodes and kill primarily through physical rather than chemical action.

DBCP

$CH_2Br{-}CHBr{-}CH_2Cl$

dibromochloropropane

METHYL BROMIDE

CH_3Br

bromomethane

TETRACHLOROTHIOPHENE (Penphene®)

Cl Cl S Cl Cl

2,3,4,5-tetrachlorothiophene

ISOTHIOCYANATES

Isothiocyanates are chemicals that contain the radical —N═C═S, which inactivates sulfhydryl groups in amino acids. Three nematicides are classified as isothiocyanates: metam-sodium, Vorlex®, and dazomet. Metam-sodium is a dithiocarbamate, to be mentioned with the fungicides in Chapter 14, but it is readily converted to an isothiocyanate, and is active against all living matter in the soil. It is highly effective in control of weeds and weed seeds, nematodes, and soil fungi.

Vorlex® is applied as a preplant soil fumigant and controls weeds, fungi, insects, and nematodes. It is also discussed in Chapter 14.

Dazomet is a diazine, slightly resembling the thiazoles (see Chapter 14) and undergoing a similar ring cleavage in the soil to produce

METAM-SODIUM (SMDC) (Vapam®)

$CH_3{-}NH{-}C({=}S){-}S{-}Na \cdot 2H_2O$

sodium *N*-methyldithiocarbamate dihydrate

Vorlex®

$CH_3{-}N{=}C{=}S$

methylisothiocyanate

DAZOMET (DMTT)

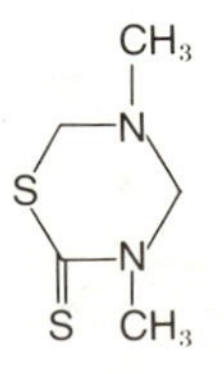

tetrahydro-3,5-dimethyl-2*H*,1,3,5-thiadiazine-2-thione

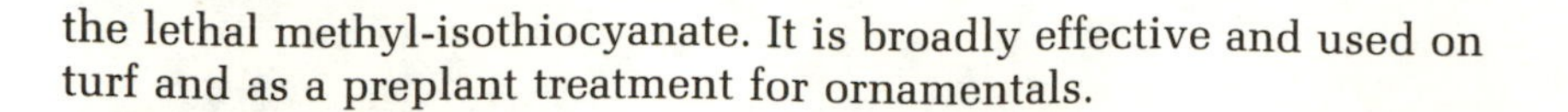
the lethal methyl-isothiocyanate. It is broadly effective and used on turf and as a preplant treatment for ornamentals.

ORGANOPHOSPHATES

Nematodes have a nervous system similar to that of insects, although they are more primitive, and are thus susceptible to the action of the organophosphate insecticides. Unfortunately, most of the organophosphates are rapidly degraded in the soil and are effective for but a short time. These do not necessarily have to be systemic in their action, but may depend on vapor pressure or water solubility to reach the target pests. Only a few of the organophosphates serve as both nematicides and insecticides, while a few are used only for their nematicidal action. All these organophosphates inhibit the neurotransmitter enzyme cholinesterase, resulting in the paralysis and ultimately the death of affected nematodes.

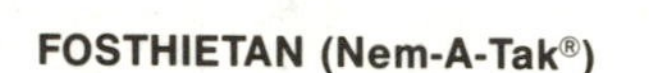
FOSTHIETAN (Nem-A-Tak®)

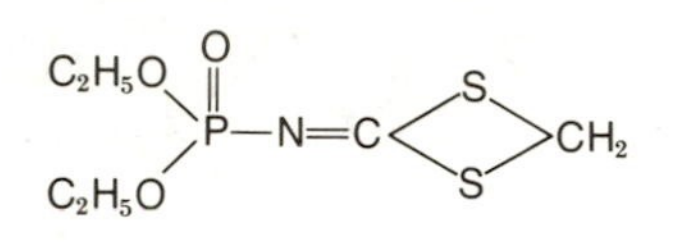

diethyl 1,3-dithietan-2-ylidenephosphoramidate

FENAMIPHOS (Nemacur®)

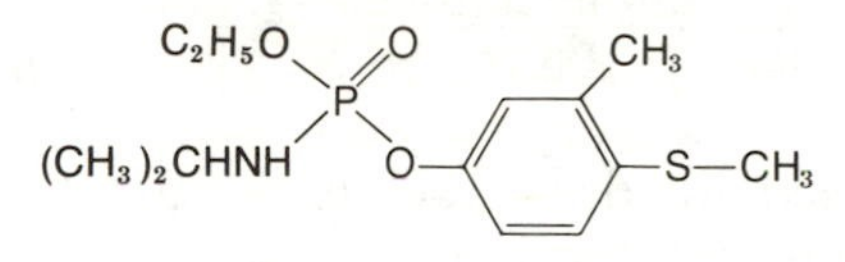

ethyl 3-methyl-4-(methylthio)-phenyl (1-methylethyl)phosphoramidate

FENSULFOTHION (Dasanit®)

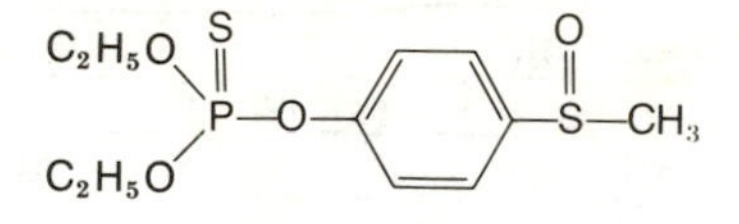

O,O-diethyl-*O*-[*p*-(methylsulfinyl)phenyl] phosphorothioate

PHORATE (Thimet®)

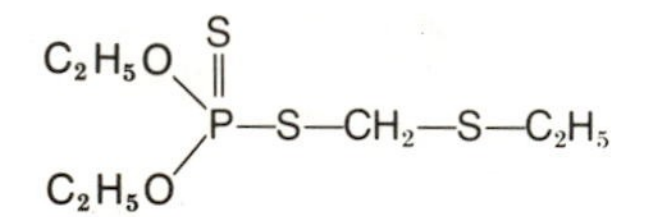

O,O-diethyl *S*-(ethylthio)methyl phosphorodithioate

DISULFOTON (Disyston®)

O,O-diethyl *S*[2-(ethylthio)ethyl] phosphorodithioate

ETHOPROP (Mocap®)

O-ethyl *S,S*-dipropyl phosphorodithioate

Some of the organophosphate nematicides that are no longer manufactured are dichlofenthion (Nemacide®, Mobilawn®), thionazin (Zinophos®), and diamidfos (Nellite®).

CARBAMATES

ALDICARB (Temik®)

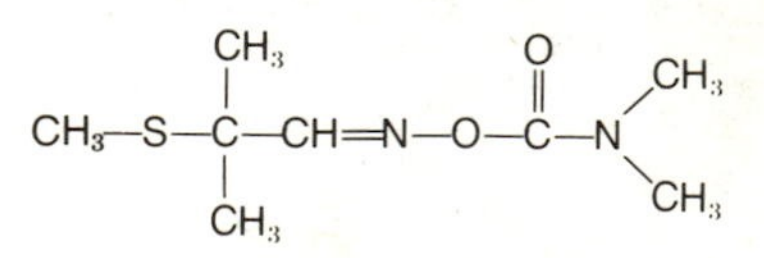

2-methyl-2-(methylthio)propionaldehyde *O*-(methylcarbamoyl) oxime

The carbamate aldicarb, mentioned in Chapter 4 as a systemic insecticide, is the only example of an oxime nematicide. It is now registered for several crops, including cotton, potatoes, sugar beets, oranges, pecans, peanuts, sweet potatoes, and ornamentals, to name a few. It shows great promise for additional crops.

Unlike most other nematicides, aldicarb is formulated only as a granular material because of the high toxicity of the parent compound. Granular formulation significantly reduces the handling hazard. It is drilled into the soil at planting or when the plants are in various stages of growth. It becomes water soluble and is absorbed by the roots and translocated throughout the plant, killing insects that pierce and suck foliage as well as nematodes in and around the roots.

Carbofuran is a most promising systemic carbamate insecticide-nematicide. It is currently registered as a nematicide for alfalfa, tobacco, peanuts, and sugarcane, and looks promising on soybeans, cotton, grapes, and grains. It has a relatively short residual life, making it useful on forage and vegetable crops.

Oxamyl is a carbamate insecticide, nematicide, and acaricide that is useful on many field crops, vegetables, fruits, and ornamentals. Its total solubility in water accounts for its effectiveness as a nematicide. Another carbamate insecticide-nematicide, carbosulfan (Advantage®), is still in the experimental stage and is described in Chapter 24.

With nematicides in general, the residue problem is relatively serious, and the EPA is examining closely the chlorinated and bromonated fumigants for subtle, long-range health effects. Their use will undoubtedly decline as long-term feeding studies on laboratory animals reveal untoward results.

Consequently, the nonvolatile carbamates and organophosphates, with their short residual effects, will become the nematicides of the future.

CARBOFURAN (Furadan®)

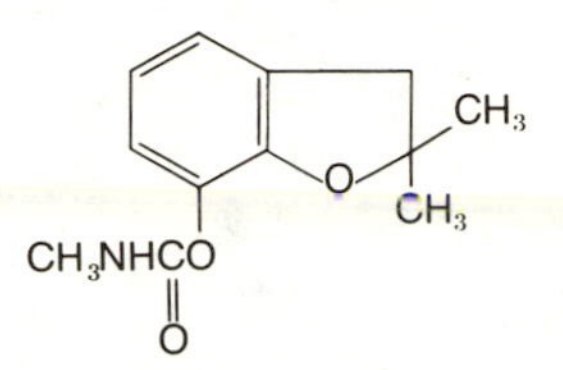

2,3-dihydro-2,2-dimethyl-7-benzofuranyl methylcarbamate

OXAMYL (Vydate®)

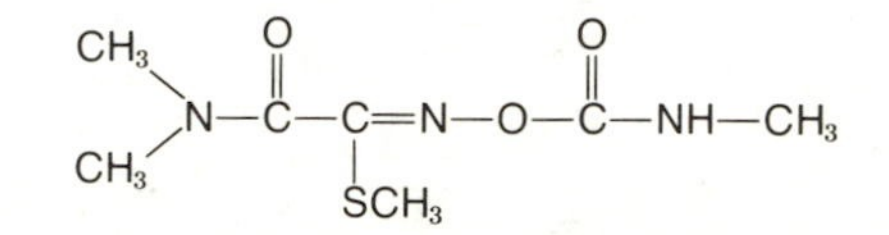

methyl N′,*N*′-dimethyl-*N*-[(methyl carbamoyl)oxy]-1-thiooxamimidate

CHEMICALS USED IN THE CONTROL OF VERTEBRATES—ANIMALS WITH BACKBONES

CHAPTER 7

Rodenticides

> Rats!
> They fought the dogs and killed the cats,
> And bit the babies in the cradles,
> And ate the cheeses out of the vats,
> And licked the soup from the cooks'
> own ladles.
>
> Robert Browning, *The Pied Piper of Hamelin*

Several small mammals, especially rodents, damage human dwellings, stored products, and cultivated crops. Among these are native rats and mice, squirrels, woodchucks, pocket gophers, hares, and rabbits. Rats are notorious freeloaders, and, in some of the underdeveloped countries, where it is necessary to store grain in the open, as much as 20 percent may be consumed by rats before people can do so.

About one-half of all mammalian species are rodent species (order Rodentia). Because they are so very highly prolific and widespread, they are continuously competing with humans for food. Most of the methods used to control rodents are aimed at destroying them. Poisoning, shooting, trapping, and fumigation are among the methods selected. Of these, poisoning is most widely used and is probably the most effective and economical. Because rodent control is in itself a diverse and complicated subject, here we mention only those more commonly used rodenticides.

Rodenticides differ widely in their chemical nature. Strange to say, they also differ widely in the hazard they present under practical conditions, even though all of them are used to kill animals that are physiologically similar to humans.

COUMARINS (ANTICOAGULANTS)

The most successful group of rodenticides are the coumarins, represented classically by warfarin. Seven compounds belong to this classification, all of which have been very successful rodenticides. Their mode of action is twofold: (1) inhibition of prothrombin formation, the material in blood responsible for clotting; and (2) capillary damage, resulting in internal bleeding. The earlier coumarins require repeated ingestion over a period of several days, leaving the unsuspecting rodents growing weaker daily. The earlier coumarins are thus considered relatively safe, since repeated accidental ingestion would be required to produce serious illness. In the case of most other rodenticides, a single accidental ingestion could be fatal. Of the several types of rodenticides available, only those with anticoag-

DICUMAROL

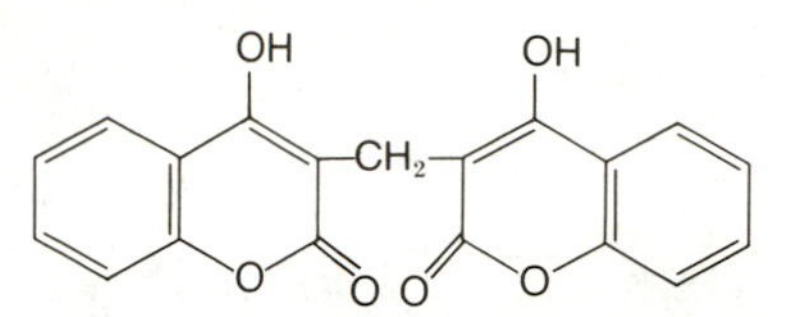

3,3′-methylene bis (4-hydroxycoumarin)

WARFARIN

3-(α′-acetonylbenzyl)-4-hydroxycoumarin

COUMACHLOR

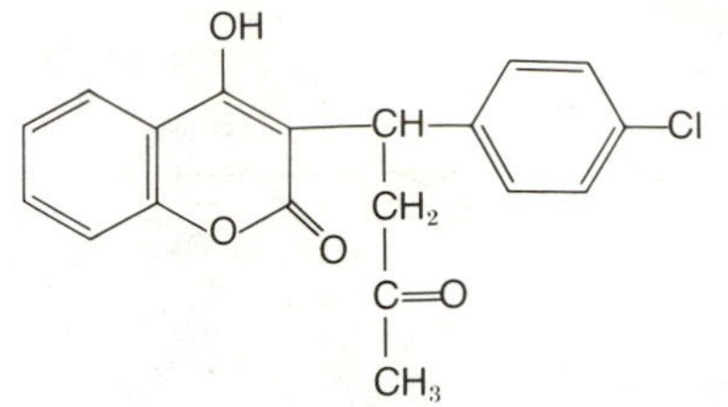

3-(α-acetonyl-4-chlorobenzyl)-4-hydroxycoumarin

COUMATETRALYL

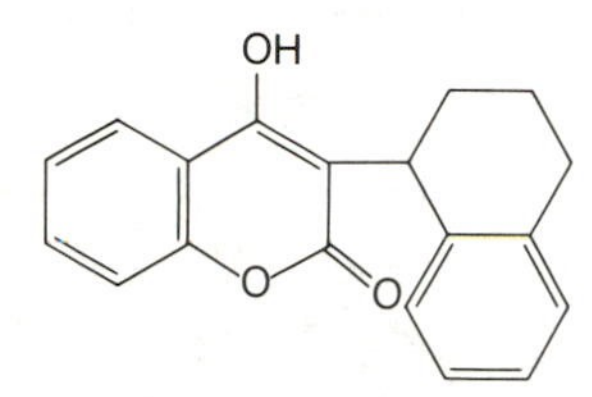

3-(d-tetralyl)-4-hydroxycoumarin

COUMAFURYL

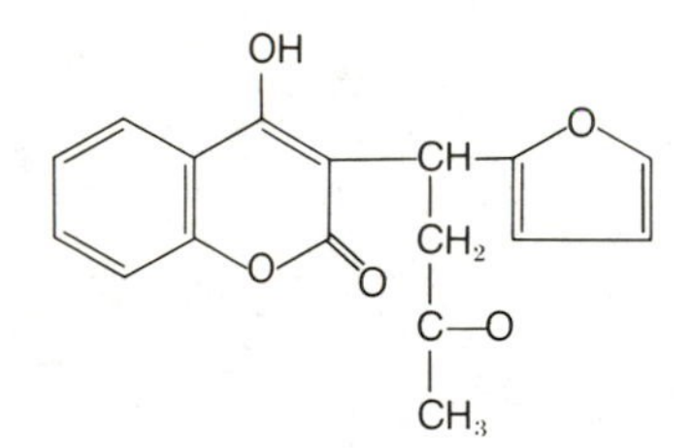

3-(α-acetonylfurfuryl)-4-hydroxycoumarin

ulant properties are safe to use around the home. Vitamin K_1 is the antidote for accidental poisoning by these and all other anticoagulant rodenticides.

The first coumarin was dicumarol, introduced in 1948 after the molecule was identified as the compound responsible for sweet clover's toxicity to cattle. As a rodenticide, however, dicumarol has been superseded by warfarin. Warfarin was released in 1950 by Wisconsin Alumni Research Foundation (thus its name *WARF coumarin*, or *WARFarin*). It was immediately successful as a rat poison, because rats did not develop "bait shyness," as they did with other baits during the required ingestion period of several days. Coumachlor was introduced in 1953, but has never been successful in the United States because of warfarin's wide acceptance. Coumatetralyl was developed in Germany and introduced in the United States in 1957 with a fair degree of success in situations where warfarin had resulted in bait shyness of rodents. The last of the early coumarins is coumafuryl, a commonly used material.

Two new coumarin rodenticides have appeared in the past three years, brodifacoum (Talon®) and bromadiolone (Maki®). These differ from the earlier coumarins in that, although they are anticoagulant in their mode of action, they require but a single feeding for rodent death to occur within 4 to 7 days. They are both effective against rodents that are resistant to conventional anticoagulants.

To protect children and pets from accidental ingestion of these highly toxic anticoagulants, the treated baits must be placed in tamper-proof bait boxes or in locations not accessible to children.

These rodenticides are relatively specific for rodents in that they are offered as rodent-attractive, premixed baits inaccessible to pets

BRODIFACOUM (Talon®)

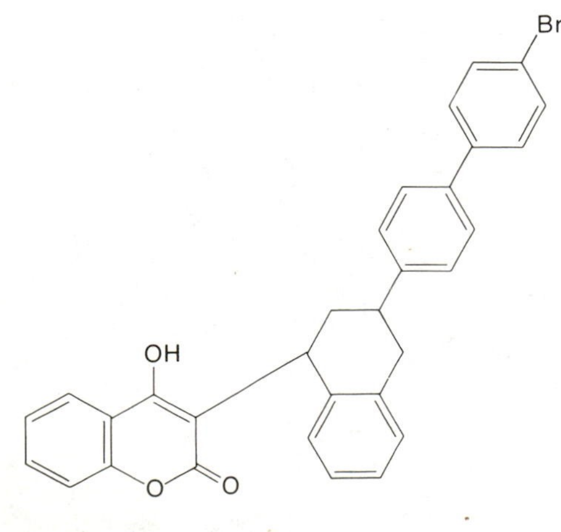

3-[3-[4′-bromo(1-1′-biphenyl)-4-yl]-
1,2,3,4-tetrahydro-1-naphthalenyl]-4-
hydroxy-2*H*-1-benzopyran-2-one

BROMADIOLONE (Maki®)

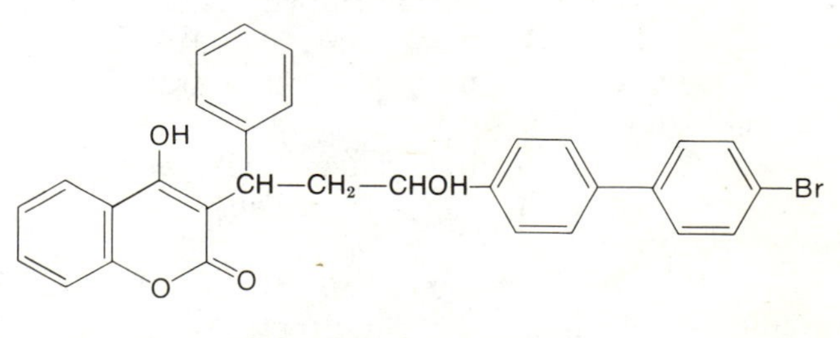

3-[3-[4′-bromo(1,1′-biphenyl)-4-yl]-
3-hydroxy-1-phenylpropyl]-4-hydroxy-
2*H*-1-benzopyran-2-one

TABLE 7-1
Grams of rodenticide baits that must be eaten by animals to equal their LD_{50}.

Species (body weight, kg)	*Zinc phosphide 2.5%*	*Diphacinone 0.005%*	*Vacor® 2.0%*	*Warfarin 0.025%*	*Talon® 0.005%*	*Maki® 0.005%*
Rat (0.25)	0.45	16–25.5	0.06	58	1.4	
Mouse (0.025)	—[a]	70	0.12	37	0.2–0.43	
Rabbit (1)	—[a]	700	15.0	3200	5.8	
Pig (50)	40–80	150,000	1250	200–1000	500–2000	
Dog (5)	4–8	88	125	400–5000	25–100	1500–2000
Cat (2)	1.6–3.2	588	6.2	48–320	1000	1000
Chicken (1)	0.8 1.2	—[a]	36	4000	200–2000	28[b]

[a] LD_{50} for the animal is not available.
[b] Consumption of 28 g/day for 20 consecutive days.
Source: National Pest Control Association, Inc. Technical release, *Talon®, a New Rodenticide,* Nov. 21, 1979.

and domestic animals, and their relative toxicity to other warm-blooded animals is low. Table 7-1 shows the actual amounts of treated baits that would have to be consumed in one feeding by different species of animals to equal their LD_{50}. For instance, the rat is 2 times more sensitive to brodifacoum than the pig and dog, 36 times more than the chicken, and 90 times more sensitive than the cat. To read the table, divide the number of grams that must be eaten to equal the LD_{50} by the weight of the animal. For instance, with the rat, $1.4 \div 0.25 = 5.6$, the LD_{50} in grams of food per kilogram of rat body weight. With the cat, $1000 \div 2 = 500$, the LD_{50} in grams of food for the cat. Divide 500 by 5.6 equals 90, or the number of times less sensitive the cat is to the rodenticide than the rat.

INDANDIONES (ANTICOAGULANTS)

The three compounds pindone, diphacinone, and chlorophacinone belong to the indandione class of anticoagulants, which differs chemically from the coumarin anticoagulants. Pindone appeared in 1942, truly the first anticoagulant rodenticide. As with the early coumarins, daily feedings by rodents are necessary to produce death. Diphacinone became available in the early 1950s, and was the first single-dose anticoagulant. In rare instances rodents will die after feeding on bait only once. However, in most circumstances two to three feedings are necessary, with death occurring in five to seven days.

The most recent addition to this group, in 1961, is chlorophacinone, which does indeed require but a single dose of a bait containing 50 mg/kg, killing rats from the fifth day. It does not induce bait shyness as do pindone and diphacinone. In addition to its anticoagulant action it also uncouples oxidative phosphorylation, partially explaining its success as a single-dose rodenticide. Again, vitamin K_1 is the antidote for accidental poisoning by these and all other anticoagulant rodenticides.

PINDONE (Pival®)

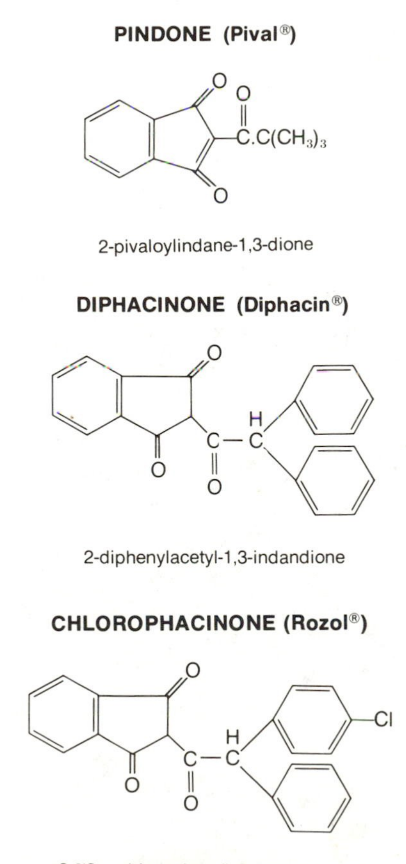

2-pivaloylindane-1,3-dione

DIPHACINONE (Diphacin®)

2-diphenylacetyl-1,3-indandione

CHLOROPHACINONE (Rozol®)

2-[(2-*p*-chlorophenyl)-2-phenylacetyl]-1,3-indandione

RED SQUILL (Scilliroside)

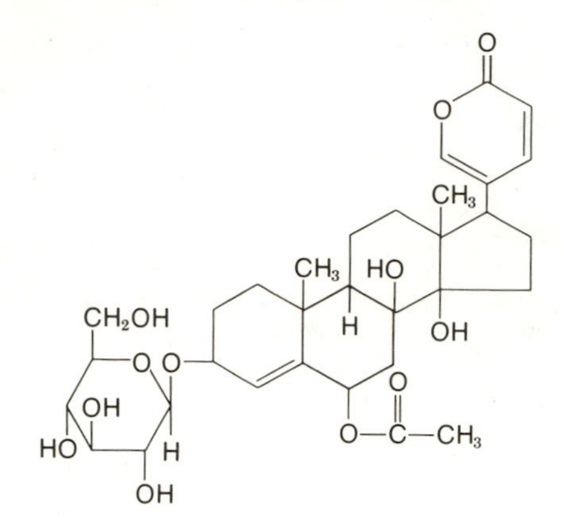

STRYCHNINE

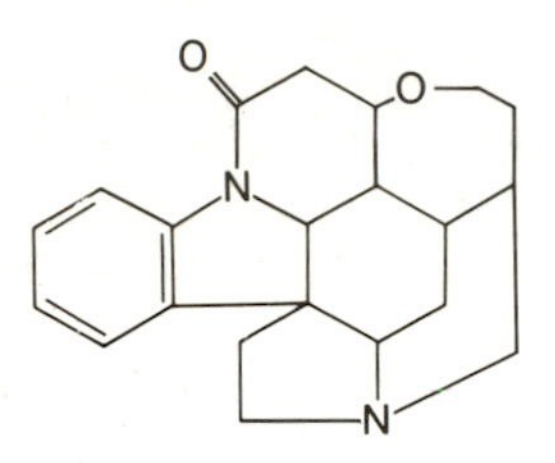

BOTANICALS

Red squill, which comes from the powdered bulbs of a plant, Mediterranean squill, was used before 1935, but was never more than a mediocre rodenticide. The active ingredient is scilliroside, classed as a cardiac glycoside. Its specific activity is due to the inability of rats to vomit—thus they must absorb the toxicant. Other animals ingesting squill do vomit, which permits them to survive accidental poisoning.

Strychnine is an alkaloid, from the Asiatic tree *Strychnos nux-vomica*, that is usually converted to strychnine sulfate for use as a rodenticide. Strychnine is highly toxic to all warm-blooded animals and acts by paralyzing specific muscles, resulting in cessation of breathing and heart action. It is only a fair rodenticide and has been replaced by the anticoagulants.

ORGANOCHLORINES

DDT 50 percent dust was used for years by structural pest control operators as a tracking powder. The dust was sprinkled in the known runs of mice, and, after tracking through the dust, the mice stopped to preen themselves and clean their feet. Death resulted from convulsions and paralysis, just as in insects. DDT was exceptionally effective against bats.

Similarly, endrin was sprayed on orchard soils and fruit tree trunks in rather heavy concentration during the fall or winter months. Field mice eating the bark of fruit trees and trailing across treated soil quickly ingested lethal quantities. Although quite effective, neither practice is acceptable today, because of persistence and hazard to other species. (Structures are given in Chapter 4, with the organochlorine insecticides.)

PHOSPHORUS

Phosphorus occurs in two common forms, the relatively harmless red and the highly toxic white or yellow phosphorus. (The striking surface of a safety match contains 50 percent red phosphorus.) Phosphorus use is almost nonexistent today, having been replaced by the anticoagulants. Yellow phosphorus attacks the liver, kidney, and heart, resulting in rapid tissue disintegration; it also causes rats to attempt to vomit, a function that, uniquely, they cannot perform.

Zinc phosphide (Zn_3P_2) is an intense poison to mammals and birds and is used against rats, mice, ground squirrels, prairie dogs, and gophers. It has an unpleasant, garliclike odor, which is evidently not offensive to rodents. Its mode of action is similar to that of phosphorus.

ORGANOPHOSPHATES

Phorazetim (Gophacide®) was a very successful rodenticide, developed in Germany, and eventually accepted in the United States. It gave excellent control of pocket gophors, rats, and mice after a 24-hour exposure to baits. It is an organophosphate with a high degree of cholinesterase inhibition as its mode of action. Its production has been discontinued.

PHORAZETIM (Gophacide®)

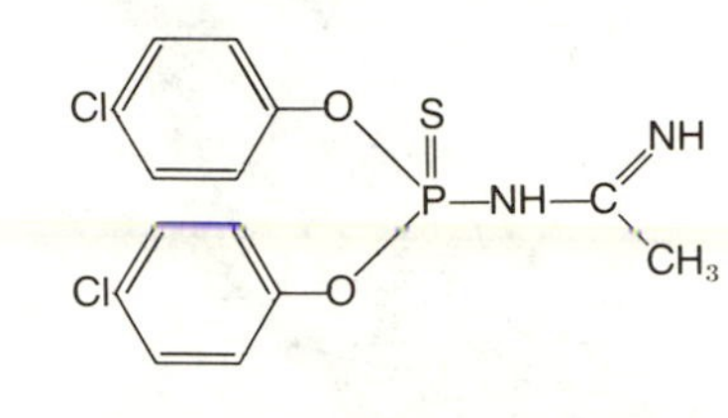

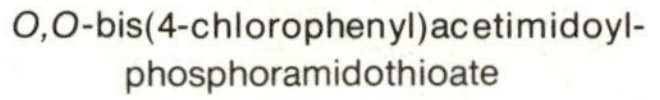
O,O-bis(4-chlorophenyl)acetimidoyl-phosphoramidothioate

MISCELLANEOUS RODENTICIDES

Compound 1080, sodium fluoroacetate, one of the most toxic poisons known to warm-blooded animals, was introduced in 1947. Its use is now restricted to authorized trained personnel because of its extreme hazard to humans and domestic animals. Sodium fluoroacetate has a strong effect on both the heart and nervous system, resulting in convulsions, paralysis, and death. More recently it has made the news in coyote control programs in the western United States.

Compound 1081 is a moderately fast-acting rodenticide closely related to sodium fluoroacetate. It possesses a lower mammalian toxicity and a longer latent period before animals become distressed and stop feeding. Its use is less likely to lead to poison shyness because of sublethal dosing.

Antu® is an acronymic form of the chemical nomenclature given it when introduced in 1946. Because rodents quickly develop a tolerance to Antu®, it has been displaced by other materials.

Thallium sulfate ($TlSO_4$) is an old and reliable rodenticide that has also been replaced by the safer anticoagulants, for the same general reasons. It is a general cellular toxin that resembles arsenic in its effects and attacks or inhibits enzymes other than those containing —SH groups.

Pyriminil (Vacor®) was introduced in 1977 as an acute toxicant that killed rodents in one feeding. It was effective against rats that are resistant to warfarin, which kills over a period of several days by causing internal hemorrhaging. Pyriminil killed in four to eight hours after ingestion of a single dose as small as 0.5 g of the 2 percent bait. Its mode of action is to inhibit niacinamide metabolism, and the rodents die from paralysis and pulmonary arrest. It was removed from the market voluntarily by its manufacturer in 1978 following several near-fatal human exposures to the toxicant. Apparently, pyriminil was surprisingly toxic to humans, a fact not revealed in the original animal toxicity studies.

SODIUM FLUOROACETATE (1080®)

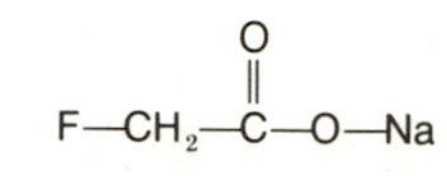

FLUOROACETAMIDE (1081, Fluorakil 100®)

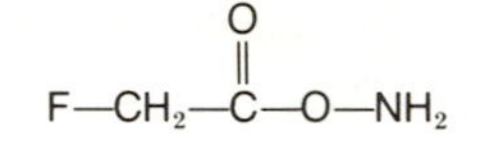

Antu®

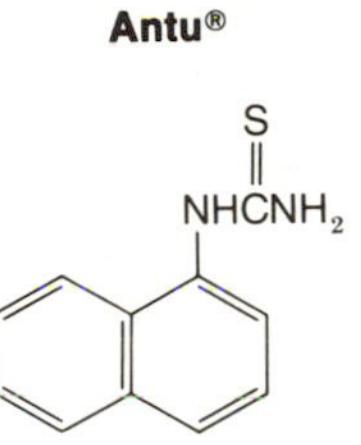

α-naphthylthiourea

PYRIMINIL (Vacor®)

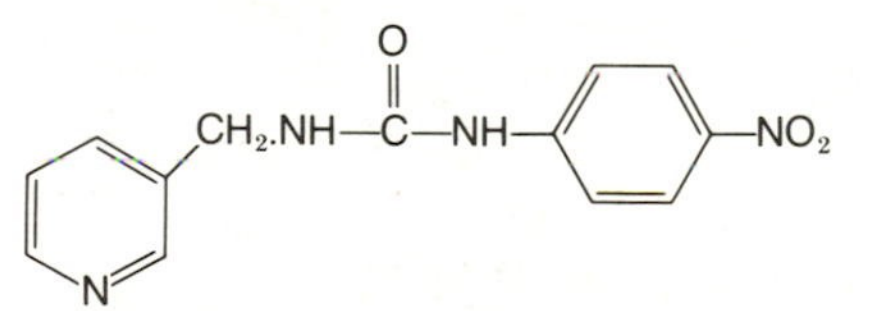

N-3-pyridylmethyl-*N'*-*p*-nitrophenyl urea

CHAPTER 8

Avicides

Sing a song of sixpence,
A pocket full of rye,
Four and twenty blackbirds,
Baked in a pie.
Old English Nursery Rhyme

All birds can, in one way or another, at times be beneficial to humans. They provide enjoyment and wholesome recreation for most of us regardless of where we live. Despite the fact that wild bird populations are for the most part beneficial, there are occasions when individuals of certain species can seriously compete with human interests. When these situations occur, control measures are necessary.

These beautiful winged creatures create pest problems singly or in small groups, but especially when in large aggregations. Most of the areas of conflict with humans are: (1) destruction of agricultural foodstuffs and predation; (2) contamination of foodstuffs or defacing of buildings with their feces; (3) transmission of diseases, directly and indirectly, to man, poultry, and dairy animals; (4) hazards at airports and freeways; and (5) as a general nuisance that affects human comfort.

Of the various methods of control, we will mention repellents, perch treatments (sticky chemicals on ledges and roosts), toxic baits, soporifics (stupefacients), and chemosterilants.

REPELLENTS

Avitrol®

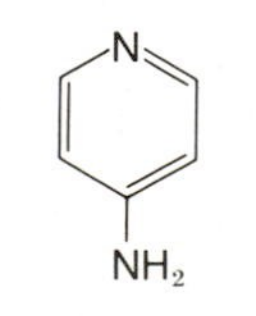

4-aminopyridine

Probably the most prominent of the avicides is Avitrol® (4-aminopyridine), which is used as a repellent; the effects result from the distress calls made by affected birds. The material has a relatively low LD_{50} of 20 mg/kg, thus mortality in some individuals is inevitable (LD_{50} is the dose lethal to 50 percent of organisms tested). It is to be used only by licensed pest control operators for driving away flocks of nuisance or feed-consuming birds from feedlots, fields, airports, warehouse premises, public buildings, and grain-processing plants. Only a small number of birds need be affected to alarm the rest of the flock. The treated grain must be eaten to be effective. After one alarming exposure, birds will not return to treated areas.

Methiocarb (Mesurol®), a carbamate insecticide, is registered for use as a bird repellent on cherries and blueberries. In other countries it is used as a repellent on corn, sorghum, rice, and grapes. (See also Chapter 10.)

PERCH TREATMENTS

The chlorinated cyclodiene insecticide endrin has been used for years as an elevated, out-of-reach perch treatment for the control of pigeons, starlings, and English sparrows. It is quite effective, if used properly, and is available only to licensed pest control operators trained in bird control.

Another effective perch treatment is the organophosphate insecticide fenthion (Entex®), used to control pigeons, starlings, and English sparrows. This compound acts quickly after absorption through the feet of perching birds. Fenthion is for use only by licensed pest control operators and trained personnel of industry and government. Queletox® is a special formulation for control of weaver birds in Africa.

TOXIC BAITS

Starlicide® is a chlorinated compound used as a slow-acting avicide against starlings and blackbirds. It is not effective against house sparrows. Since the material kills slowly, requiring from 1 to 3 days, large numbers of dead birds do not appear in the treated area but rather die in flight or at their roosts. It is formulated as 0.1 and 1.0 percent baits and is available only to pest control operators trained in bird control.

The old, general-purpose poison strychnine is registered by the EPA as an avicide for the control of pigeons and English sparrows when used as a 0.6 percent grain bait. Strychnine acts very quickly, leaving the treated area strewn with dead birds, which should be removed at regular intervals. Prebaiting for several days is necessary before distributing the treated bait. For bird protection, treated grain should be colored before distribution, and uneaten bait should be removed.

SOPORIFIC

Glucochloralose (Alfamat®) is registered as a bird repellent and hypnotic, leaving some of the affected pests stunned and unable to fly for several hours.

CHEMOSTERILANT

Ornitrol® (SC-12937) is a new and apparently humane approach to bird control. It is a chemosterilant, a birth control agent for pigeons, designed to control population growth rather than to eradicate. It causes temporary sterility in pigeons after a 10-day feeding exposure by inhibiting egg production. It has little, if any, effect on mammals and is selective of pigeons when impregnated on whole corn grains too large for other species to feed on. It is formulated as a 0.1 percent bait.

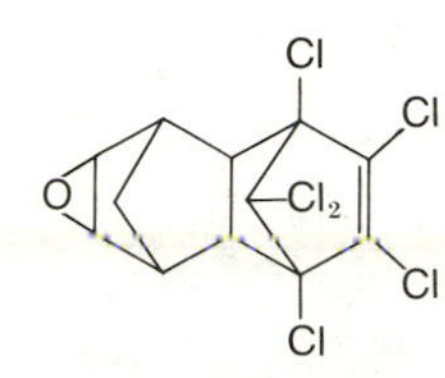

1,2,3,4,10,10-hexachloro-6,7-epoxy-1,4,4*a*,5,6,7,8,8*a*-octahydro-1,4-*endo*-*endo*-5,8-dimethanonaphthalene

FENTHION (Entex®)

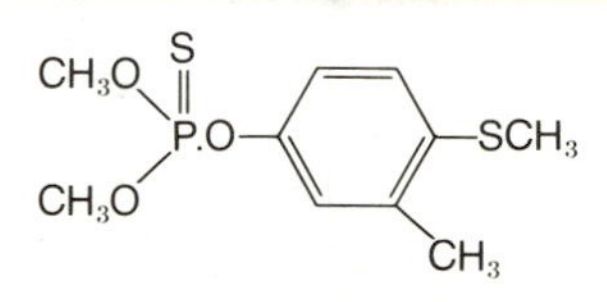

O,O-dimethyl *O*-[(4-methylthio)-*m*-tolyl] phosphorothioate

Starlicide®, DRC 1339

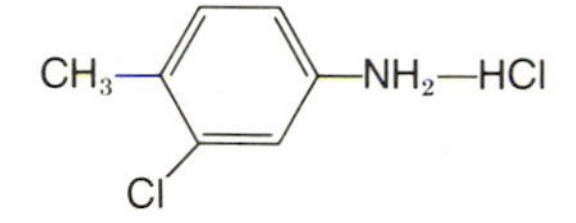

3-chloro-*p*-toludine hydrochloride

STRYCHNINE

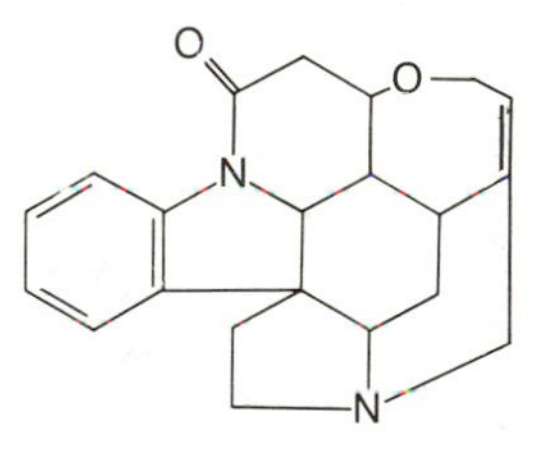

Ornitrol®, SC-12937

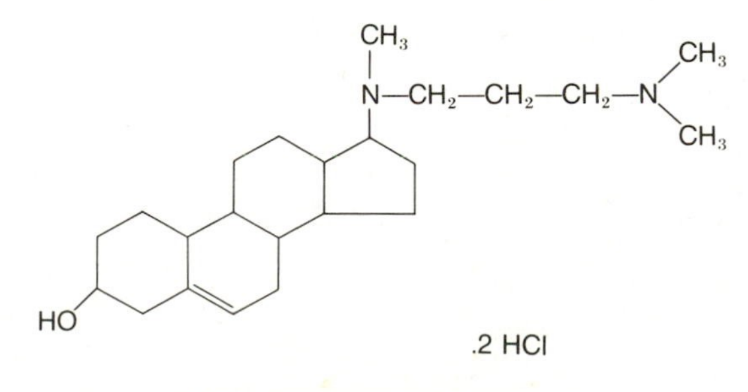

20,25-diazacholesterol, hydrochloride

Third Fisherman: Master, I marvel how the fishes live in the sea.

First Fisherman: Why, as men do aland; the great ones eat up the little ones.

William Shakespeare, *Pericles*

Piscicides

Piscicides are a small, heterogeneous assortment of chemicals that are rather nonspecific for fish. They are used to remove all fish from a body of water, for restocking with desirable game fish after some safe waiting interval.

Most pest fish conflicts involve undesirable species or rough or trash fish competing with more desirable game fish. There are also aesthetic problems: Fish may muddy the water or create odor problems as ponds and agricultural ditches dry up, and fish are even known to choke municipal water intake pipes. Introduced species, such as minnows, carp, suckers, crappie, catfish, and bass, can on occasion create rough fish–sport fish conflicts, frequently requiring the poisoning of ponds, lakes, and streams by fisheries personnel and sportsmen, so that more desirable species can be established.

Pest fish species include the destructive sea lamprey, parasitic on commercial fish species, which gained access to the upper Great Lakes through the Welland Ship Canal built in 1829 to bypass Niagara Falls; the alewife, a native of the North Atlantic, which entered the Great Lakes in the same way as the sea lamprey; wild goldfish, members of the carp family, established as a result of the home-aquarist and bait-fish trades; the walking catfish, which escaped from a tropical fish dealer in Florida; and the carp, introduced into the United States more than 100 years ago from Germany. These are just a few examples of fish that, for one reason or another, must be removed by poisoning. Unfortunately, with the exception of the lamprey eel, the piscicides are nonspecific with regard to fish.

More than 25 years ago, formal investigations were initiated to find and perfect control chemicals specific to fish. Previous to that, our fish poisons were borrowed from the agricultural insecticides and used without regard for the unique complexities of aquatic environments.

CHLORINATED INSECTICIDES

Many of the organochlorine insecticides were, early in their use history, found to be extremely toxic to fish and were in a few cases used as piscicides. Among these were toxaphene, beginning in 1948, endrin in 1952, and endosulfan (Thiodan®) in 1960. Their use was usually catastrophic, since they are toxic to all forms of wildlife and very persistent. Consequently, none of the three were ever registered for use. (Endrin gained a bad reputation in the Mississippi River fish kills of the mid 1960s, as did endosulfan in the accidental fish kill in the Rhine River in 1969.) Needless to say, none of the organochlorine insecticides should be used as piscicides.

ROTENONE

Rotenone is the most useful piscicide available for reclaiming lakes for game fishing. It eliminates all fish, closing the lake to reintroduction of rough species. After treatment, the lake can be restocked with the desired species. Rotenone is a selective piscicide, in that it kills all fish at dosages that are relatively nontoxic to fish food organisms. It also breaks down quickly, leaving no residues harmful to fish used for restocking. The recommended rate is 0.5 ppm or 5.1 kg per hectare-meter of water (1.36 pounds per acre-foot).

Rotenone is a strong inhibitor of the respiratory chain in fish, and the site of action is located in the flavoprotein region of this chain. The specialized structure of the gills favors passage of rotenone into the blood, which is then transported to vital organs for respiration inhibition.

ROTENONE

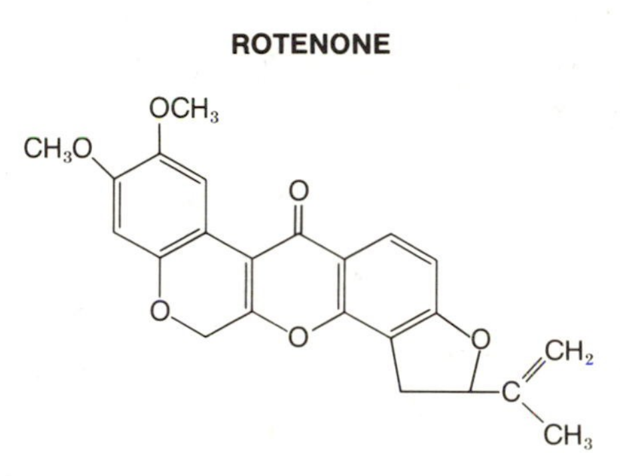

ANTIMYCIN®

Antimycin® is actually an antibiotic, produced by the fungus *Streptomyces*, whose fish poison characteristics were discovered in 1963. In the dosages used, it is specific for fish, leaving unharmed other aquatic life, waterfowl, and mammals. It is lethal in small concentrations to all stages of fish, egg through adult. It passes into the blood through the gills and irreversibly blocks cellular respiration at the cytochromes, thus inhibiting oxidative phosphorylation. Because it is not repellent to fish, it is the first piscicide to be successfully used as spot treatments in lakes. It degrades rapidly in water, usually within a few days, and even faster if the pH of the water is high. Antimycin®, also sold under the name of Fintrol®, is formulated as a 1 and 5 percent wettable powder and as a 10 percent solution, and must be used under technical supervision of state and federal fish and game agencies.

CLONITRALID

CLONITRALID (Bayluscide®)

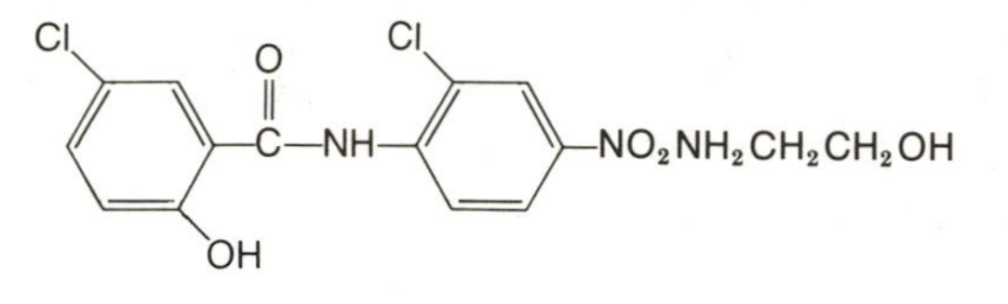

2′,5-dichloro-4′-nitrosalicylanilide,
2-aminoethanol salt

Clonitralid (Bayluscide®) is registered for the control of larval stages of the sea lamprey but is also outstanding against other fish and as a molluscicide. It is used primarily in combination with TFN (see next paragraph) to kill sea lamprey larvae in tributary streams of the Great Lakes. The liquid toxicant mixture is metered with precision into streams to kill lamprey residing in bottom muds, resulting in little harm to other aquatic organisms.

TFN

TFN (Lamprecide®)

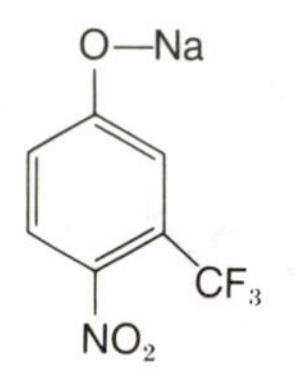

3-trifluoromethyl-4-nitrophenol, sodium salt

TFN (Lamprecide®) is a selective toxicant against the larval stages of the sea lamprey and resulted from research by biologists of the U.S. Fish and Wildlife Service. The later discovery of synergistic action between it and clonitralid (see preceding paragraph) gave the Great Lakes Fishery Commission a highly effective piscicide for eliminating most of the larval stages of the lamprey harbored in hundreds of miles of tributary streams.

Despite the successes of reducing lamprey populations in Lake Superior especially, the lamprey continues to cause great losses to the Great Lakes fishing industry. In large bodies of water, even when the target pest is confined to tributary streams for breeding, the size of this breeding ground may be too large for chemical control.

Repellents

> Do what we can, summer will have its flies. If we walk in the woods, we must feed mosquitoes.
>
> Ralph Waldo Emerson, *Essays*

Most of us think of repellents strictly in the sense of driving away mosquitoes, biting flies, and gnats—insect repellents, which are discussed in Chapter 4. These are the best known. There are, however, repellents for birds, dogs, deer, moles, rabbits, and rodents. Have you ever wished for some way to prevent the dog next door from defecating on your lawn or urinating on your shrubs? Well, perhaps you may find a solution here.

Animal repellents have been used for centuries, but with little success. One of the oldest known general repellents still in use is the plant resin asafetida, used as a medicinal and hung around the neck by the superstitious to ward off contagious diseases. In this respect, it may very well have been effective. It is so foul smelling that it undoubtedly serves also as a human repellent.

BIRD REPELLENTS

Bird repellents can be divided into three categories: (1) olfactory (odor), (2) tactile (touch), and (3) gustatory (taste). In the first category, only naphthalene granules or flakes are registered by the EPA. It should come as no surprise that naphthalene is repellent to all domestic animals as well.

Tactile repellents are made of various gooey combinations of castor oil, petrolatum, polybutane, resins, diphenylamine, pentachlorophenol, quinone, zinc oxide, and aromatic solvents applied as thin strips or beads to roosts, window ledges, and other favorite resting places. Roost-No-More® is a successful bird repellent containing several ingredients just mentioned.

The taste repellents are varied and somewhat surprising in certain instances, since they have other uses. The fungicides captan and copper oxalate are examples and are used as seed treatments to repel seed-pulling birds. Two other popular seed treatments are anthraquinone and glucochloralase. Turpentine, an old standby with multiple uses, can also be used as a seed treatment.

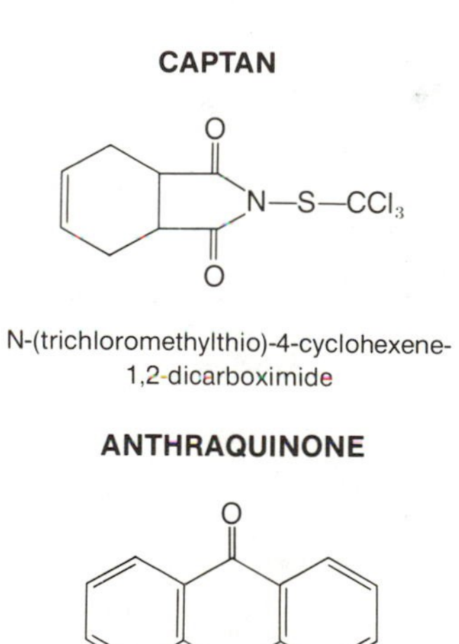

GLUCOCHLORALOSE (Alfamat®)

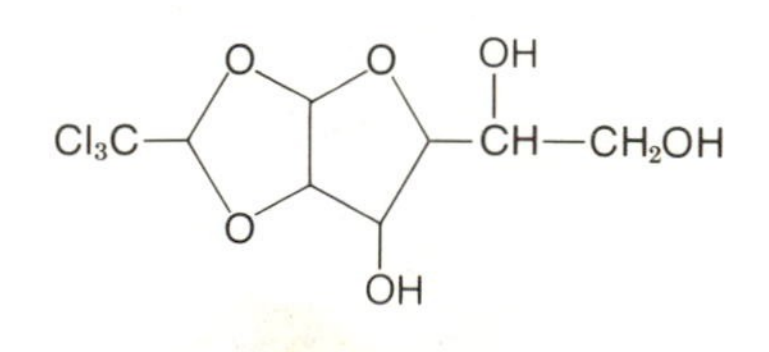

METHIOCARB (Mesurol®)

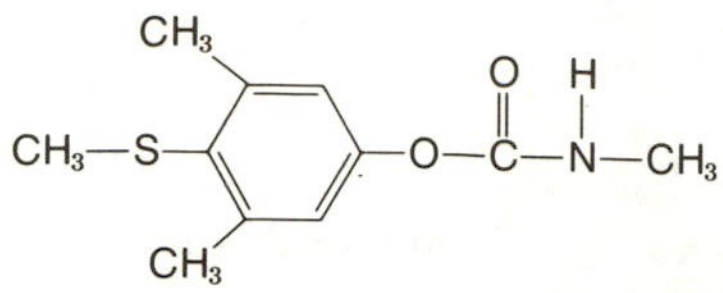

3,5-dimethyl-4-(methylthio) phenol methylcarbamate

Methiocarb (Mesurol®) is registered as a bird repellent on cherries and blueberries, and shows great promise for other crops including corn, sorghum, rice, and grapes. Avitrol®, a highly effective bird repellent, is discussed in Chapter 8.

DOG AND/OR CAT REPELLENTS

The materials registered as dog and cat repellents are almost too numerous to describe. Both forms of moth crystals, naphthalene and paradichlorobenzene, are readily available and very effective both indoors and outdoors.

Other materials found in commercial mixtures are: allyl isothiocyanate, amyl acetate, anethole, bittrex, bone oil, capsaicin, citral, citronella, citrus oil, creosote, cresylic acid, eucalyptus, geranium oil, lavender oil, lemongrass oil, menthol, methyl nonyl ketone, methyl salicylate, nicotine, pentanethiol, pyridine, sassafras oil, and thymol. Some of these substances are readily detectable and may be highly annoying to humans if used indoors.

DEER REPELLENTS

As rural dwellers know, deer can prune leaves and limbs from fruit and ornamental trees up to 8 feet above the ground by standing on their hind legs. Materials registered as effective deer repellents include two fungicides—thiram and ziram—and bone oil. Hinder® is a new deer and rabbit repellent approved for use on a full range of food crops, ornamentals, and nursery stock. It is exempt from a tolerance on food crops.

MOLE REPELLENTS

THIRAM

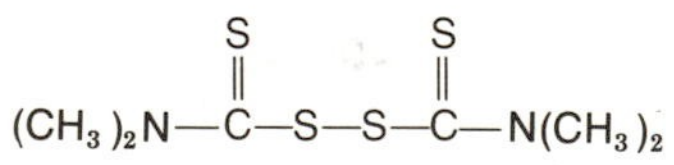

$(CH_3)_2N{-}C({=}S){-}S{-}S{-}C({=}S){-}N(CH_3)_2$

tetramethylthiuramdisulfide

Only the fungicide thiram and the liquid form of paradichlorobenzene are registered as mole repellents.

RABBIT REPELLENTS

Rabbit repellents include blood meal, naphthalene, nicotine, and the two fungicides thiram and ziram. (See also Hinder® under Deer Repellents.)

ZIRAM

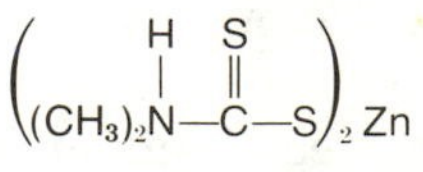

$((CH_3)_2N{-}C({=}S){-}S)_2Zn$

zinc dimethyldithiocarbamate

NICOTINE

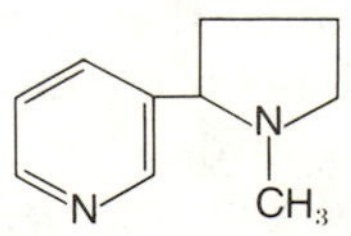

3-(1-methyl-2-pyrrolidyl) pyridine

RODENT REPELLENTS

The term *rodent* is rather all-inclusive and perhaps a bit deceptive, because not all rodent repellents are registered for all rodents. Users must make an accurate identification of the particular pest and select one of the following materials or combinations that clearly indicates that pest on its label: Biomet-12 (tri-n-butyltin chloride), naphthalene, paradichlorobenzene, polybutanes, polyethylene, R-55® (tert-butyl dimethyltrithioperoxycarbamate), and thiram.

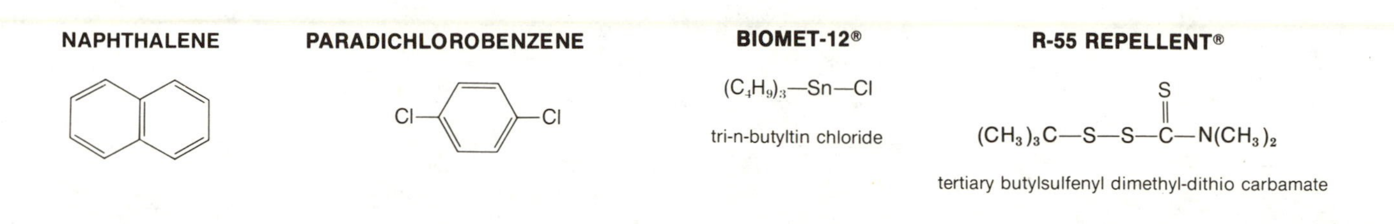

PART IV

CHEMICALS USED IN THE CONTROL OF PLANTS

More energy is expended for the weeding of man's crops than for any other single human task.

LeRoy Holm (1971)

Herbicides

In the past three decades, herbicides, or chemical weed killers, have largely replaced mechanical methods of weed control in countries where intensive and highly mechanized agriculture is practiced. Herbicides provide a more effective and economical means of weed control than cultivation, hoeing, and hand pulling. Together with fertilizers, other pesticides, and improved plant varieties, they have made an important contribution to the increased yields we now have and serve to combat rising costs and shortages of agricultural labor. The heavy use of herbicides is confined to North America, western Europe, Japan, and Australia. Without the use of herbicides, it would have been impossible to mechanize fully the production of cotton, sugar beets, grains, potatoes, and corn.

Herbicides are also used extensively away from the farm, in areas such as industrial sites, roadsides, ditch banks, irrigation canals, fence lines, recreational areas, railroad embankments, and power lines. Herbicides remove undesirable plants that might cause damage, present fire hazards, or impede work crews. They also reduce costs of labor for mowing.

Herbicides are classed as *selective* when they are used to kill weeds without harming the crop and as *nonselective* when the purpose is to kill all vegetation. Both selective and nonselective materials can be applied to weed foliage or to soil containing weed seeds and seedlings, depending on the mode of action. The term *true selectivity* refers to the capacity of an herbicide, when applied at the proper dosage and time, to be active only against certain species of plants but not against others. But selectivity can also be achieved by placement, as when a nonselective herbicide is applied in such a way that it reaches the weeds but not the crop.

Herbicide classification would be a simple matter if only the selective and nonselective categories existed. However there are multiple-classification schemes that may be based on selectivity, contact versus translocation, timing, area covered, and chemical classification.

Herbicides are also classified as contact or translocated. *Contact herbicides* kill the plant parts to which the chemical is applied and are most effective against annuals, those weeds that germinate from seeds and grow to maturity each year. Complete coverage is essential in weed control with contact materials. *Translocated herbicides* are absorbed either by roots or above-ground parts of plants and then circulate within the plant system to distant tissues. Translocated herbicides may be effective against all weed types; however, their greatest advantage is in the control of established perennials, those weeds that continue their growth from year to year. Uniform application is needed for the translocation materials, whereas complete coverage is not required.

Another method of classification is the timing of herbicide application with regard to the stage of crop or weed development. Timing depends on many factors, including the chemical classification of the material and its persistence, the crop and its tolerance to the herbicide, weed species, cultural practices, climate, and soil type and condition. The three categories of timing are preplanting, preemergence, and postemergence.

Preplanting applications for control of annual weeds are made to an area before the crop is planted, within a few days or weeks of planting. Preemergence applications are completed prior to emergence of the crop or weeds, depending on definition, after planting. Postemergence applications are made after the crop or weed emerges from the soil.

Herbicide application based on the area covered involves four categories: band, broadcast, spot treatments, and directed spraying. A band application treats a continuous strip, as along or in a crop row. Broadcast applications cover the entire area, including the crop. Spot treatments are confined to small areas of weeds. Directed sprays are applied to selected weeds or to the soil to avoid contact with the crop.

In this chapter, we emphasize the chemical classification of herbicides. The two major classifications are inorganic and organic herbicides.

INORGANIC HERBICIDES

The first chemicals used in weed control were inorganic compounds. Brine and a mixture of salt and ashes were both used to sterilize the land as early as biblical times, by the Romans. In 1896, copper sulfate was used selectively to kill weeds in grain fields. From about 1906 until 1960, sodium arsenite solutions were the standard herbicides of commerce. Arsenic trioxide has been used at rates of 450 to 900 kg/ha for soil sterilization, whereas 1 or 2 kg of selected organic herbicides would easily produce the same result. As with the arsenical insecticides, the trivalent arsenicals were nonspecific inhibitors of those enzymes containing sulfhydryl groups. They also uncouple oxidative phosphorylation.

Ammonium sulfamate ($NH_4SO_3NH_2$), another salt, was introduced in 1942 for brush control. Other salts have been used over the years. These include ammonium thiocyanate, ammonium nitrate, ammonium sulfate, iron sulfate, and copper sulfate, each applied heavily as a foliar spray. Their mechanisms of action are desiccation and plasmolysis (shrinkage of cell protoplasm away from its wall due to removal of water from its large central vacuole).

The borate herbicides were another family of inorganics, for example, sodium tetraborate ($Na_2B_4O_7 \cdot 5H_2O$), sodium metaborate ($Na_2B_2O_4 \cdot 4H_2O$), and amorphous sodium borate ($Na_2B_8O_{13} \cdot 4H_2O$). The amount of boron or boric acid determined their effectiveness. Borates are absorbed by plant roots, are translocated to above-ground parts, and are nonselective, persistent herbicides. Boron accumulates in the reproductive structures of plants, but its mechanism of toxicity is unclear. The borates are still used to give a semipermanent form of sterility to areas where no vegetation of any sort is wanted.

A nonselective herbicide used extensively for the last 40 years is sodium chlorate ($NaClO_3$). It acts as a soil sterilant at rates of 225 kg/ha, but it can be used as a foliar spray at 5.7 kg/ha as a defoliant of cotton. Caution must be taken with sprays of sodium chlorate to be certain the formulation contains flame retardants, usually sodium metaborate. Sulfuric acid has also been used as a foliar herbicide, but its use is limited considerably by its corrosiveness to metal spray rigs. Their mechanisms of action are those described for the miscellaneous salts, namely, desiccation and plasmolysis.

Several of the inorganic herbicides are still useful in weed and brush control, but are gradually being replaced by organic materials. Although organic herbicides are not superior to inorganic ones, the EPA has placed intensive restrictions on some inorganic herbicides because of their persistence in soils. The inorganics are not wise choices for use around the home except by experts aiming to remove all vegetation from an area.

ORGANIC HERBICIDES

Petroleum Oils

The earliest organic herbicides were the petroleum oils, which are a complex mixture of long-chain hydrocarbons containing traces of nitrogen- and sulfur-linked compounds. These mixtures include alkanes, alkenes, and often alicyclics and aromatics, produced by distillation and refinement of crude oils. The petroleum oils are effective contact herbicides for all vegetation. Many homeowners today make use of old crankcase oil, gasoline, kerosene, and diesel oil for spot treatments. This is not an ecological choice in view of the drive to save energy and avoid further release of hydrocarbons in our already overburdened atmosphere. Although only temporary in effect, they are fast-acting and very safe to use around the home. The petroleum oils exert their lethal effect by penetrating and disrupting plasma membranes. Other herbicidal oils used in agriculture include naphtha, Stoddard solvent, fuel oils, and mineral spirits. Petroleum oils are used mostly for nonselective, but occasionally selective, weed control.

Organic Arsenicals

Widely used as agricultural herbicides are the organic arsenicals, namely the arsinic and arsonic acid derivatives. Cacodylic acid (dimethylarsinic acid) and its sodium salt are the only derivatives of

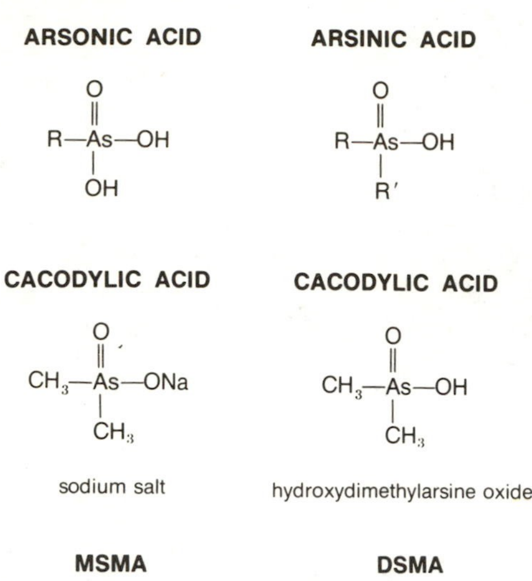

MSMA

CH_3—As(=O)—ONa, —OH

monosodium methanearsonate

DSMA

CH_3—As(=O)—ONa, —ONa

disodium methanearsonate

arsinic acid. Disodium methanearsonate (DSMA) and monosodium methanearsonate (MSMA) are salts of arsonic acid.

The organic arsenicals are much less toxic to mammals than the inorganic forms, are crystalline solids, and are relatively soluble in water. The arsonate, or pentavalent arsenic, acts in a way different from the trivalent form of the inorganic arsenicals just described. The arsonates upset plant metabolism and interfere with normal growth by entering into reactions in place of phosphate. They not only substitute for essential phosphate but also are absorbed and translocated in a manner similar to that of phosphates.

The trivalent arsenates are exclusively contact herbicides, whereas the pentavalent arsonates are translocated to underground tubers and rhizomes, making them extremely useful against Johnsongrass and nut sedges. They are usually applied as spot treatments.

Phenoxyaliphatic Acids

An organic herbicide introduced in 1944, later to be known as 2,4-D, was the first of the "phenoxy herbicides," "phenoxyacetic acid derivatives," or "hormone" weed killers. These were highly selective for broad-leaf weeds and were translocated throughout the plant. 2,4-D provided most of the impetus in the commercial search for other organic herbicides in the 1940s. Several compounds belong to this group, of which 2,4-D and 2,4,5-T are the most familiar. Other important compounds in this group are 2,4-DB, MCPA, and silvex.

The phenoxy herbicides have complex mechanisms of action resembling those of auxins (growth hormones) in some. They affect cellular division, activate phosphate metabolism, and modify nucleic acid metabolism.

2,4-D, MCPA, and 2,4,5-T have been used for years in very large volume worldwide with no adverse effects on human or animal health. However, 2,4,5-T, used mainly to control woody perennials, became the subject of extended investigation, particularly because of its use in Vietnam in combination with 2,4-D as Agent Orange. Certain samples were found to contain excessive amounts of a highly toxic impurity, 2,3,7,8-tetrachlorodibenzo-*p*-dioxin, commonly referred to as tetrachlorodioxin, or dioxin. Subsequent slight alterations in manufacturing procedures have brought the dioxin content down to tolerable levels.

2,4-D has been and continues to be one of the most useful herbicides developed. More than 32 million kg, manufactured by six basic manufacturers in the United States, is used each year in 1500 formu-

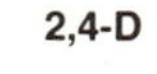

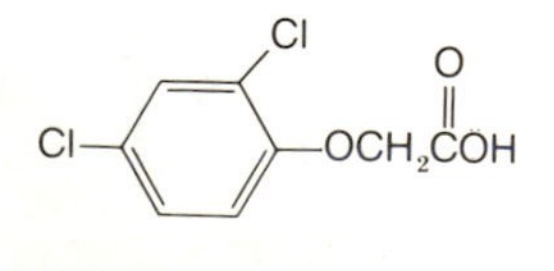

(2,4-dichlorophenoxy)acetic acid

2,4,5-T

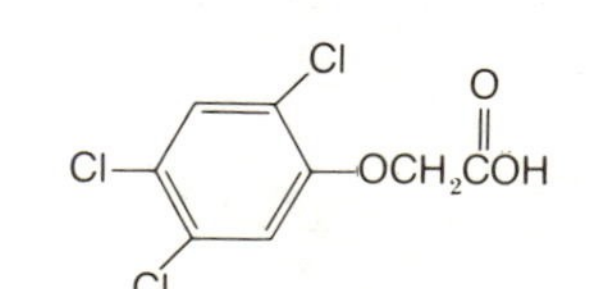

(2,4,5-trichlorophenoxy)acetic acid

SILVEX

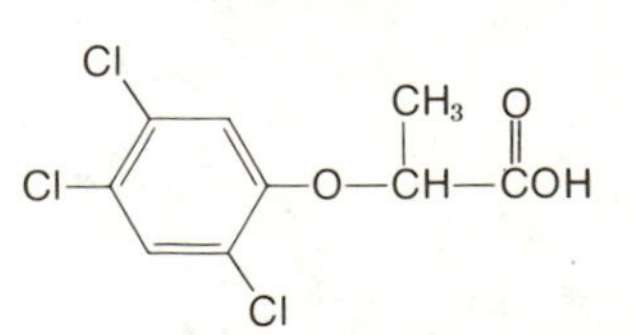

2-(2,4,5-trichlorophenoxy)propionic acid

MCPA

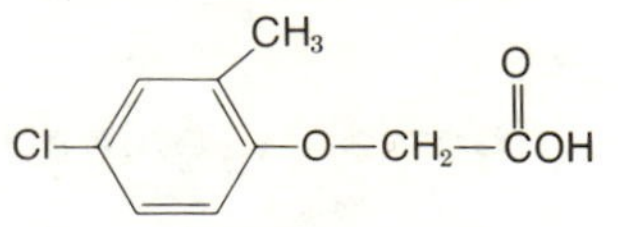

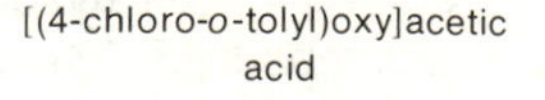
[(4-chloro-*o*-tolyl)oxy]acetic acid

lated products and in 35 ester and salt forms. In agriculture it is used on cereal and grain crops for the control of broad-leaf weeds, and on rights-of-way, turf and lawns, and in forest conservation programs. The manufacturing process for 2,4-D used in the United States does not result in any level of tetrachlorodioxin contamination.

Acifluorfen (sodium salt) is a recent addition to the phenoxys, as is diclofop methyl. Acifluorfen controls annual broad-leaf and grass weeds in soybeans, peanuts, and other legumes. Diclofop methyl is sold as Illoxan® in other countries and is used for postemergence in wheat, barley, soybeans, sugar beets, and vegetables.

ACIFLUORFEN (Blazer®)

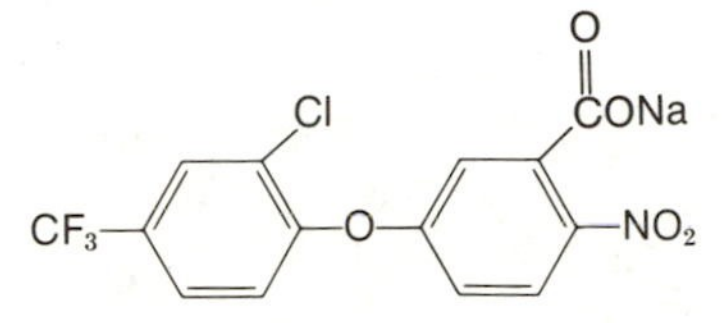

sodium 5-[2-chloro-4-(trifluoro-methyl)-phenoxy]-2-nitrobenzoate

DICLOFOP METHYL (Hoelon®)

2-[4-(2,4-dichlorophenoxy)-phenoxy]-methyl-propanoate

Substituted Amides

The amide herbicides have diverse biological properties. These are simple molecules that are easily degraded by plants and soil. One of the earliest was allidochlor, which is selective for the grasses. Allidochlor inhibits the germination or early seedling growth of most annual grasses probably by alkylation of the —SH groups of proteins.

Diphenamid is used as a preemergence soil treatment and has little contact effect. Most established plants are tolerant to diphenamid, because it affects only seedlings. It persists from 3 to 12 months in soil.

Propanil has been used extensively on rice fields as a selective postemergence control for a broad spectrum of weeds. Napropamide is used in the control of grass and broad-leaf weeds in vineyards and orchards, and in direct-seeded tomatoes, strawberries, ornamentals, and tobacco.

There is no consistent pattern in the mechanism of action of amide herbicides. Some are applied only to the soil and are active through the root system or seeds, while others are applied only to foliage.

ALLIDOCHLOR or CDAA

N,N-diallyl-2-chloroacetamide

DIPHENAMID

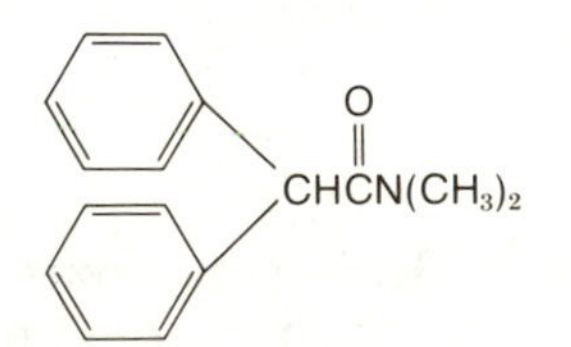

N,N-dimethyl-2,2-diphenylacetamide

PROPANIL

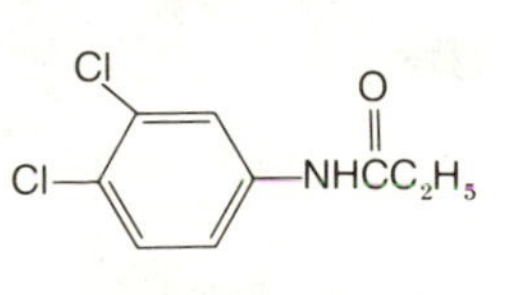

3′,4′-dichloropropionanilide

NAPROPAMIDE (Devrinol®)

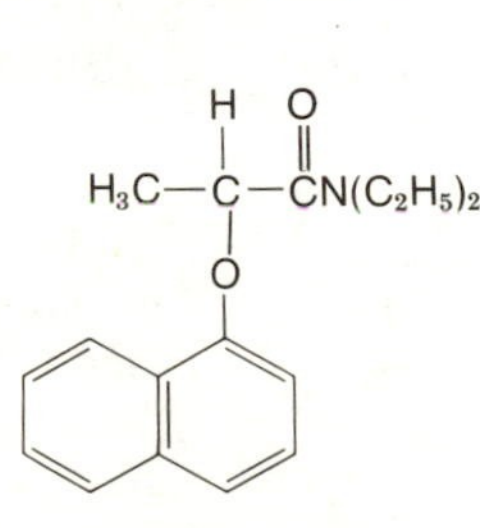

2-(α-naphthoxy)-*N,N*-diethylpropionamide

TRIFLURALIN (Treflan®)

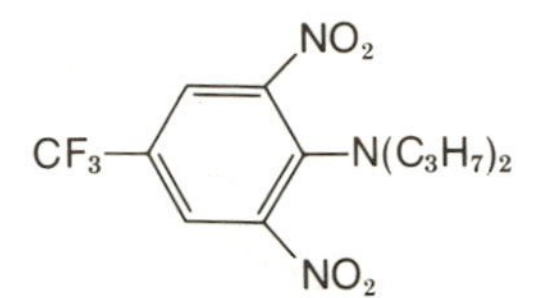

α,α,α-trifluoro-2,6-dinitro-*N,N*-dipropyl-*para*-toluidine

BENEFIN

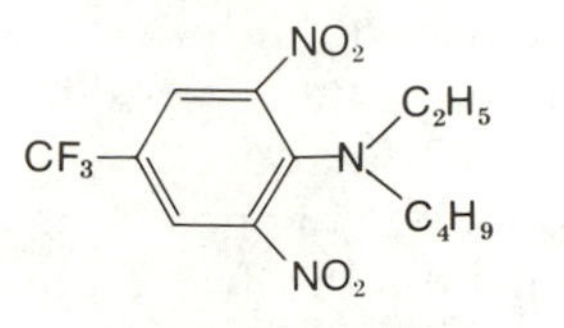

N-butyl-*N*-ethyl-α,α,α-trifluoro-2,6-dinitro-*p*-toluidine

ISOPROPALIN

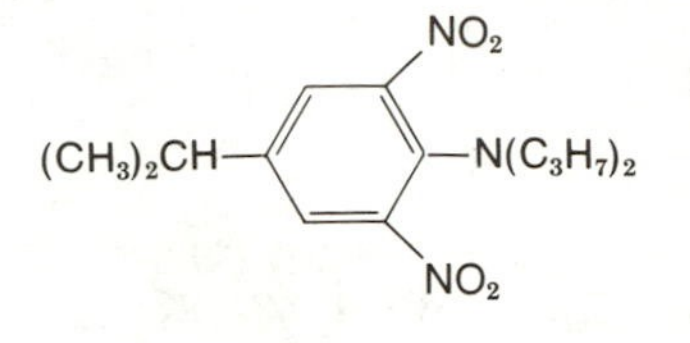

2,6-dinitro-*N,N*-dipropylcumidine

NITRALIN

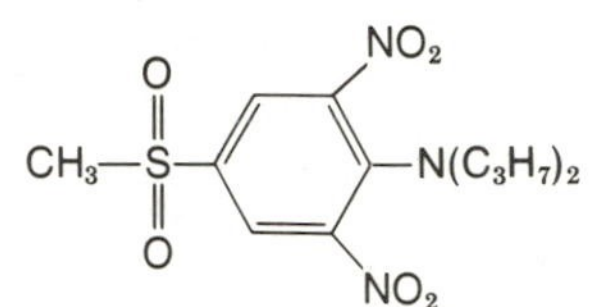

4-(methylsulfonyl)-2,6-dinitro-*N,N*-dipropylaniline

ORYZALIN (Surflan®)

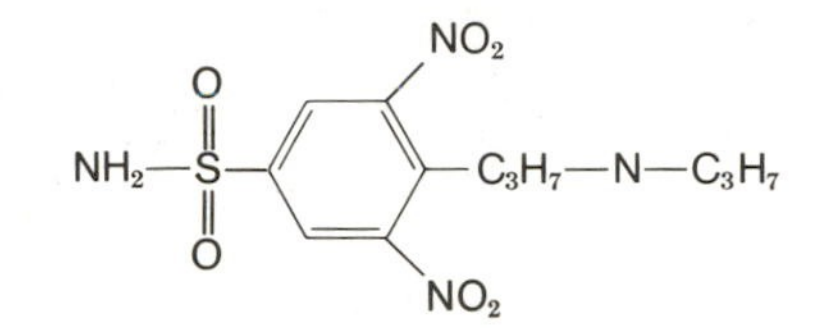

3,5-dinitro-N^4,N^4-dipropylsulfanilamide

PENDIMETHALIN (Prowl®)

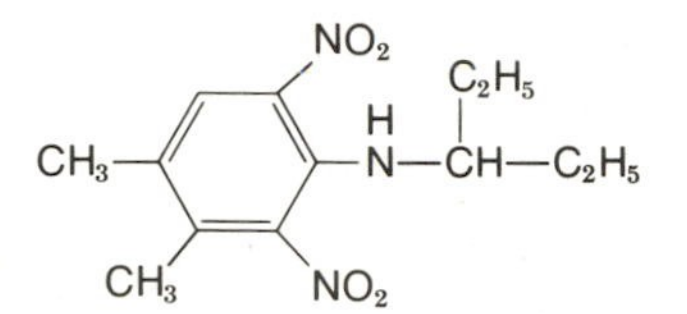

N-(1-ethylpropyl)-3,4-dimethyl-2,6-dinitrobenzenamine

FLUCHLORALIN (Basalin®)

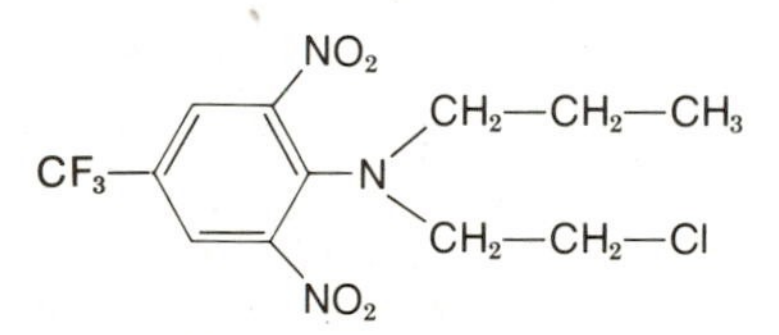

N-(2-chloroethyl)-α,α,α-trifluoro-2,6-dinitro-*N*-propyl-*p*-toluidine

Nitroanilines

The nitroanilines are probably the group of herbicides most heavily used in agriculture. They are used almost exclusively as soil-incorporated, preemergence selective herbicides in many field crops. Examples of the substituted dinitrotoluidines are trifluralin, benefin, nitralin, and isopropalin. Trifluralin has very low water solubility, which minimizes leaching and movement from the target. The nitroanilines inhibit both root and shoot growth when absorbed by roots, but they have an involved mode of action, which includes inhibiting the development of several enzymes and the uncoupling of oxidative phosphorylation.

Oryzalin controls annual grasses and broad-leaf weeds in cotton, soybeans, nonbearing fruit trees, nut trees, vineyards, and ornamentals. Pendimethalin controls annual grass and certain broad-leaf weeds in corn and is preplant incorporated in cotton, soybeans, and tobacco. Fluchlorin is preplant and preemergence incorporated for control of annual grasses and certain broad-leaf weeds in soybeans and cotton.

Substituted Ureas

The substituted ureas, $H_2N\overset{O}{\overset{\|}{C}}NH_2$, have their hydrogen atoms replaced (substituted) by various carbon chains and rings, yielding a utilitarian group of compounds. Thousands of substituted ureas have been tested as herbicides and many are in use today. Illustrated are monuron, diuron, fenuron-TCA, linuron, siduron, and tebuthiuron. Most of the ureas are relatively nonselective and are usu-

MONURON

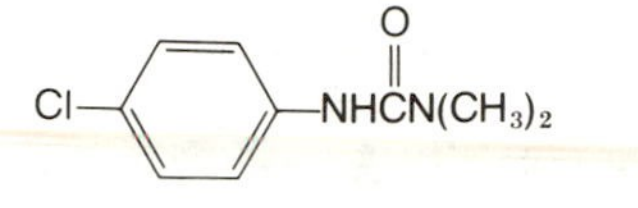

3-(*p*-chlorophenyl)-1,1-dimethylurea

DIURON

Cl, Cl, O, NHCN$(CH_3)_2$

3-(3,4-dichlorophenyl)-1,1-dimethylurea

FENURON-TCA

O, O, NHCN$(CH_3)_2$ · Cl_3COH

1,1-dimethyl-3-phenylurea mono(trichloroacetate)

LINURON

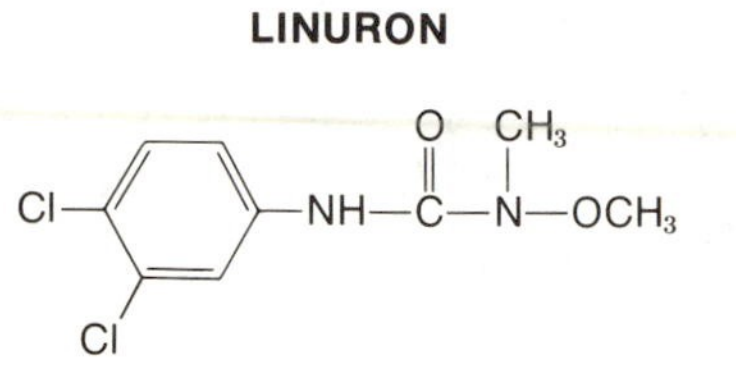

3-(3,4-dichlorophenyl)-1-methoxy-1-methylurea

SIDURON

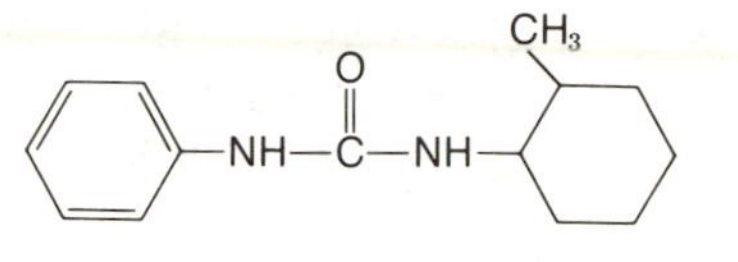

1-(2-methylcyclohexyl)-3-phenylurea

TEBUTHIURON (Spike®)

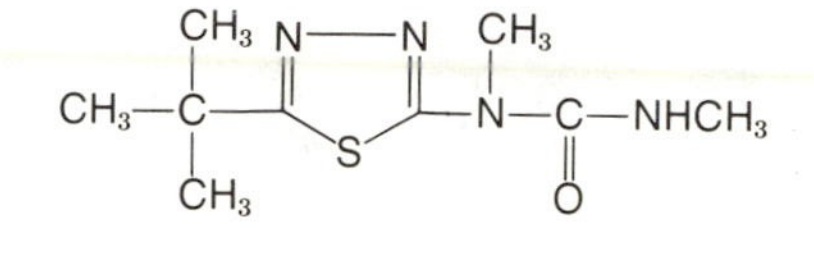

N-[5-(1,1-dimethyl)-1,3,4-thiadiazol 2-yl]-N,N′-dimethylurea

ally applied to the soil as preemergence herbicides; some have post-emergence uses, while others are active when foliar applied. The ureas are strongly adsorbed by the soil, then absorbed by roots. Their mechanism of action is to inhibit photosynthesis—the production of plant sugars—and, indirectly, to inhibit the Hill reaction (see p. 162).

Carbamates

The esters of carbamic acid are physiologically quite active. As we have seen, some carbamates are insecticidal, and, as we shall see later, others are fungicidal. Still other carbamates are herbicidal, and we treat that group in this section. Discovered in 1945, the carbamates are used primarily as selective preemergence herbicides, but some are also effective in postemergence use.

PROPHAM **BARBAN**

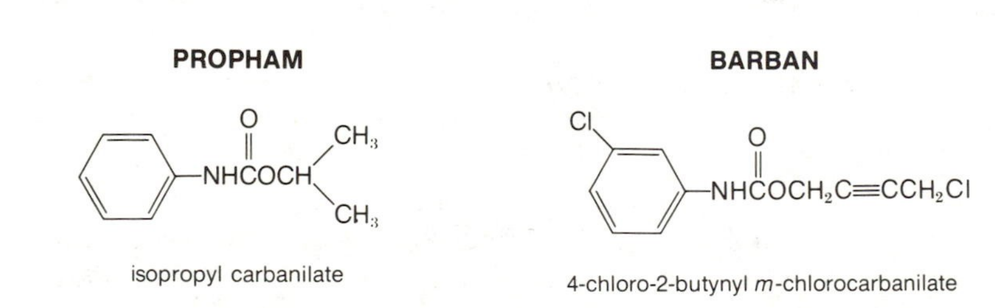

isopropyl carbanilate

4-chloro-2-butynyl *m*-chlorocarbanilate

The first of the herbicidal carbamates was propham; it was followed first by chlorpropham, then by barban and terbucarb. These carbamates kill plants by stopping cell division and plant tissue growth. Two effects are noted: cessation of protein production and shortening of chromosomes undergoing mitosis (duplication).

Propham and barban are aryl carbamates, because they all contain an aryl group, the six-member carbon ring, normally the phenyl ring; the alkyl group is the carbon chain or hydrocarbon radical with one

PHENMEDIPHAM

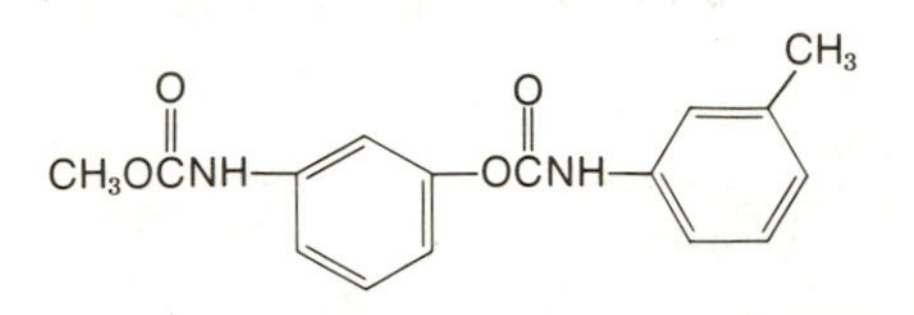

methyl *m*-hydroxycarbanilate *m*-methylcarbanilate

hydrogen displaced by the attachment of the alkyl group to the remainder of the molecule.

Phenmedipham contains two carbamate radicals in its molecule and is used as a postemergence herbicide in sugar beet production. Chlorpropham has largely replaced propham for field crops and is particularly useful for alfalfa, clovers, tomatoes, safflower, and soybeans. Potatoes placed in storage may be treated with chlorpropham to inhibit sprouting. Asulam is applied in most instances for grass control, such as crabgrass, Johnsongrass, goosegrass, foxtail, and barnyardgrass in sugarcane. It is also effective in reforestation and Christmas-tree plantings.

CHLORPROPHAM

isopropyl *m*-chlorocarbanilate

ASULAM (Asulox®)

methyl sulfanilylcarbamate

Thiocarbamates

The thiocarbamates contains sulfur (from the Greek word *theion*) and are derived from the hypothetical thiocarbamic acid, $HS{-}\overset{O}{\overset{\|}{C}}{-}NH_2$. The thiocarbamates are selective herbicides marketed for weed control in croplands. The thio- and dithiocarbamates have exceptional vapor pressures, are quite volatile, and must be incorporated in the soil at time of application. They inhibit the development of seedling shoots and roots as they emerge from the weed seeds. Consequently, they are all used as either preplant or preemergence, soil-incorporated herbicides.

Examples of thiocarbamates are EPTC and pebulate. EPTC is used mostly to control annual and perennial grassy weeds, including nutgrass, Johnsongrass, and quackgrass. Pebulate is effective against both grassy and broad-leaf weeds in sugar beets, tobacco, and tomatoes. Also included among the thiocarbamates, but not illustrated, are butylate, cycloate, diallate, molinate, triallate, and vernolate.

Among the dithiocarbamates, CDEC is an exceptionally good herbicide for preemergence weed control in vegetable and field crops. It is effective in muck as well as sandy soils. Metham is a general-purpose soil fumigant for weeds, weed seeds, nematodes, insects, and soil-borne disease microorganisms. Molinate is effective for watergrass control in rice. Cycloate is used mostly in sugar beets, table beets, and spinach to control annual and perennial grasses and broad-leaf weeds. Vernolate is effective for control of most grass weeds and some broad-leafs in soybeans and peanuts. Butylate is incorporated preplant to control grass weeds in corn.

EPTC

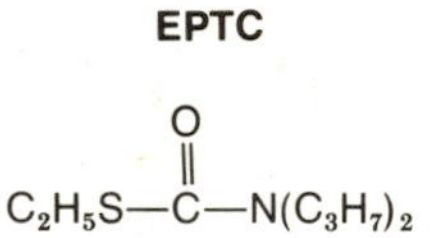

S-ethyl dipropylthiocarbamate

PEBULATE

S-propyl butylethylthiocarbamate

CDEC

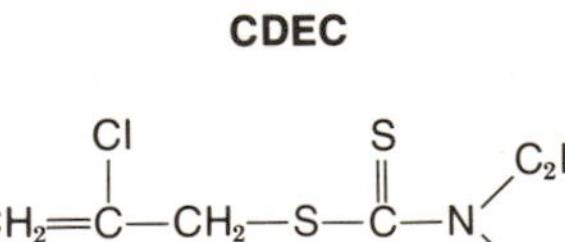

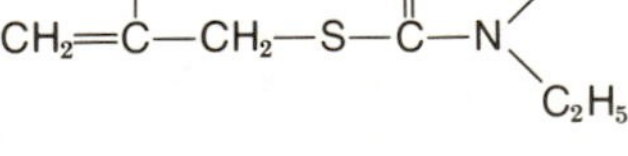

2-chloroallyl-diethyldithiocarbamate

METHAM

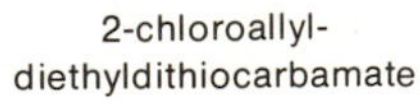

sodium *N*-methyl-dithiocarbamate

MOLINATE (Ordram®)

S-ethyl hexahydro-1 *H*-azepine-1-carbothioate

CYCLOATE (Ro-Neet®)

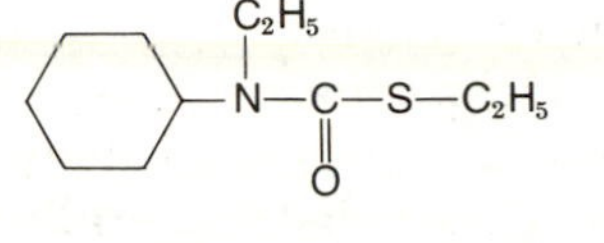

S-ethylcyclohexylethylthiocarbamate

VERNOLATE (Vernam®)

S-propylidpropylthiocarbamate

BUTYLATE (Sutan®)

S-ethyl diisobutylthiocarbamate

Heterocyclic Nitrogens

The *triazines,* which are six-member rings containing three nitrogens (the prefix *tri-* means "three") and azine (a nitrogen-containing ring) make up the heterocyclic nitrogens. The fundamental triazine nucleus is illustrated, showing the placement of the three nitrogens.

Probably the most familiar group of heterocyclic nitrogens, because of their heavy use and notoriety, are the triazines, which are strong inhibitors of photosynthesis. Their selectivity depends on the ability of tolerant plants to degrade or metabolize the parent compound (the susceptible plants do not). Triazines are applied to the soil primarily for their postemergence activity. There are many triazines on the market today, seven of which are illustrated.

TRIAZINE NUCLEUS

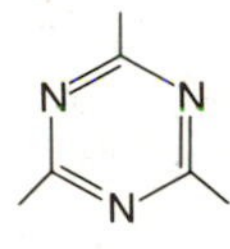

ATRAZINE

2-chloro-4-(ethylamino)-6-(isopropylamino)-*s*-triazine

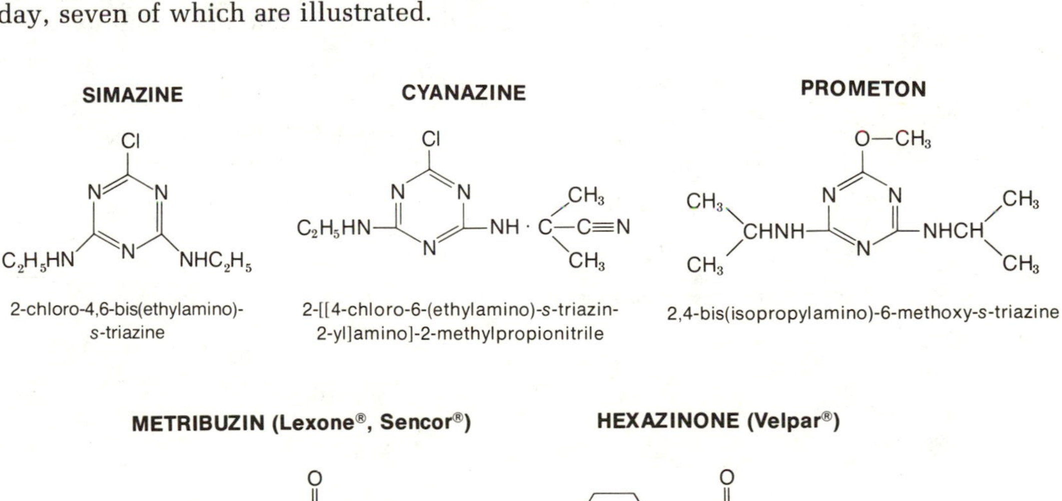

SIMAZINE

2-chloro-4,6-bis(ethylamino)-*s*-triazine

CYANAZINE

2-[[4-chloro-6-(ethylamino)-*s*-triazin-2-yl]amino]-2-methylpropionitrile

PROMETON

2,4-bis(isopropylamino)-6-methoxy-*s*-triazine

PROPAZINE

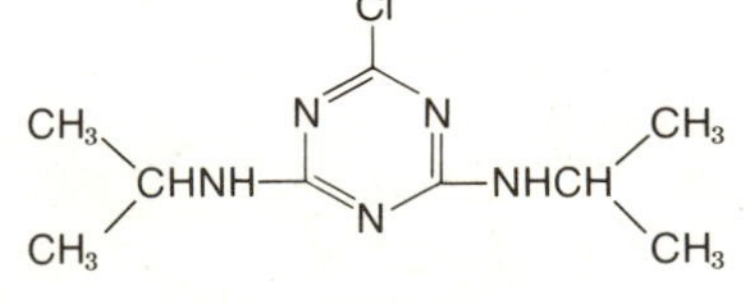

2-chloro-4,6-bis(isopropylamino)-*s*-triazine

METRIBUZIN (Lexone®, Sencor®)

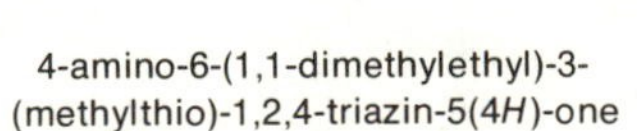

4-amino-6-(1,1-dimethylethyl)-3-(methylthio)-1,2,4-triazin-5(4*H*)-one

HEXAZINONE (Velpar®)

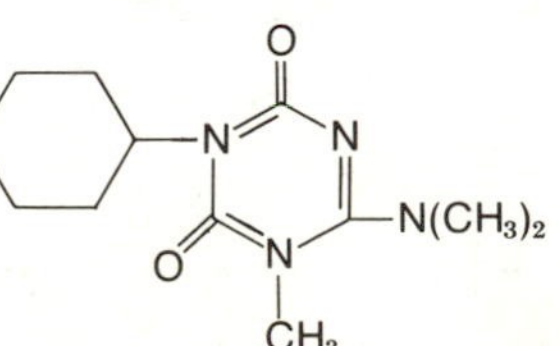

3-cyclohexyl-6-(dimethylamino)-1-methyl-1,3,5-triazine-2,4-(1*H*,3*H*)-dione

AMITROLE

3-amino-*s*-triazole

OXADIAZON (Ronstar®)

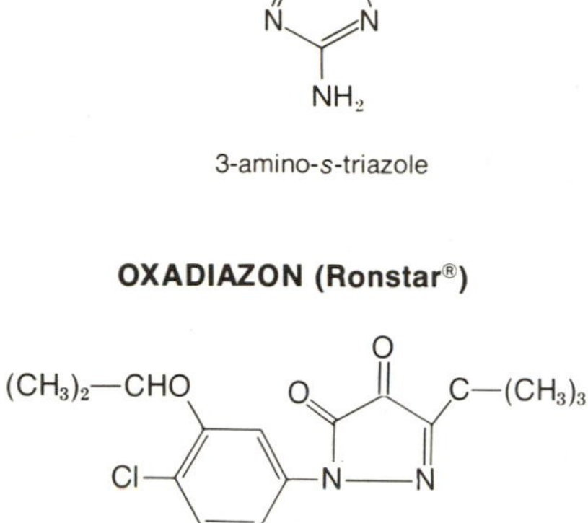

2-tert-butyl-4-(2,4-dichloro-5-isopropoxy-phenyl)-Δ²-1,3,4-oxadiazolin-5-one

BENTAZON (Basagran®)

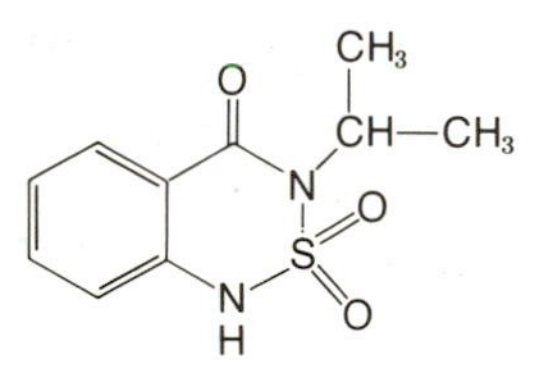

3-(1-methylethyl)-1*H*-2,1,3-benzo-thiadiazin-4(3*H*)-one 2,2-dioxide

PYRIDINE

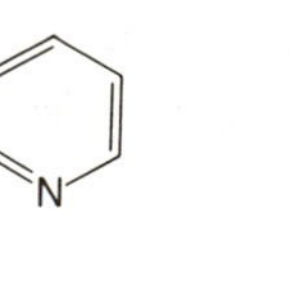

PICLORAM

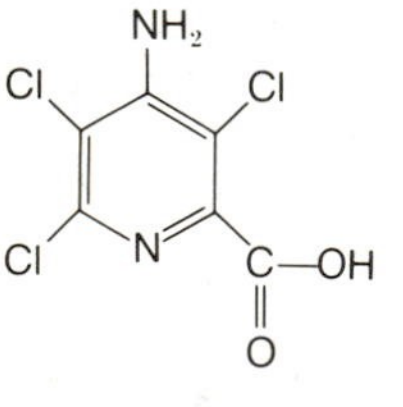

4-amino-3,5,6-trichloropicolinic acid

The triazines are used selectively in certain crops and nonselectively in others. They are used in greatest quantity in corn production and nonselectively on industrial sites.

Metribuzin is used on soybeans, wheat, sugarcane, and a few vegetables. Hexazinone is used in the control of annual, biennial, and perennial weeds and woody plants on noncropland. Sometimes it is applied as pellets or gridballs to control woody plants in conifer plantings.

The *triazoles* are five-member rings containing three nitrogens. The outstanding member of this group is amitrole, which acts in about the same way as the triazines, by inhibiting photosynthesis. In the form of aminotriazol, it became embroiled in the historic "cranberry incident" of 1959. This event culminated in the addition of the Delaney Amendment to the Pure Food and Drug Act in 1960 (see Chapter 23). In essence, this amendment stated that no residue of any carcinogen (cancer-producing agent) in a food crop would be tolerated.

Oxadiazon is used for weed control in dry-seeded rice, and for turf and ornamentals. Bentazon is effective in controlling broad-leaf weeds in soybeans, rice, corn, peanuts, beans, and peas.

A popular herbicide derived from the pyridine molecule is picloram. Picloram is a readily translocated herbicide used against broad-leafed and woody plants; it may be taken up from either the roots or the foliage. It has also been used experimentally as a growth regulator at extremely low dosages on apricots, cherries, and figs. Its mechanism of action is probably the regulation of protein and enzyme synthesis in cells through its effects on nucleic acid synthesis and metabolism.

The *substituted uracils* give a wide range of grass and broad-leaf control for an extended period by inhibition of photosynthesis. Illustrated are terbacil, bromacil, and lenacil.

URACIL NUCLEUS

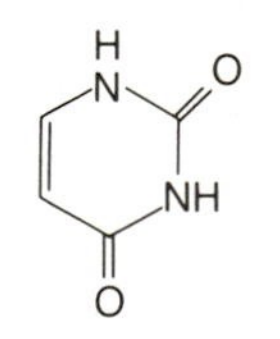

LENACIL

3-cyclohexyl-6,7-dihydro-1*H*-cyclopentapyrimidine-2,4-(3*H*,5*H*)-dione

BROMACIL

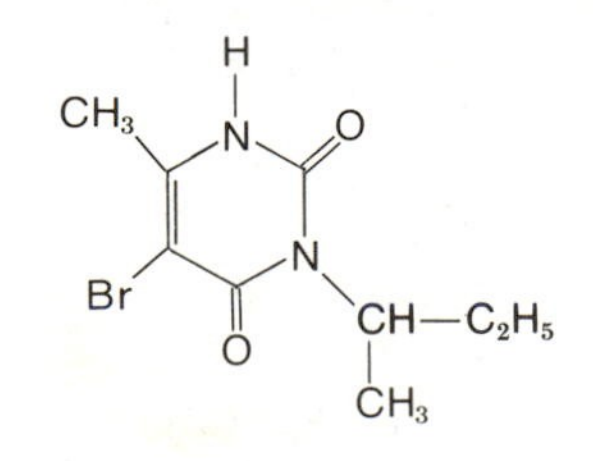

5-bromo-3-*sec*-butyl-6-methyluracil
[5-bromo-6-methyl-3-(1-methylpropyl)uracil]

TERBACIL

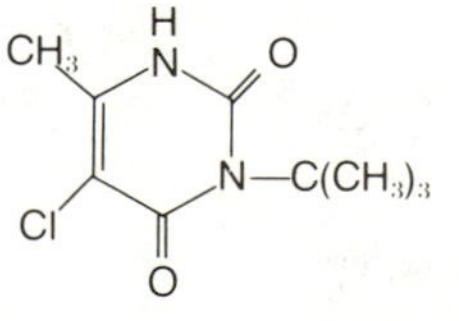

3-*tert*-butyl-5-chloro-6-methyluracil

Aliphatic Acids

Two aliphatic acids heavily used as herbicides are TCA and dalapon, used against grasses, particularly quackgrass and Bermudagrass. They both act by precipitation of protein within the cells. Dalapon is widely used around homes to control Bermudagrass.

TCA

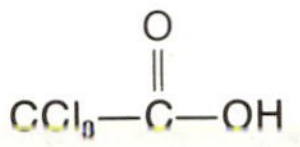

$CCl_3—\overset{O}{\overset{\|}{C}}—OH$

trichloroacetic acid

DALAPON

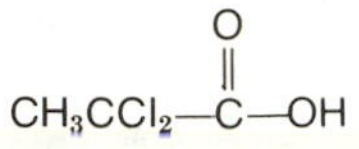

$CH_3CCl_2—\overset{O}{\overset{\|}{C}}—OH$

2,2,2-dichloropropionic acid

Arylaliphatic Acids (Substituted Benzoic Acids or "Benzoics")

The arylaliphatic acids are aryls, or six-member rings, attached to aliphatic acids. A number of these materials are employed as herbicides and are applied to the soil against germinating seeds and seedlings. The mechanism of action for dicamba and fenac is not completely understood; however, it is probably similar to that of the phenoxy herbicides (2,4-D, etc.), which interfere with the metabolism of nucleic acids.

DCPA and chloramben produce auxinlike growth effects in plants but the available research data are insufficient to propose a mechanism of action.

DICAMBA **FENAC**

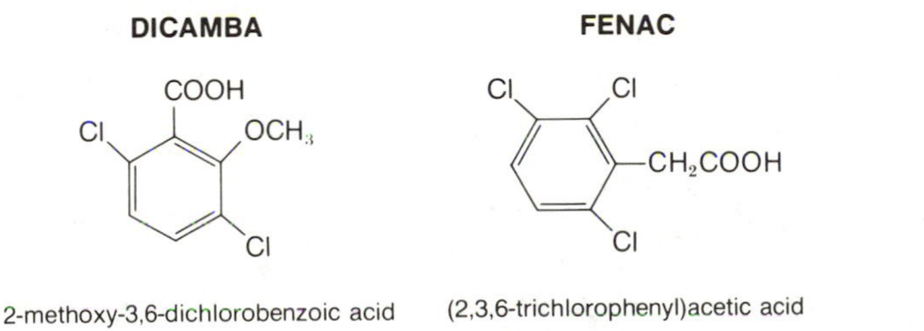

2-methoxy-3,6-dichlorobenzoic acid

(2,3,6-trichlorophenyl)acetic acid

DCPA

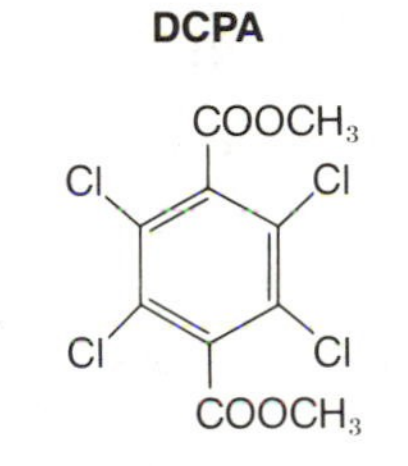

dimethyl tetrachloroterephthalate

CHLORAMBEN

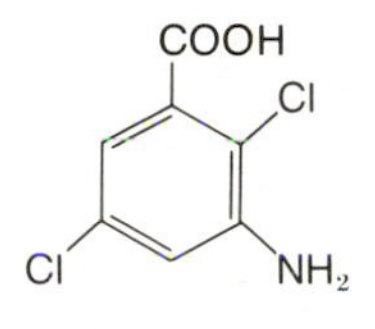

3-amino-2,5-dichlorobenzoic acid

Phenol Derivatives

Though rather old, the phenols are still used in great quantity as general contact herbicides, usually against broad-leaf weeds, and selectively in cereal crops.

Phenol derivatives are highly toxic to humans by every route of entry into the body and are nonselective foliar herbicides that are most effective in hot weather. The nitrophenols, represented by DNOC, were first introduced as herbicides in 1932. Dinoseb is the best-known representative of the nitrophenols, which act by uncoupling oxidative phosphorylation.

The dinitrophenols have also been used as ovicides, insecticides (see Chapter 4), fungicides, and blossom-thinning agents.

Nitrofen is a somewhat different relative of the phenols, with two phenyl rings, as pictured. It has broad-spectrum use as both a pre- and postemergence herbicide in vegetable crops, sugar beets, rice, and a few ornamentals. It is a contact rather than a translocated material.

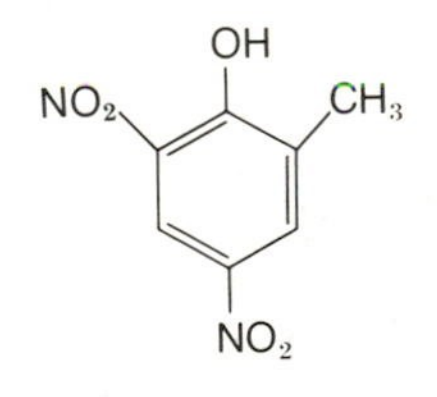

4,6-dinitro-*o*-cresol

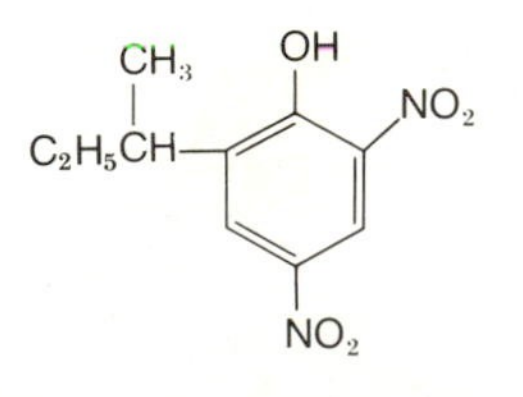

2-*sec*-butyl-4,6-dinitrophenol

NITROFEN

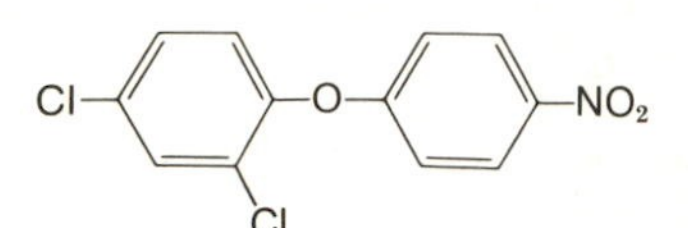

2,4-dichlorophenyl *p*-nitrophenyl ether

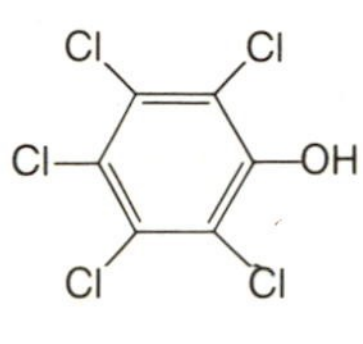

Another subgroup of the phenol derivatives is the chlorinated phenols, of which only one member is recognized as a herbicide, namely PCP or pentachlorophenol, which is also used for termite protection and wood treatment for fungal rots, and as a nonselective herbicide and preharvest defoliant. Its mechanism is a combination of plasmolysis, protein precipitation, and desiccation. Because of its wide effectiveness and multiple routes of action, PCP is recognized as being destructive to all living cells.

Substituted Nitriles

Nitriles are organic compounds containing the C≡N or cyanide grouping. There are three substituted nitrile herbicides, which have a wide spectrum of uses against grasses and broad-leaf weeds. Their mechanisms of action are broad, involving seedling growth inhibition, potato sprout inhibition, and gross disruption of tissues by inhibiting oxidative phosphorylation and preventing the fixation of CO_2. These effects, however, do not explain their rapid action. Ioxynil is not registered for use in the United States.

DICHLOBENIL

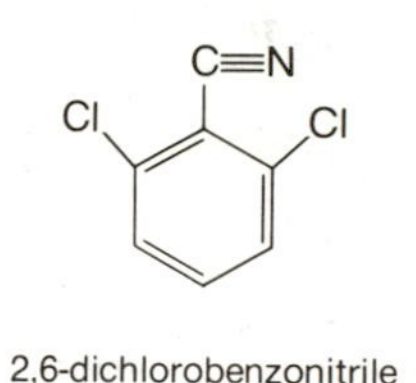

2,6-dichlorobenzonitrile

BROMOXYNIL

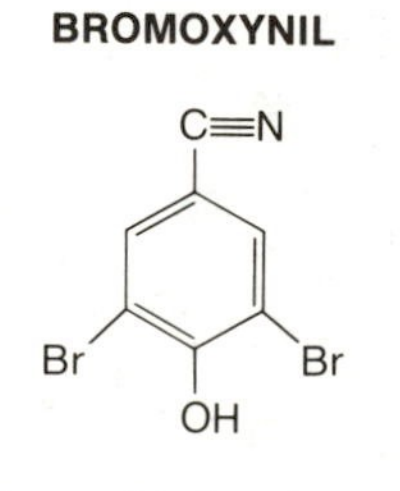

3,5-dibromo-4-hydroxybenzonitrile

IOXYNIL

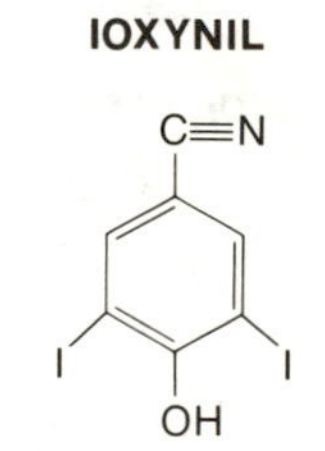

4-hydroxy-3,5-diiodobenzonitrile

Bipyridyliums

The name *bipyridylium* suggests the attachment of two pyridyl rings. There are two important herbicides in this group, diquat and paraquat. Both are contact herbicides that damage plant tissues quickly, causing the plants to appear frostbitten because of cell membrane destruction. This rapid wilting and desiccation occur within hours, making these novel herbicides also useful as preharvest desiccants for seed crops, cotton, soybeans, sugarcane, and sunflowers. Diquat is also used in aquatic weed control. Neither material is active in soils. They are available only to professional weed control specialists, who can achieve spectacular results for homeowners.

DIQUAT

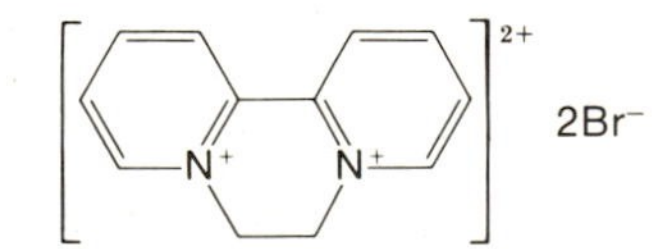

6,7-dihydrodipyrido(1,2-α:2′,1′-c)pyrazidinium (dibromide)

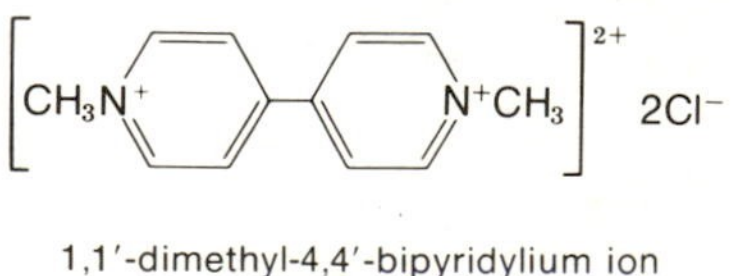

1,1′-dimethyl-4,4′-bipyridylium ion (dichloride)

Myco-Herbicides

A new concept in weed control is the use of disease microorganisms as useful, sometimes self-perpetuating pathogens. The first to be registered in this field is *Phytophthora palmivora* (Devine®), a naturally occurring, highly selective disease of milkweed vine (*Morrenia odorata*), a serious pest in citrus groves. Properly applied to the soil beneath citrus trees, *P. palmivora* kills existing milkweed vines. The microorganism persists in the vine root debris and continues to control germinating vines for more than one year after a single treatment. This pathogen is selective and does not infest citrus tree roots, fruit, or foliage. However, because certain other ornamental crops are susceptible to *P. palmivora*, caution is required in its use.

Miscellaneous Herbicides

Within the miscellaneous herbicides belongs methyl bromide (CH_3Br), which is used as a fumigant for any known organism in soil—nematodes, fungi, seed, insects, and other plant parts. Allyl alcohol ($CH_2{=}CHCH_2OH$) is a volatile, water-soluble fumigant used for the same purposes as methyl bromide.

Endothall is used as an aquatic weed and a selective field crop herbicide. It acts by interfering in RNA synthesis. Endothall has one distinct advantage over most aquatic weed killers—its low toxicity to fish, an outstanding example of environmental protection through pesticide selectivity.

One of the better turf herbicides, especially for the control of crabgrass, is bensulide. It is an organophosphate, but is considered one of the less toxic herbicidal materials. Bensulide acts by inhibiting cell division in root tips. It is used as a preemergence herbicide in lawns to control certain grasses and broad-leaf weeds, but it fails as a foliar spray because it is not translocated.

Acrolein requires application only by licensed operators because of its frightening tear-gas effect, but it is an extremely useful aquatic weed herbicide. Plants exposed to acrolein disintegrate within a few hours and float downstream. It is a general plant toxicant, destroying plant cell membranes and reacting with various enzyme systems. Spot treatment in lakes will effectively kill weeds without destroying the fish population.

Metolachlor was introduced in 1974 and belongs to the chloroacetamide herbicides (or acetanilides). It is soil-incorporated as a preplant and preemergence herbicide for annual weed grasses and some broad-leaf weeds in corn, soybeans, and peanuts. Its presumed mode of action is the inhibition of nucleic acid metabolism and protein synthesis.

Glyphosate was discovered in 1971 and belongs to the glyphosphate herbicide classification. It is a nonselective, nonresidual, postemergence material. Glyphosate is recognized for its effectiveness against perennial, deep-rooted, grass and broad-leaf weeds, as well as woody brush problems in crop and noncrop areas. It is a translocated, foliar-applied herbicide that can be applied at any stage of plant growth or at any time of year, with most types of application equipment, including the new wick, roller, and wiper devices. Its mechanism of action appears to be the inhibition of the synthesis of aromatic amino acids, which results in the inhibition of nucleic acid metabolism and protein synthesis.

ENDOTHALL

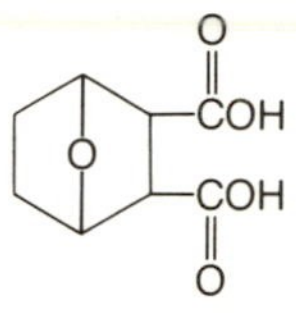

7-oxabicyclo (2,2,1)heptane-2,3-dicarboxylic acid (sodium salt)

BENSULIDE

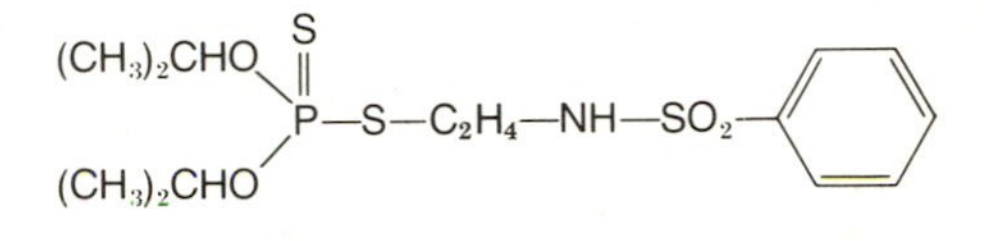

O,O-diisopropyl phosphorodithioate *S*-ester with *N*-(2-mercaptoethyl)benzenesulfonamide

ACROLEIN

$CH_2{=}CH{-}CHO$

METOLACHLOR (Dual®)

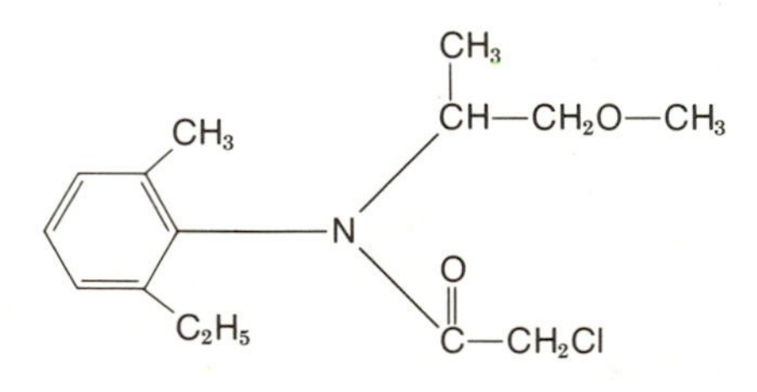

2-chloro-*N*-(2-ethyl-6-methylphenyl)-*N*-(2-methoxy-1-methylethyl) acetamide

GLYPHOSATE (Roundup®)

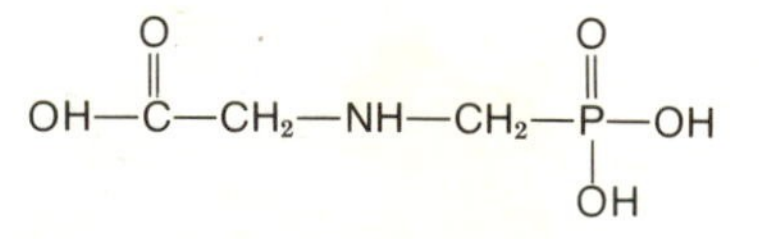

N-(phosphonomethyl)glycine (isopropylamine salt)

OXYFLUORFEN (Goal®)

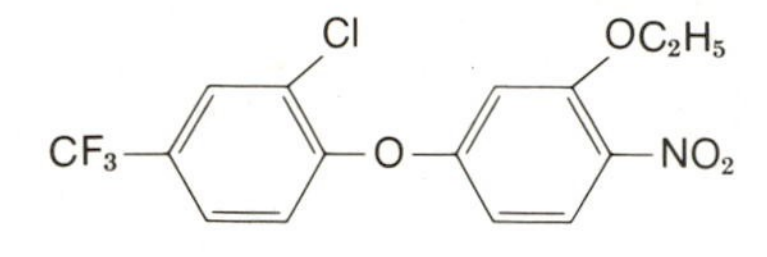

2-chloro-1-(3-ethoxy-4-nitrophenoxy)-4-(trifluoromethyl) benzene

FOSAMINE AMMONIUM (Krenite®)

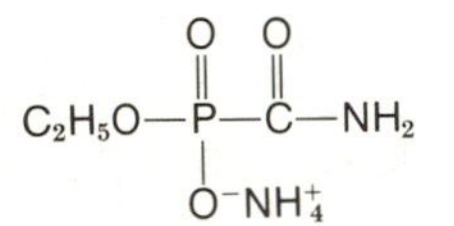

ammonium ethyl carbamoylphosphonate

Oxyfluorfen belongs to the diphenyl ether herbicides and controls annual broad-leaf and grass weeds in soybeans, corn (witchweed control), cotton, deciduous fruits, grapes, and nuts.

Fosamine ammonium belongs to the organophosphate herbicides and is used as a foliar spray for brush control in non-cropland, principally on rights-of-way.

Many chemical and use classes of herbicides are available, and some members of the same chemical class have different mechanisms of action. Those discussed and illustrated here are but a cross section of the existing herbicides. We can expect that different classes will be developed in the future and their mechanisms of action will be more clearly delineated, following intensive research on this complex subject. The wealth of materials precludes a brief summary. Rather, the reader is urged to frequently review the groups, to establish generalizations regarding herbicide use and action classes.

CHAPTER 12

Plant Growth Regulators

They came ... and cut down from there a branch with a single cluster of grapes, and they carried it on a pole between two of them.

Numbers 13:23

Plant growth regulators (PGRs) are chemicals used in some manner to alter the growth of plants, blossoms, or fruits. Legally considered pesticides, plant growth regulators are also referred to as *plant regulating substances, growth regulants, plant hormones,* and *plant regulators.* Hormones are the natural substances produced by plants that control growth, initiate flowering, cause blossoms to fall, set fruit, cause fruit and leaves to fall, control initiation and termination of dormancy, and stimulate root development.

The use and development of PGRs began in 1932, when it was discovered that acetylene and ethylene would promote flowering in pineapples. In 1934, auxins were found to enhance root formation in cuttings. Since then, outstanding developments have occurred as a consequence of the use of PGRs. Some of the major discoveries include the development of seedless fruit; prevention of berry and leaf drop in holly; prevention of early, premature drop of fruit; promotion of heavy setting of fruit blossoms; thinning of blossoms and fruit; prevention of sprouting in stored potatoes and onions; and the inhibition of buds in nursery stock and fruit trees to prolong dormancy. These are just a few of the hundreds of achievements made possible by these growth-controlling chemicals. As a result, we eat better-quality foods, eat fruits and vegetables "out-of-season" or literally year-round, and pay less for our food because of the reduced need for hand labor in thinning and harvesting.

Six classes of PGRs are recognized by the American Society for Horticultural Science: auxins, gibberellins, cytokinins, ethylene generators, inhibitors, and growth retardants.

AUXINS

Auxins are compounds that induce elongation in shoot cells. Some occur naturally, whereas others are manufactured. Auxin precursors are materials that are metabolized to auxins in the plant. Antiauxins are chemical compounds that inhibit the action of auxins.

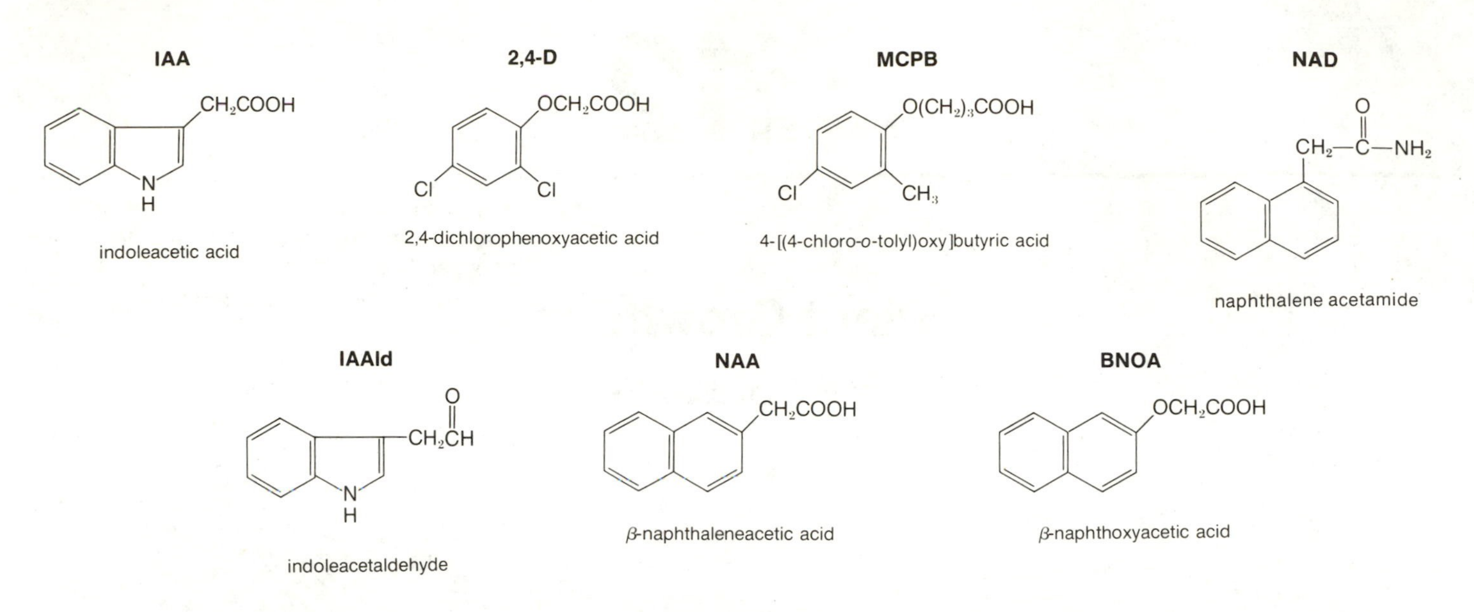

Auxins are used to thin apples and pears, increase yields of potatoes, soybeans, and sugar beets, to assist in the rooting of cuttings, and to increase flower formation, among other things. The mechanism of action is not completely understood, but they may work by controlling the type of enzyme produced in the cell. In any event, with the addition of auxin the individual cells become larger by a loosening of the cell wall, which is followed by the increased uptake of water and expansion of the cell wall.

Seven auxins are illustrated here. Among them, 2,4-D is classed as a herbicide with auxinlike characteristics. In citrus culture, 2,4-D is used to prevent preharvest fruit drop in older trees (6 years and older), to prevent leaf and fruit drop following pesticide oil sprays, to delay fruit maturity, and to increase fruit size. To achieve these effects 2,4-D is applied at precise times, to the entire tree, in the form of water sprays containing the herbicide at 8 to 16 ppm. It is effective on grapefruit, lemons, and navel and Valencia oranges.

GIBBERELLINS

Gibberellins are compounds that stimulate cell division or cell elongation or both, and have a gibbane skeleton. In 1957, gibberellin, or gibberellic acid, was introduced to the horticultural world. It caused incredible growth in many types of plants. Although originally isolated from a fungus, gibberellins were later found to be natural constituents in all plants. Since then, more than 57 gibberellins have been isolated and are identified as GA_1, GA_2, etc. The gibberellic acid most commonly used is GA_3. The mechanism of action for gibberellins is the induction or manufacture of more enzyme(s) in the cells, resulting in cell growth, particularly by elongation. The most striking effect of treatment with gibberellin is the stimulation of growth, expressed as long stems.

Examples of gibberellins' uses are: to increase stalk length and yields of celery, to break dormancy of seed potatoes, to increase grape size, to induce seedlessness in grapes, to improve the size of

GIBBANE

GA₃

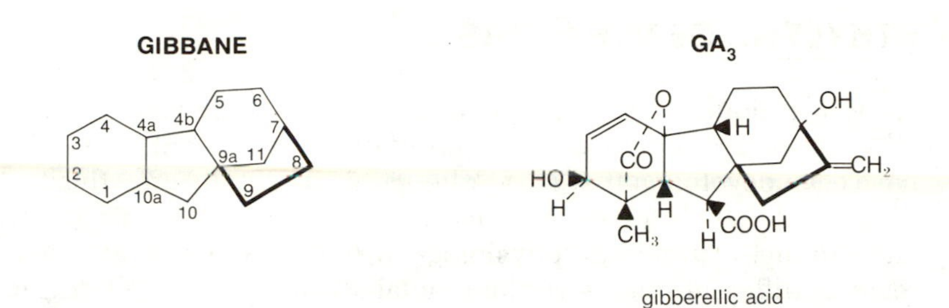

gibberellic acid

greenhouse-grown flowers, to delay fruit maturity, to extend fruit harvest, and to improve fruit quality.

CYTOKININS

Cytokinins (also referred to as *phytokinins*) are naturally occurring or manufactured compounds that induce cell division in plants. Most of the cytokinins are derivatives of adenine. These useful materials were discovered in 1955. Their practical potential lies in prolonging the storage life of green vegetables, cut flowers, and mushrooms.

Cytokinins cause two outstanding effects in plants—the induction of cell division and the regulation of differentiation in removed plant parts. Their mechanism of action is not known, but they apparently act at the gene level, becoming incorporated in the nucleic acids of the cell, where they influence cell division.

Naturally Occurring Cytokinins

ZEATIN

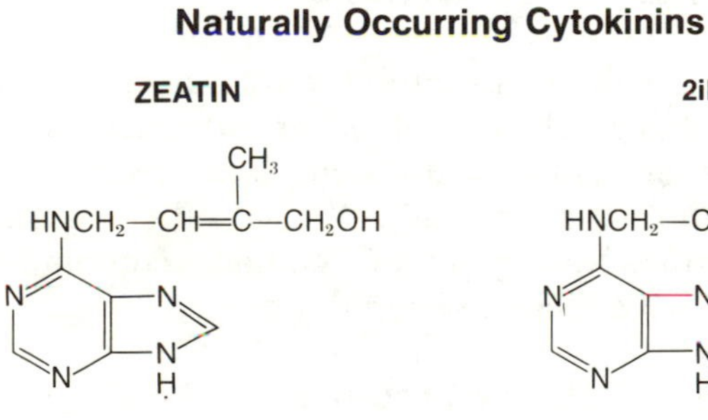

2iP

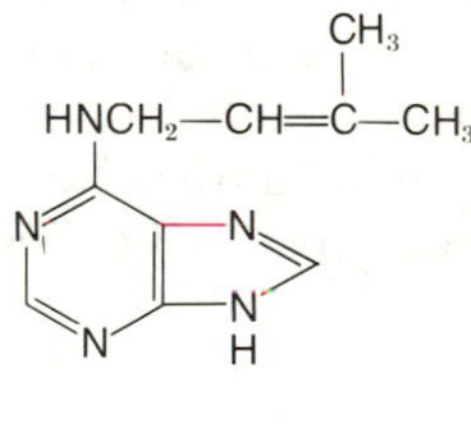

6-(γ,γ-dimethylallylamino)purine

Synthetic Cytokinins

BA

PBA

ADENINE

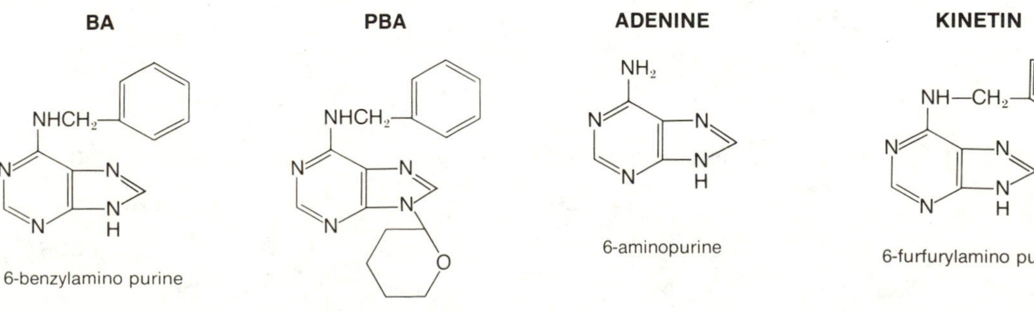

6-benzylamino purine

6-(benzylamino)-9-(2-tetrahydro-pyranyl)-9*H*-purine

6-aminopurine

KINETIN

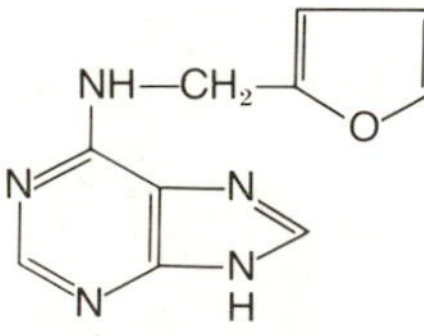

6-furfurylamino purine

ETHYLENE GENERATORS

As we mentioned, the first use of a plant growth regulator was that of ethylene in promotion of pineapple flowering. Recently, materials have been developed that are applied as sprays to the growth sites of plants and that stimulate the release of ethylene ($H_2C{=}CH_2$). Ethylene produces numerous physiological effects and can be used to regulate different phases of plant metabolism, growth, and development.

The ethylene generators also have many uses. They can be used to accelerate pineapple maturity; to induce uniform fig ripening; to facilitate mechanical harvest of peppers, cherries, plums, and apples; and to induce uniform flowering and fruit thinning, among other possibilities.

Ethylene can be used to stimulate seed germination and sprouting; abscission of flowers, leaves, and fruit; regulation of growth; and ripening of fruit, when introduced at the proper time. None of these biological effects results from a clearly defined mechanism of action. Generally, however, ethylene apparently has its greatest influence on the dominant enzymes of the particular physiological state of the absorbing tissue. The ethylene apparently serves as a trigger or synergist, resulting in a chain of premature biochemical events, expressed in the final and usually desirable result. Ethephon gradually releases ethylene as a degradation product when applied to plant surfaces.

ETHEPHON (Ethrel®)

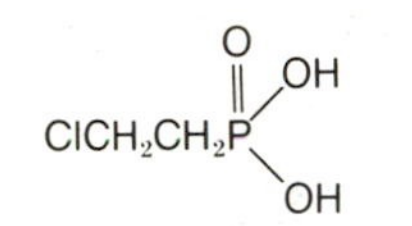

2-(chloroethyl)phosphonic acid

INHIBITORS AND RETARDANTS

An assorted group of substances inhibit or retard certain physiological processes in plants. Those that occur naturally in plants are usually hormones and inhibit different functions; for instance, growth or seed germination, or the action of other hormones, gibberellins, and auxins. New types of synthetic compounds, plant growth retardants, have recently been discovered.

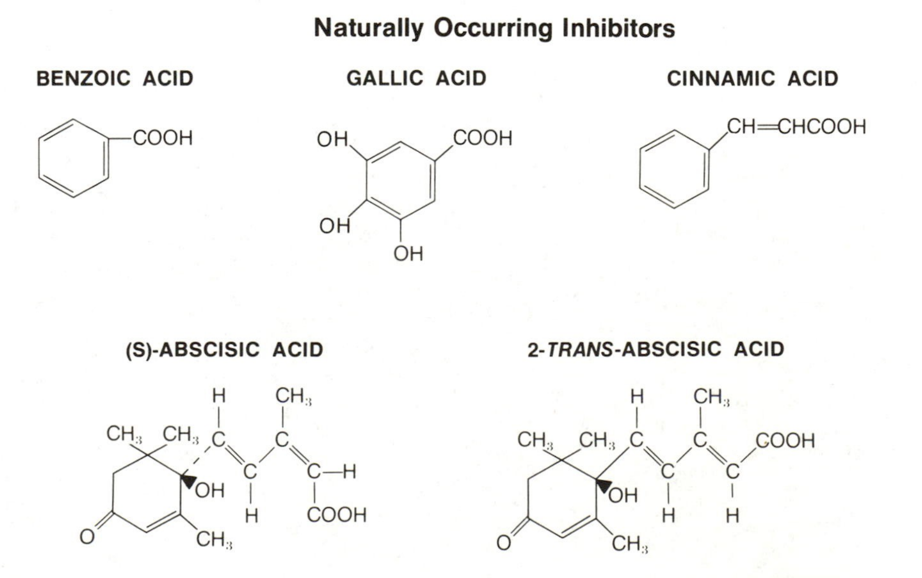

The inhibitors and retardants have many uses. They prevent sprouting of stored onions, potatoes, and other root crops; they retard sucker development on tobacco plants; they induce shortening of stems; they favor redistribution of dry matter; they prevent lodging of grain; and they permit controlled growth of flower crops and ornamentals.

Synthetic inhibitors include mepiquat chloride (Pix®), a recent experimental growth regulator for cotton. It is applied at early bloom or when the plants are 2 feet high, whichever comes first, to reduce vegetative growth. In some instances it is reported to increase cotton maturity and yields. Mepiquat chloride in combination with ethephon is used as a growth regulator (Terpal®) for winter and spring barleys, rye, and oats to speed heading and early harvest. The mode of action for mepiquat chloride has not yet been identified.

Chlormequat chloride, another synthetic inhibitor, has a wide range of uses, including promoting flower production by shortening internodes and thickening stems. It is also used in Europe to prevent lodging of small grains and in vegetable production to produce stockier plants with thicker stems. Ancymidol reduces internode elongation and is used in greenhouse flower production, applied either to soil or foliage. Dikegulac serves as a pinching agent for azaleas and as a growth retardant for shrubs, hedges, and ground covers. Daminozide has a broad range of effects from restricting vegetative growth to increasing fruit set. Mefluidide suppresses seedhead formation and regulates growth of turf grasses.

These materials are a diverse group of compounds and thus have different biological effects on plants. They are antagonistic to the growth-promoting hormones, such as auxins, gibberellins, and cytokinins, through a multitude of biochemical actions.

MEPIQUAT CHLORIDE (Pix®)

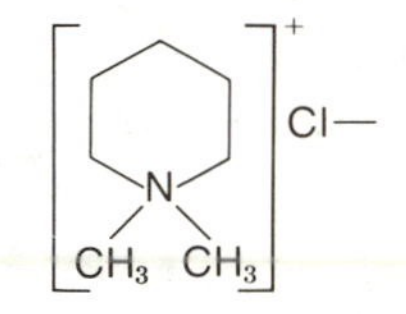

1,1-dimethyl-piperidiniumchloride

CHLORMEQUAT CHLORIDE (CeCeCe®, Cycocel®)

$ClCH_2CH_2N^+(CH_3)_3\ Cl^-$

2-chloroethyltrimethylammonium ion

MH—MALEIC HYDRAZIDE

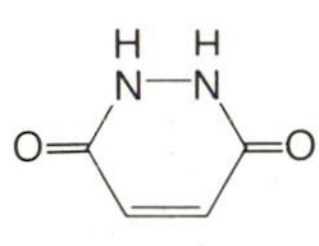

1,2-dihydro-3,6-pyridazinedione

DAMINOZIDE (Alar®)

$(CH_3)_2N{-}NHC(=O)CH_2CH_2COOH$

butanedioic acid mono-(2,2-dimethyl hydrazide)

ANCYMIDOL (A-Rest®)

a-cyclopropyl-*a*-(*p*-methoxyphenyl)-5-pyrimidine methanol

DIKEGULAC SODIUM (Atrinal®)

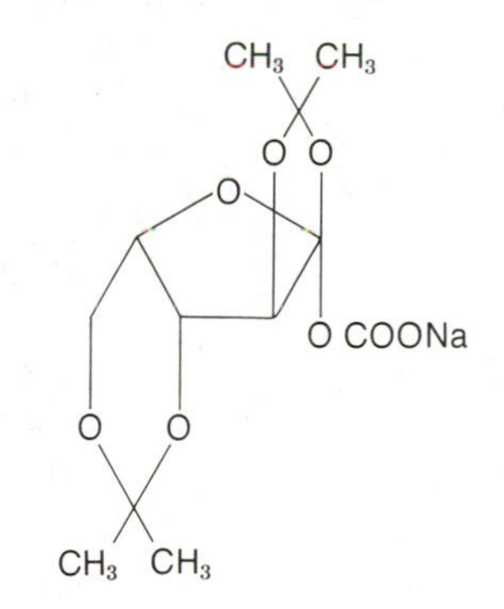

sodium salt of 2,3:4,6-bis-*O*(1-methylethylidene)-*O*-(-L-xylo-2-hexulofuranosonic acid)

MEFLUIDIDE (Embark®)

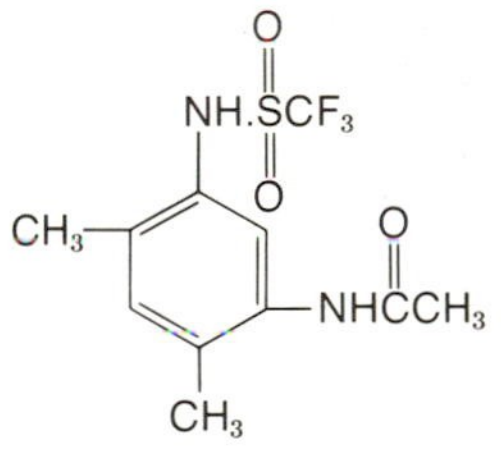

N-(2,4-dimethyl-5-(((trifluoromethyl) sulfonyl)amino)phenyl)acetamide

CHAPTER 13

Defoliants and Desiccants

Certain trees are more susceptible to ... arsenic than others. After each rain the poison ... is absorbed more, or is more active when wet, and ... it acts by dehydrating thereafter.

C. V. Riley, *U.S. Agricultural Report* (1886)

Chemical defoliation is the premature abscission of leaves (that is, separation of leaves from the plant), brought on by the formation of the abscission layer at the point where the leaf petiole joins the plant stem. Defoliants facilitate harvest operations by accelerating leaf fall from crop plants such as cotton, soybeans, or tomatoes. For example, the premature removal of leaves from the cotton plant permits earlier harvesting, and results in higher grades of cotton because few leaves remain to clog the mechanical picker, add trash, or stain the fibers. Defoliation often helps lodged (fallen-over) plants to straighten up, increasing the plants' exposure to sun and air. This enables the plants to dry quickly and thoroughly, and the mature bolls open faster, reducing boll rots that damage fiber and seed. And, finally, defoliation of the cotton plant reduces populations of fiber-staining insects, particularly aphids and whiteflies, which deposit honeydew on the open bolls.

Two conditions must exist before defoliation of any crop is effective: First, the plant must be in a state of maturity in which growth has stopped, and second, the temperatures must exceed 27°C during the day and 10°C at night.

Chemicals that are used to speed the drying of crop plant parts such as cotton leaves and potato vines are called *desiccants*. They usually kill the leaves rapidly, freezing them to the plant and thus mimicking the effect of a light frost. Desiccants cause foliage to lose water. Leaves, stems, and even branches of plants are sometimes killed so rapidly by desiccants that an abscission layer has insufficient time to develop, and the drying leaves thus remain attached to the plant. The advantage of desiccants over defoliants is that they can be applied at a later date than defoliants. During this additional time, for example, a cotton plant's leaves continue to function and contribute to seed and fiber quality.

DEFOLIANTS

For several reasons, only a small number of compounds are registered as defoliants. First, only a few crops require defoliation. Second, the materials must act rapidly to bring about abscission within a minimum time after application. Third, the compounds must break down rapidly, leaving no undesirable residue in the target portion of the crop. Defoliants are not for lawn and garden use.

Inorganic Salts

The inorganic salts are the oldest defoliants. Sodium chlorate ($NaClO_3$), magnesium chlorate ($Mg(ClO_3)2 \cdot 6H_2O$), disodium octaborate tetrahydrate ($Na_2B_8O_{14} \cdot 4H_2O$), and the other sodium polyborates are good examples of these old standbys, which are still used, primarily on cotton. These are contact materials that, by virtue of their high acidity, bring about rapid destruction of the delicate protoplasmic structures, resulting in the formation of the petiole abscission layer. The exact nature and sequence of the chemical reactions are unknown. The chlorates cause chlorosis of leaves and a starch depletion in stems and roots when applied in less than lethal doses.

Aliphatic Acids

As indicated in Chapter 11, an aliphatic acid is a carbon-chain acid. Their sodium salts are also included in this group. Neodecanoic acid is a readily degradable compound of moderate effectiveness for use on most crops requiring defoliation.

Prep® is the sodium salt of an aliphatic acid, is used as a defoliant for cotton and potatoes, and is rapidly absorbed but not translocated.

NEODECANOIC ACID

$$(CH_3)_3C(CH_2)_5\overset{\overset{\displaystyle O}{\|}}{C}OH$$

Prep®

$$Cl(CH_2)_2\overset{\overset{\displaystyle O}{\|}}{C}—ONa$$

sodium *cis*-3 chloroacrylate

Paraquat

Paraquat—also mentioned in Chapter 11 in the bipyridylium class—damages plant tissue very rapidly. Its swift action results from the breakdown of plant cells responsible for photosynthesis, giving the leaves a waterlogged appearance within a few hours of treatment. Paraquat results in the formation of OH^- radicals or hydrogen peroxide (H_2O_2) as the primary toxicant(s). Because most leaves drop off, paraquat is considered a defoliant.

PARAQUAT

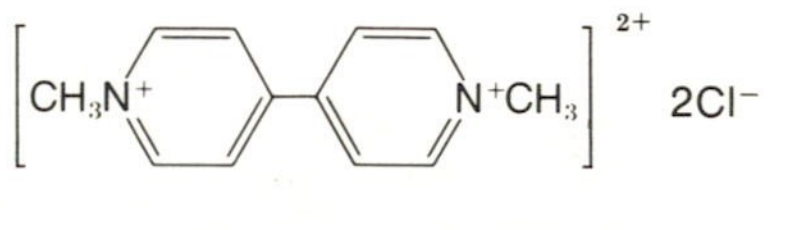

1,1′-dimethyl-4,4′-bipyridylium ion (dichloride)

Organophosphates

Merphos and DEF® are two organophosphate defoliants that have proved extremely useful in cotton production. Neither of these compounds is hormone acting. Rather, they induce abscission by injur-

MERPHOS (Folex®)

$(C_4H_9S)_3P$

S,S,S-tributyl phosphorotrithioite

DEF®

$(C_4H_9S)_3P{=}O$

S,S,S-tributyl phosphorotrithioate

THIDIAZURON (Dropp®)

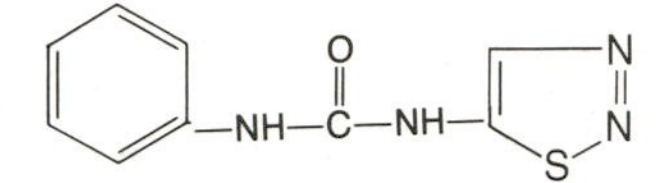

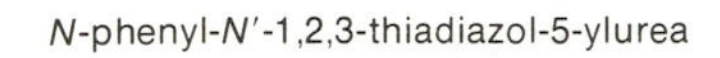
N-phenyl-*N′*-1,2,3-thiadiazol-5-ylurea

Harvade®

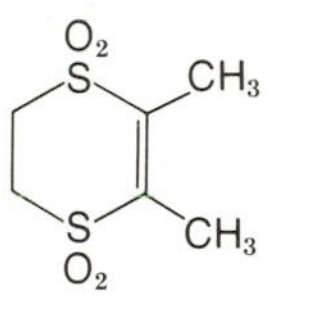

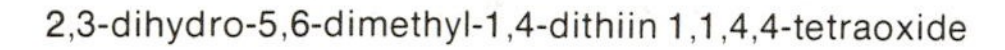
2,3-dihydro-5,6-dimethyl-1,4-dithiin 1,1,4,4-tetraoxide

ARSENIC ACID

H_3AsO_4

ing the leaf, causing changes in the levels of naturally occurring plant hormones that induce the early formation of the leaf abscission layer. Defoliation follows treatment by 4 to 7 days.

Miscellaneous Defoliants

Two very new defoliants, thidiazuron and Harvade® are still in the testing and registration stages at this writing. Thidiazuron is a defoliant for cotton that stimulates the formation of the abscission layer, causing shedding of the leaves, and also inhibits regrowth of leaves on treated cotton plants. Harvade® is an effective cotton defoliant even at temperatures of 21°C and below. It also defoliates nursery stock, grapes, and natural rubber. With rice and sunflower it has been shown to enhance maturation and subsequently reduce seed moisture content.

DESICCANTS

The Inorganics

The number of desiccants available is much larger than the number of defoliants. Ammonia and ammonium nitrate are both used as desiccants for cotton; they also ultimately add nutrients to the soil after the crop residue is returned to the soil. Petroleum solvents are applied to alfalfa and clover seed crops as well as to potatoes to enhance leaf drying and harvest.

Two of the inorganic salts mentioned in earlier chapters are used as desiccants for cotton: sodium borate(s) and sodium chlorate.

Arsenic acid is probably the oldest of the cotton desiccants, and it is still used in quantity. The material penetrates the leaf cuticle and the rapid contact injury precludes any extensive translocation. Arsenic acid uncouples oxidative phosphorylation and forms complexes with —SH-containing enzymes.

Phenol Derivatives

Dinoseb and pentachlorophenol, both phenol derivative herbicides of universal effectiveness, are used as desiccants for cotton. Pentachlorophenol has additional uses on seed crops of alfalfa, clovers, lespedeza, and vetch.

DINOSEB

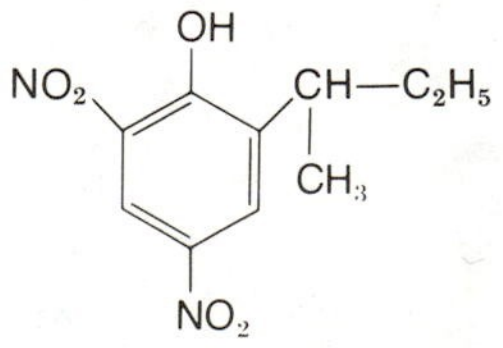

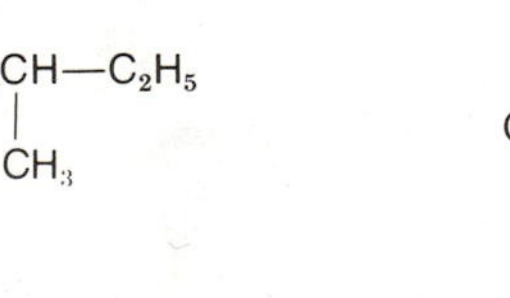

2-sec-butyl-4,6-dinitrophenol

PCP

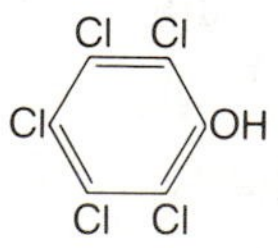

Cl Cl
Cl OH
Cl Cl

pentachlorophenol

Bipyridyliums

The bipyridylium herbicides, diquat dibromide and paraquat, because of their frostlike effect on green foliage, are outstanding desiccants. Diquat dibromide is used on the seed crops only of alfalfa, clover, soybeans, and vetch, while paraquat is used on cotton, potatoes, and soybeans.

DIQUAT **PARAQUAT**

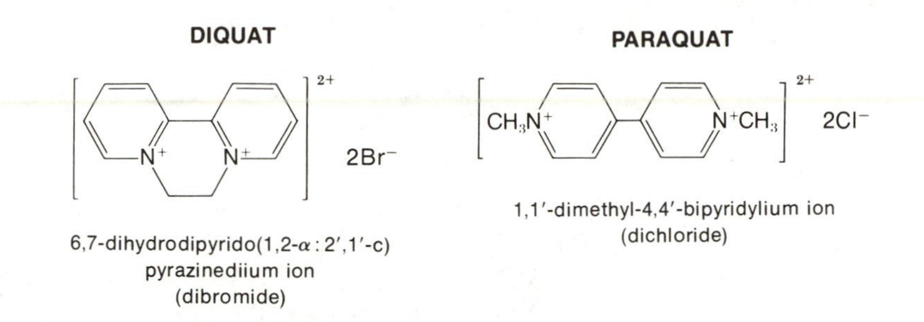

6,7-dihydrodipyrido(1,2-α : 2′,1′-c) pyrazinediium ion (dibromide)

1,1′-dimethyl-4,4′-bipyridylium ion (dichloride)

Miscellaneous Desiccants

Endothall, examined briefly toward the end of Chapter 11, is also a desiccant. It is used on cotton and—for the seed crops only—alfalfa, clovers, soybeans, trefoil, and vetch. It kills the leaves by contact through rapid penetration of the cuticle and results in desiccation and browning of the foliage. Its mode of action is not understood.

Ametryn, one of the very old and popular triazine herbicides, is also used as a potato vine desiccant prior to digging the potatoes. Ametryn penetrates the leaves rapidly and is a strong inhibitor of photosynthesis, causing the leaves to desiccate within 72 hours after application. (Newer, experimental desiccants are described in Chapter 24.)

ENDOTHALL

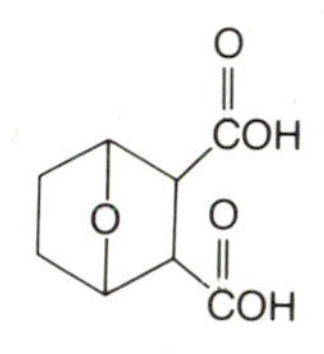

7-oxabicyclo (2,2,1)heptane-2,3-dicarboxylic acid

AMETRYN

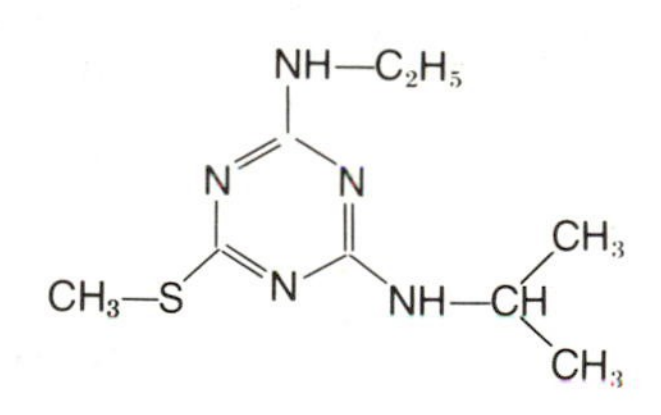

2-(ethylamino)-4-(isopropylamino)-6-(methylthio)-S-triazine

PART V

CHEMICALS USED IN THE CONTROL OF MICROORGANISMS

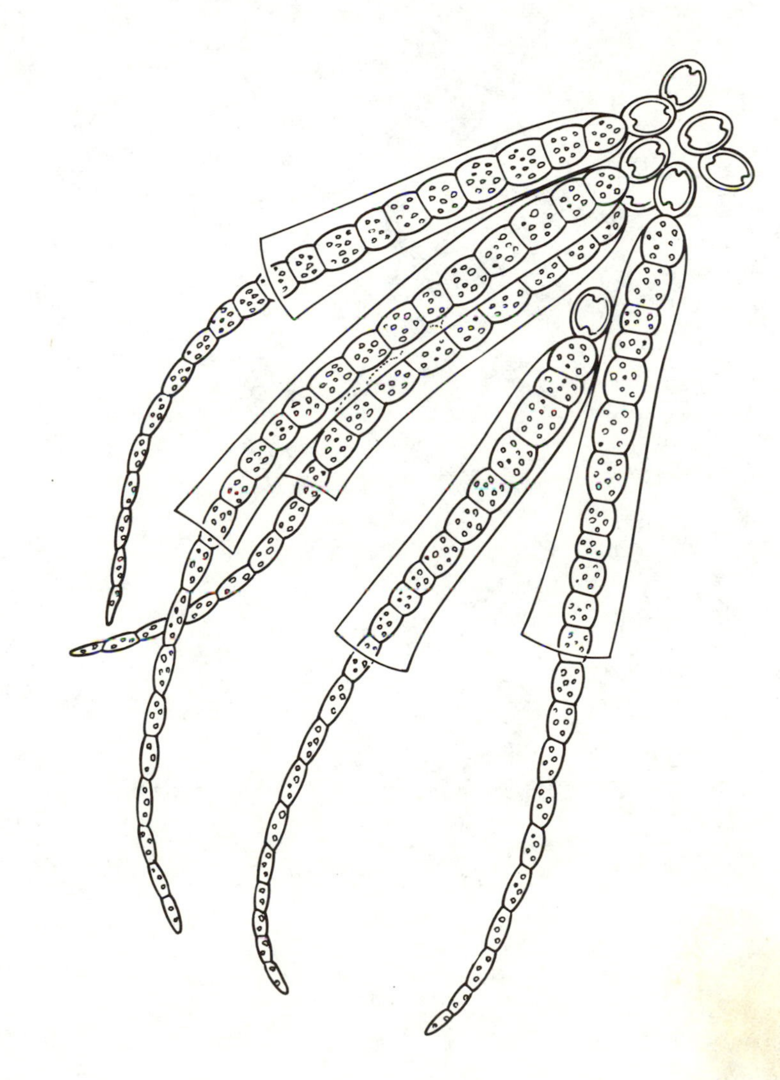

CHAPTER 14

Fungicides and Bactericides

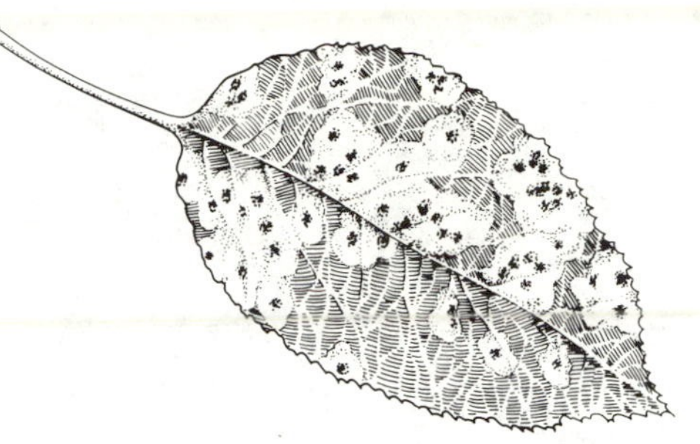

The Lord shall smite thee ...
with blasting and with mildew.

Deuteronomy 28:22

I have smitten you with blasting
and with mildew; I laid waste
your gardens and your vineyards;
and your fig and your olive trees.

Amos 4:9

Fungicides, strictly speaking, are chemicals used to kill or halt the development of fungi. However, for our purposes we shall consider them as chemicals used to control bacterial as well as fungal plant pathogens, the causal agents of most plant diseases. Other organisms that cause plant disease are viruses, rickettsias, algae, nematodes, mycoplasmalike organisms, and parasitic seed plants. In this chapter we discuss only those chemicals used for the control of fungi and bacteria.

There are hundreds of examples of plant disease. These include storage rots, seedling diseases, root rots, gall diseases, vascular wilts, leaf blights, rusts, smuts, mildews, and viral diseases. These can, in many instances, be controlled by the early and continued application of selected fungicides that either kill the pathogens or inhibit their development.

Most plant diseases can be controlled to some extent with today's fungicides. Among those that are only beginning to be controlled with chemicals are *Phytophthora* and *Rhizoctonia* root rots, *Fusarium*, *Verticillium* and bacterial wilts, and the viruses. The difficulties with these diseases are that they either occur below ground, and thus beyond the reach of fungicides, or they are systemic within the plant.

Fungal diseases are basically more difficult to control with chemicals than are insects because the fungus is a plant living in close quarters with its host. This explains the difficulty of finding selective chemicals that kill the fungus without harming the plant. Also, fungi that can be controlled by fungicides may undergo secondary cycles rapidly and produce from 12 to 25 "generations" during a 3-month growing season. Consequently, repeated applications of protective fungicides may be necessary, due to plant growth dilution and removal by rain and other weathering.

Fungicides must be applied to plants during stages when they are vulnerable to inoculation by pathogens, before there is any evidence of disease. Those fungicides referred to as *chemotherapeutants* can help to control certain diseases after the symptoms appear. Also, protective fungicides are commonly used, even after symptoms of disease have appeared. Eradicant fungicides are usually applied directly to the pathogen during its overwintering stage, long before disease has begun and symptoms have appeared. In the case of crops whose sale depends on appearance, such as lettuce and celery, however, the fungicide must be applied as a protectant spray, in advance of the pathogens, to prevent the disease.

There are about 225 fungicidal materials in our present arsenal, most of which are recently discovered organic compounds. Most of these act as protectants, preventing spore germination and subsequent fungal penetration of plant tissues. Protectants are applied repeatedly to cover new plant growth and to replenish the fungicide that has deteriorated or has been washed off by rain.

The application principle for fungicides differs from that of herbicides and insecticides. Only that portion of the plant that has a coating of dust or spray film of fungicide is protected from disease. Thus a good uniform coverage is essential. Fungicides, with several new exceptions, are not systemic in their action. They are applied as sprays or dusts, but sprays are preferable, since the films stick more readily, remain longer, can be applied during any time of the day, and result in less off-target drift.

Thanks to modern chemistry, many of the serious diseases of grain crops are controlled by treating the seeds with selective materials. Others are controlled with resistant varieties. Diseases of fruit and vegetables are often controlled by sprays or dusts of fungicides.

Historically, fungicides have relied on sulfur, copper, and mercury compounds, and even today most of our plant diseases could be controlled by these groups. However, the sulfur and copper compounds can retard growth in sensitive plants, and therefore the organic fungicides were developed. These sometimes have greater fungicidal activity and usually have less phytotoxicity.

The general-purpose fungicides for agriculture include inorganic forms of copper, sulfur (and mercury, until recently), and metallic complexes of cadmium, chromium, and zinc, along with a variety of organic compounds. The general-purpose lawn and garden fungicides are few in number and are usually organic compounds.

INORGANIC FUNGICIDES

Sulfur

Sulfur in many forms is probably the oldest effective fungicide known and is still a very useful garden fungicide. There are three physical forms or formulations of sulfur used as fungicides. The first is finely ground sulfur dust that contains 1 to 5 percent clay or talc to assist in dusting qualities. The sulfur in this form may be used as a carrier for another fungicide or an insecticide. The second is flotation or colloidal sulfur, which is so very fine that it must be formulated as a wet paste in order to be mixed with water. In its original, dry, microparticle size, it would be impossible to mix with water,

TABLE 14-1
A sampling of the vast number of inorganic copper compounds used as fungicides.

Name	Chemical formula	Uses
Cupric sulfate	$CuSO_4 \cdot 5H_2O$	Seed treatment and preparation of Bordeaux mixture
Cupric hydroxide (Kocide®)	$Cu(OH)_2$	Seed treatment, foliage spray; many fungal diseases
Copper oxychloride	$3Cu(OH)_2 \cdot CuCl_2$	Powdery mildews
Copper oxychloride sulfate	$3Cu(OH)_2 \cdot CuCl_2 +$ $3Cu(OH)_2 \cdot CuSO_4$	Many fungal diseases
Copper ammonium carbonate (Copper-Count-N®)	Chemical complex (formula not known)	Many citrus, deciduous fruit, and vegetable diseases
Cuprous oxide	Cu_2O	Powdery mildews
Basic copper sulfate	$CuSO_4 \cdot Cu(OH)_2 \cdot H_2O$	Seed treatment and preparation of Bordeaux mixture
Cupric carbonate (Malachite®)	$Cu(OH)_2 \cdot CuCO_3$	Many fungal diseases
Copper resinate (Citcop®)	Salts of fatty and rosin acids	Bacterial and fungal diseases of grapes, citrus, vegetables

but would merely float. Wettable sulfur is the third form; it is finely ground with a wetting agent so that it will mix readily with water for spraying. The easiest to use, of course, is dusting sulfur, applied when plants are slightly moist with the morning dew.

For most effective disease control the particle size of wettable sulfurs should be no larger than 7 μm. A good grade of dusting sulfur should pass through a 325-mesh or finer screen.

Copper

The majority of inorganic copper compounds are practically insoluble in water and are pretty blue, green, red, or yellow powders sold as fungicides. The various forms include Bordeaux mixture, named after the Bordeaux region in France, where it originated. Bordeaux is a chemically undefined mixture of copper sulfate and hydrated lime, which was accidentally discovered when sprayed on grapes in Bordeaux to scare off "freeloaders." It was soon observed that downy mildew, a disease of grapes, disappeared from the treated plants. From this unique origin began the commercialization of fungicides. The copper ion, which becomes available from both the highly soluble and relatively insoluble copper salts, provides the fungicidal as well as phytotoxic and poisonous properties. A few of the many inorganic copper compounds used over the years are presented in Table 14-1.

The EPA has determined that no tolerance level need be set for a large number of copper compounds: Bordeaux mixture, copper acetate, copper carbonate–basic, copper-lime mixtures, copper oxychloride, copper silicate, copper sulfate, copper sulfate–basic, copper-zinc-chromate, cuprous oxide, cupric oxide, and copper hydroxide.

A comment on solubility is appropriate at this point. In general, protective fungicides have low ionization constants, but in water some toxicant does go into solution. That small quantity absorbed by the fungal spore is then replaced in solution from the residue. The spore continues to accumulate the toxic ion in sublethal doses, whose cumulative effect is lethal. Except for powdery mildews, water in the penetration court—the place on plant tissue softened by fungal mycelium, which permits it to penetrate the plant cuticle—thus permits spore germination and makes soluble the toxic portion of the fungicidal residue.

The copper ion is toxic to all plant cells and must be used in discrete dosages or in relatively insoluble forms, to prevent killing all or portions of the host plant. This is the basis for the use of relatively insoluble or "fixed" copper fungicides, which release only very low levels of copper, adequate for fungicidal activity, but not enough to affect the host plant.

Copper compounds are not easily washed from leaves by rain, since they are relatively insoluble in water, and thus give longer protection against disease than do most of the organic materials. They are relatively safe to use and require no special precautions during spraying. Although copper is an essential element for plants, there is some danger in an accumulation of copper in agricultural soils resulting from frequent and prolonged use. In fact, certain citrus growers in Florida have experienced a serious problem of copper toxicity after using fixed copper for disease control.

The currently accepted theory for the mode of action of copper's fungistatic action is its nonspecific denaturation of protein. The Cu^{++} ion reacts with enzymes having reactive sulfhydryl groups—which would explain its toxicity to all forms of plant life.

Mercury

The inorganic mercurial fungicides are probably the most toxic of the fungicides. Mercury's fungicidal properties and toxicity to animals are due in part to the degree of association of divalent mercury ions, which are toxic to all forms of life. As a result, no mercury residues are permitted in foods or feed.

Over the past 30 years, many organic mercury compounds were developed, but they have been replaced by other organic fungicides. Ceresan is typical of those used as seed treatment, whereas phenylmercury acetate (PMA) was useful for turf diseases, as a seed treatment, and as a dormant spray for fruit trees. The mode of action for

CERESAN

$CH_3OCH_2CH_2HgCl$

2-methoxyethylmercuric chloride

PMA

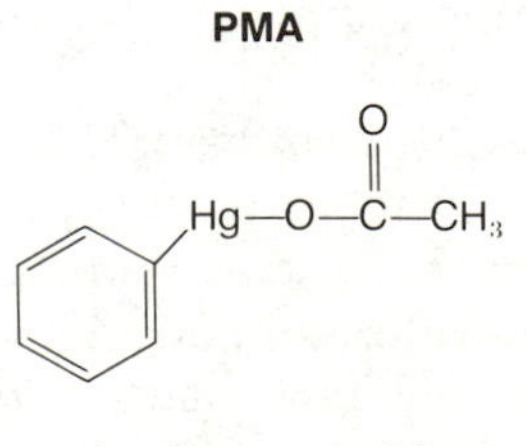

phenylmercury acetate

the mercurials is the nonselective or nonspecific inhibition of enzymes, especially those containing iron and sulfhydryl sites.

Both organic and inorganic mercurial fungicides, with one or two exceptions, have been banned from home and agricultural use by the EPA. This decision hinged on their toxicity to warm-blooded animals and accumulation of mercury in the environment.

ORGANIC FUNGICIDES

Many synthetic sulfur and other organic fungicides have been developed over the past 35 years to replace the more harsh, less selective inorganic materials. Most of them have had no measurable buildup effect on the environment after many years of use. Thiram, the first of the organic sulfur fungicides, was discovered in 1931; it was followed by many others. Then came other new classes, the dithiocarbamates and dicarboximides (zineb and captan) introduced in 1943 and 1949, respectively. Now more than 200 fungicides of all classes are in use or in various stages of development.

The newer organic fungicides possess several outstanding qualities. They are extremely efficient—that is, smaller quantities are required than of those used in the past; they usually last longer; and they are safer for crops, animals, and the environment. Most of the newer fungicides also have very low phytotoxicity, many being at least ten times safer than the copper materials. And most of them are readily degraded by soil microorganisms, thus preventing their accumulation in soils.

Dithiocarbamates

Among the dithiocarbamates we find the "old reliables"—thiram, maneb, ferbam, ziram, Vapam® (SMDC), and zineb—all developed in the early 1930s and 1940s. Such fungicides probably have greater popularity and use than all other fungicides combined. Except for systemic action, they are employed collectively in every use known for fungicides. The dithiocarbamates probably act by being metabolized to the isothiocyanate radical (—N═C═S), which inactivates the —SH groups in amino acids contained within the individual pathogen cells.

Thiazoles

The thiazoles, a class of compounds that offers a surprising chemical disposition, contains among others, ethazol. The five-membered ring of the thiazoles is cleaved rather quickly under soil conditions to form either the fungicidal —N═C═S or a dithiocarbamate, depending on the structure of the parent molecule. Ethazol is used only as a soil fungicide and, as such, is exposed to the ring cleavage just mentioned. The probable mode of action is similar to that of the dithiocarbamates.

THIRAM

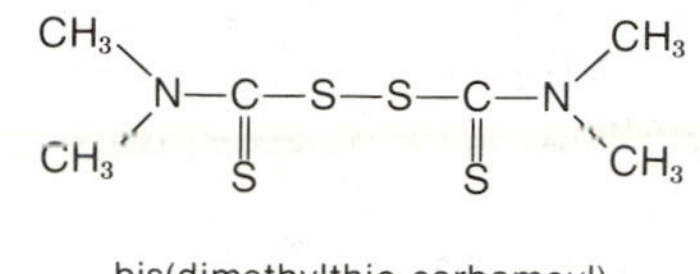

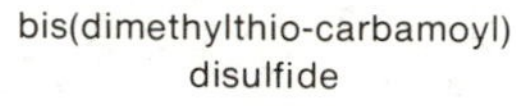
bis(dimethylthio-carbamoyl) disulfide

MANEB

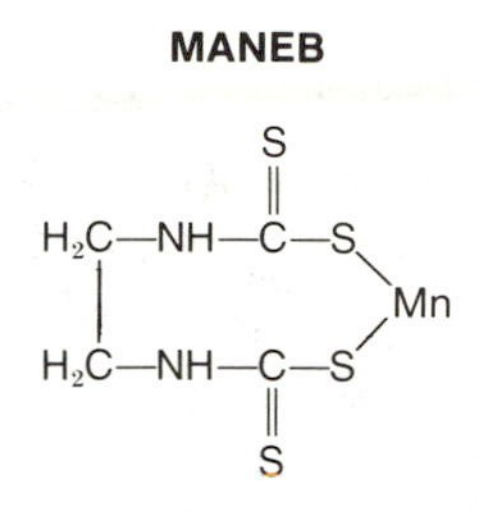

manganese ethylenebisdithiocarbamate

FERBAM

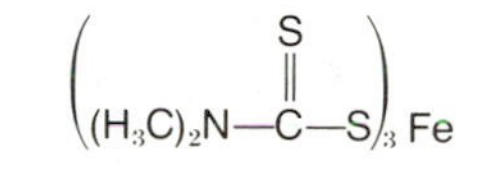

ferric dimethyldithiocarbamate

ZINEB

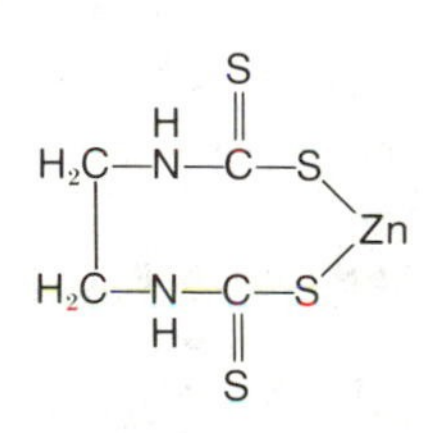

zinc ethylenebisdithiocarbamate

TRICYCLAZOLE (Beam®)

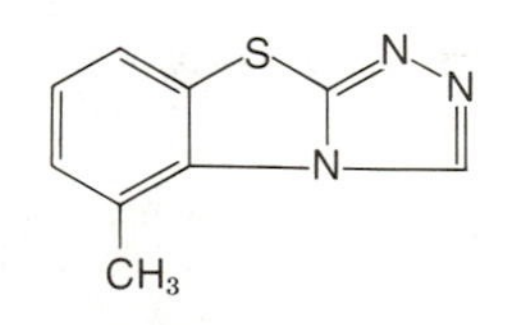

5-methyl-1,2,4-triazolo(3,4-b)-benzothiazole

ETHAZOL (Terrazole®)

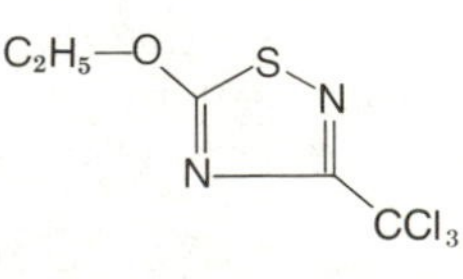

5-ethoxy-3-trichloromethyl-1,2,4-thiadiazole

Triazines

The triazine structure, seen frequently in herbicides, is found in only one fungicide. Anilazine was introduced in 1955 and has received wide use for control of potato and tomato leaf spots and turf-grass diseases.

ANILAZINE

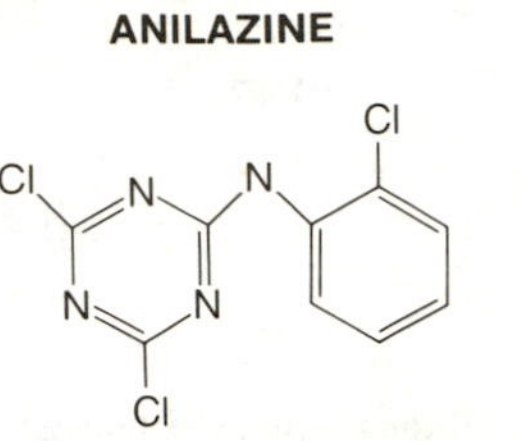

2,4-dichloro-6-(*o*-chloroanilino)-*s*-triazine

Substituted Aromatics

The substituted aromatics belong in a somewhat arbitrary classification assigned to the simple benzene derivatives that possess long-recognized fungicidal properties.

Hexachlorobenzene, introduced in 1945, is used as a seed treatment and as a soil treatment to control stinking smut of wheat. Pentachlorophenol (PCP) has been used since 1936 as a wood preservative, as a seed treatment, and, as pointed out in Chapter 11, as a herbicide. Pentachloronitrobenzene (PCNB) was introduced in the 1930s as a fungicide for seed treatment and selected foliage applications. It is also used as a soil treatment to control the pathogens of certain damping-off diseases of seedlings. Chlorothalonil is a very useful, broad-spectrum foliage-protectant fungicide made available in 1964. And chloroneb, developed in 1965, is heavily used for cotton seedling and turf diseases. Dicloran (DCNA) is a highly useful fungicide against *Botrytis, Monilinia, Rhizopus, Sclerotinia,* and *Sclerotium* species, on a wide range of fruits and vegetables.

Substituted aromatics are diverse in their modes of action. Being generally fungistatic, they reduce growth rates and sporulation of fungi, probably by combining with $—NH_2$ or —SH groups of essential metabolic compounds.

HEXACHLOROBENZENE

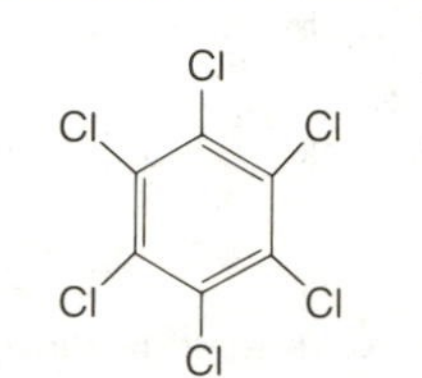

1,2,3,4,5,6-hexachlorobenzene

PCNB

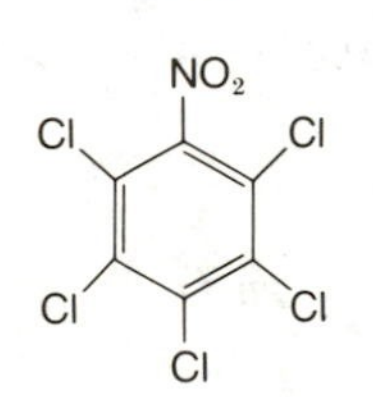

pentachloronitrobenzene

DICLORAN (DCNA, Botran®)

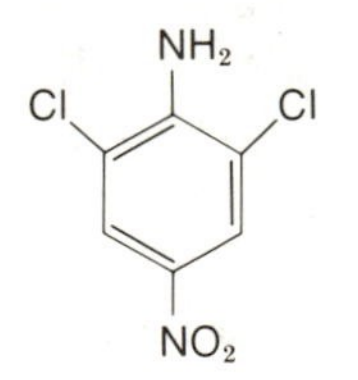

2,6-dichloro-4-nitroaniline

PCP | **CHLOROTHALONIL** | **CHLORONEB**

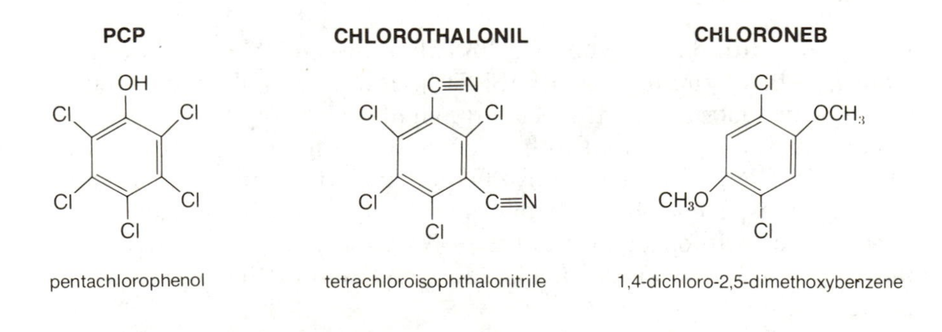

pentachlorophenol | tetrachloroisophthalonitrile | 1,4-dichloro-2,5-dimethoxybenzene

Dicarboximides (Sulfenimides)

Three extremely useful foliage protectant fungicides are dicarboximides. Captan appeared in 1949 and is undoubtedly the most heavily used fungicide around the home of all classes; folpet appeared in 1962; and captafol (Difolatan®) appeared in 1961. They are used primarily as foliage dusts and sprays on fruits, vegetables, and ornamentals.

The dicarboximides are some of the safest of all pesticides available and are recommended for lawn and garden use, as seed treatments, and as protectants for mildews, late blight, and other diseases. Remember the garden center adage, "When in doubt, use captan."

Many compounds containing the —$SCCl_3$ moiety are fungitoxic, a fact that indicates that group as a toxophore (molecular unit that accounts for the molecule's toxicity). Fungitoxicity of the dicarboximides is apparently nonspecific and is not a result of a single mode of action. The dicarboximides' lethal effect on disease organisms is probably due to the inhibition of the synthesis of amino compounds and enzymes containing the —SH radical.

Systemic Fungicides

Only in recent years have successful systemic fungicides been marketed, and very few are yet available. Systemics are absorbed by the plant and carried by translocation through the cuticle and across leaves to the growing points. Most systemic fungicides have eradicant properties that stop the progress of existing infections. They are therapeutic in that they can be used to cure plant diseases. A few of the systemics can be applied as soil treatments and are slowly absorbed through the roots to give prolonged disease control.

These systemics offer much better control of diseases than is possible with a protectant fungicide that requires uniform application and remains essentially where it is sprayed onto the plant surfaces. There is, however, some redistribution of protective fungicidal residues on the surfaces of sprayed or dusted plants, giving them longer residual activity than would be expected.

Systemics constitute the perfect method of disease control by attacking the pathogen at its site of entry or activity, and they reduce the risk of contaminating the environment by frequent broadcast fungicidal treatments. Undoubtedly, as newer and more selective systemic molecules are synthesized, they will gradually replace the protectants that compose the bulk of our fungicidal arsenal.

Those systemics currently in commercial use are mentioned in order of their appearance. None are available for home use.

Oxathiins. The oxathiins, represented by carboxin and oxycarboxin, and introduced in 1966, were the first of the systemics to succeed in practice. They are used as seed treatments for the cereal crops, particularly those affected by embryo-infecting smut fungi, and they have potential for other uses. They are selectively toxic to the smuts, rusts, and to *Rhizoctonia* (*Thanatephorus*), organisms belonging to the Basidiomycetes. The apparent mode of action of the oxathiins begins with their selective concentration in the fungal cells, followed by the inhibition of succinic dehydrogenase, an important enzyme to respiration in the mitochondrial systems.

Benzimidazoles. The benzimidazoles, represented by benomyl and thiabendazole (TBZ), were introduced in 1968 and have received wide acceptance as systemic fungicides against a broad spectrum of diseases. Benomyl has the widest spectrum of fungitoxic activity of all the newer systemics, including the *Sclerotinia*, *Botrytis*, and *Rhizoctonia* species, and the powdery mildews and apple scab.

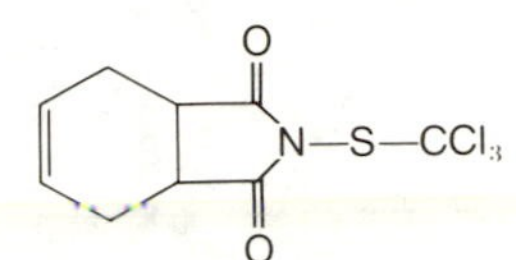

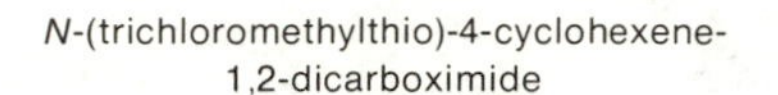
N-(trichloromethylthio)-4-cyclohexene-1,2-dicarboximide

FOLPET

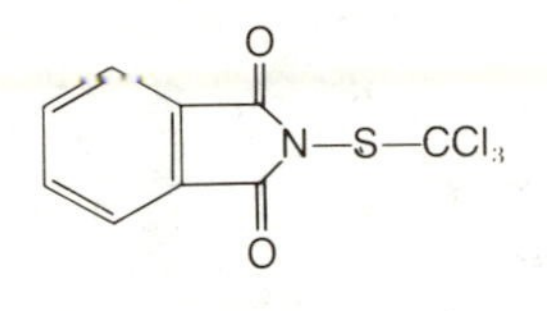

N-trichloromethylthiophthalimide

CAPTAFOL (Difolatan®)

O
N—S—CCl_2—$CHCl_2$
O

tetrachloroethylmercaptocyclohexenedicarboximide

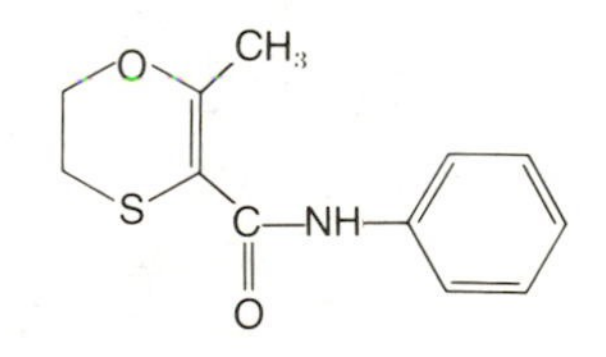

2,3-dihydro-5-carboxanilido-6-methyl-1,4-oxathiin

OXYCARBOXIN

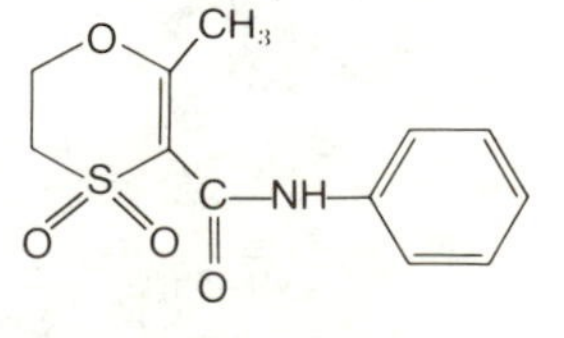

2,3-dihydro-5-carboxanilido-6-methyl-1,4-oxathiin-4,4-dioxide

BENOMYL

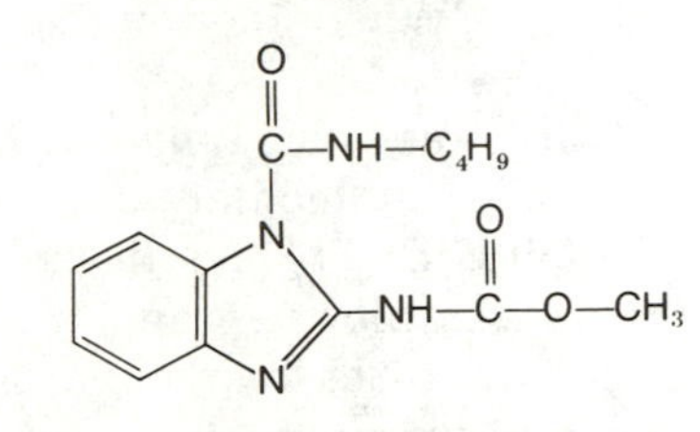

methyl 1-(butylcarbamoyl)-2-benzimidazolecarbamate

THIABENDAZOLE

2-(4′-thiazoyl) benzimidazole

THIOPHANATE

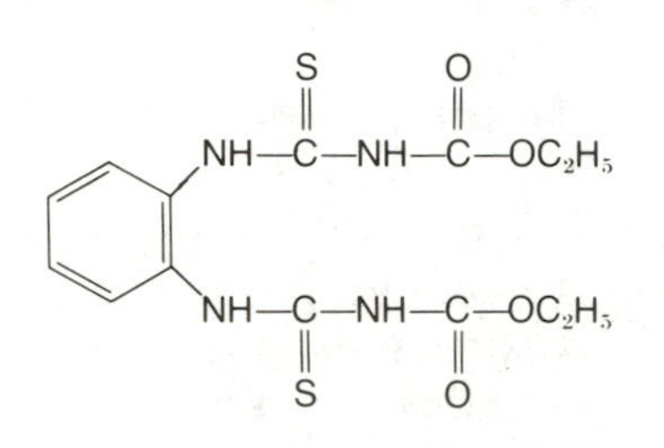

1,2-bis(3-ethoxycarbonyl-2-thioureido)benzene

CARBENDAZIM (Lignasan®)

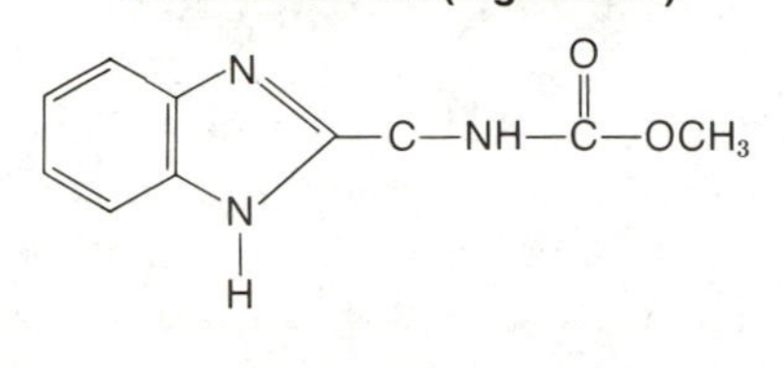

2-(methoxycarbonylamino)-benzimidazole

Thiabendazole has a similar spectrum of activity to that of benomyl. Introduced in 1969, thiophanate, although not a benzimidazole in its original structure, is converted to that group by the host plant and the fungus through their metabolism. Thiophanate has a fungitoxicity similar to that of benomyl. All three compounds have been used in foliar applications, seed treatment, dipping of fruit or roots, and soil application. Their mode of action appears to be the induction of abnormalities in spore germination, cellular multiplication, and growth, as a result of their interference in the synthesis of that vital nucleic material, DNA.

A later addition to the benzimidazoles is carbendazim, introduced in 1973. The interesting quality of this systemic is its proved usefulness in the control of the formidable Dutch elm disease. In the formulation of Lignasan®, a hydrochloride salt, it is injected into the trunks of diseased trees and results in a slow curative action.

Pyrimidines. The pyrimidine systemic fungicides appeared in the late 1960s, and include dimethirimol, ethirimol, and bupirimate.

DIMETHIRIMOL (Milcurb®)

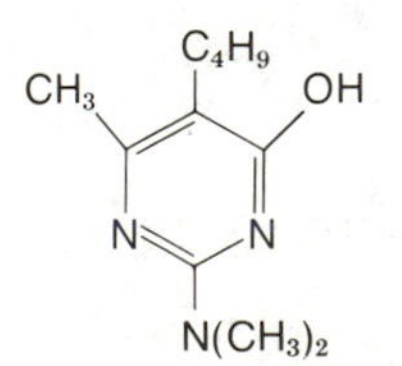

5-*n*-butyl-2-dimethylamino-4-hydroxy-6-methylpyrimidine

ETHIRIMOL (Milcurb® Super)

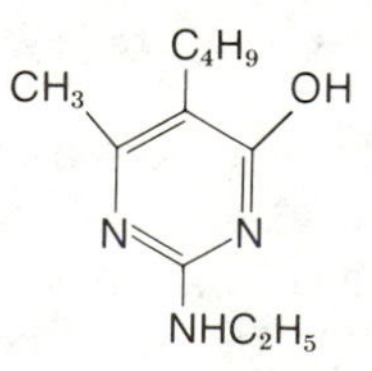

5-butyl-2-ethylamino-4-hydroxy-6-methylpyrimidine

BUPIRIMATE (Nimrod®)

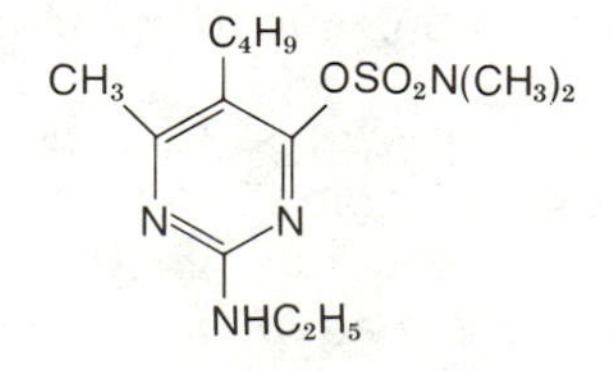

5-butyl-2-ethylamino-6-methyl-pyrimidin-4-yl-dimethylsulfamate

They are very active against specific types of powdery mildews; for instance, dimethirimol works well on cucurbits. Ethirimol is for cereals and other field crops, and bupirimate controls powdery mildews on apples and greenhouse roses.

Organophosphates. Among the newer systemic fungicides are two organophosphorous materials, IBP and Conen®. Developed in Japan and introduced in 1965, both are effective against rice blast, stem rot, and rice sheath blight.

Conen® **IBP (Kitazin®)**

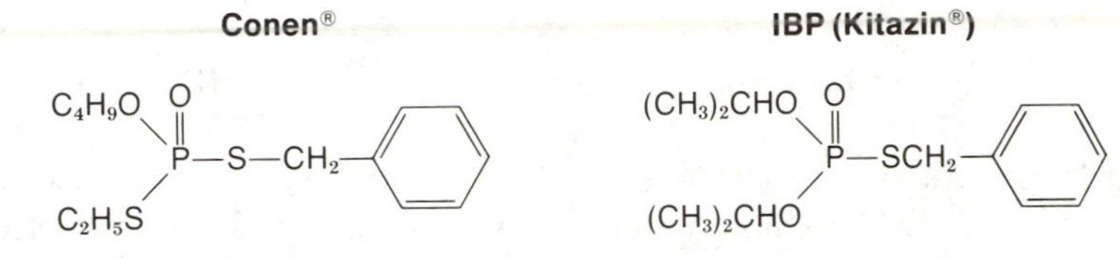

O-butyl-*S*-benzyl-*S*-ethyl phosphorodithioate *O,O*-diisopropyl-*S*-benzyl thiophosphate

Acylalanines. Another new group of systemic fungicides are the acylalanines, which include metalaxyl and furalaxyl. They are effective against soil-borne diseases caused by *Pythium* and *Phytophthora* and foliar diseases caused by the Phycomycetes (downy mildews). They show promise as foliar, soil, and seed treatments for agricultural crops. They offer systemic and corrective or curative activity as well as residual-protectant activity. However, they have little or no activity against the Ascomycetes, Basidiomycetes, and Fungi Imperfecti.

METALAXYL (Ridomil®, Dual®)

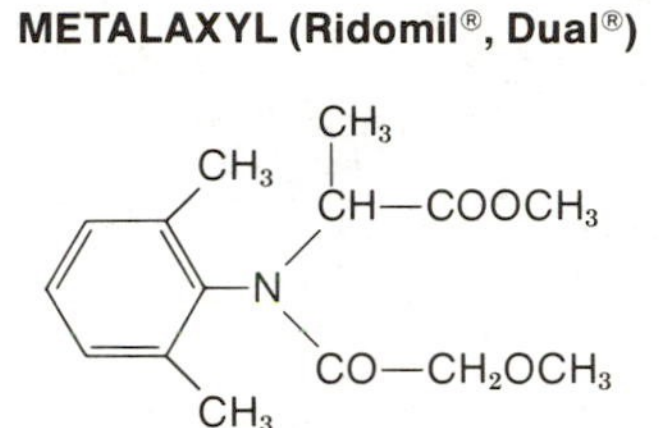

N-(2,6-dimethylphenyl)-*N*-(methoxy-acetyl)-alanine methyl ester

Triazoles. Triadimefon is the sole systemic fungicide of the triazole group. It carries both protective and curative actions, and is effective against mildews and rusts on vegetables, cereals, coffee, deciduous fruit, grapes, and ornamentals.

FURALAXYL (Fongarid®)

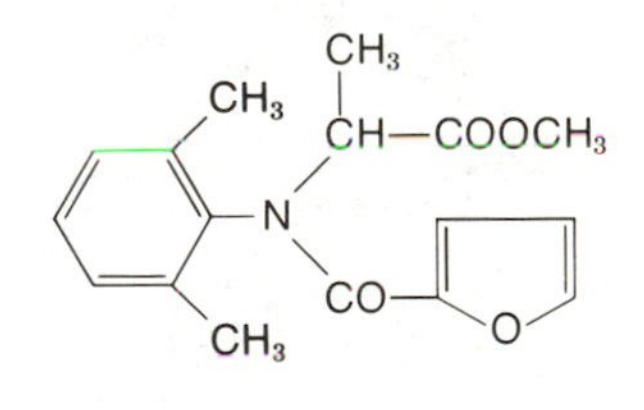

methyl *N*-2,6-dimethylphenyl-*N*-furoyl-(2)-alaninate

Piperazines. The piperazine systemic fungicides include only one member at this time, triforine, which is amazingly effective against powdery mildew on any host. Additionally it is active against scab and other diseases of fruit and berries, rust and black spot on ornamentals, powdery mildew and other leaf diseases on cereals, rust on cereals, several diseases of vegetables, and storage diseases of fruit.

TRIADIMEFON (Bayleton®)

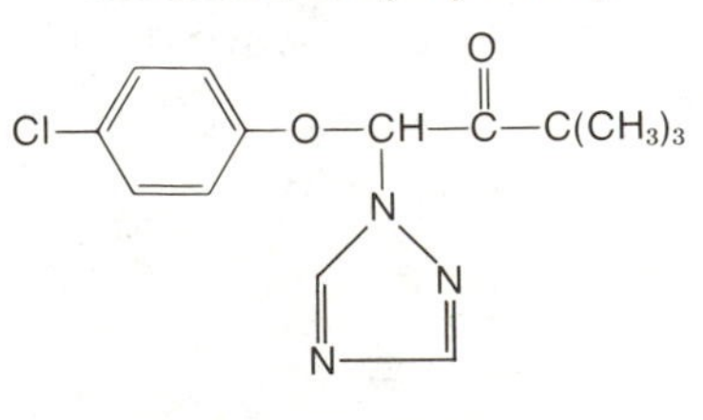

1-(4-chlorophenoxy)-3,3-dimethyl-1-(1H01,2,4-triazol-1-yl)-2-butanone

TRIFORINE

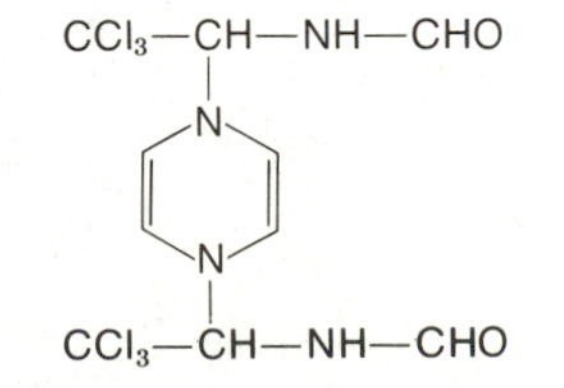

N-N′-(1,4-piperazinediyl-bis(2,2,2-trichloroethylidene))-bis-(formamide)

PROCYMIDONE (Sumilex®)

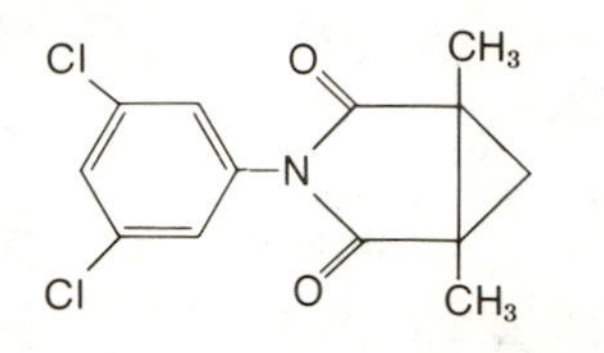

N-(3,5-dichlorophenyl)-1,2-dimethyl-cyclopropane-1,2-dicarboximide

IPRODIONE (Rovral®)

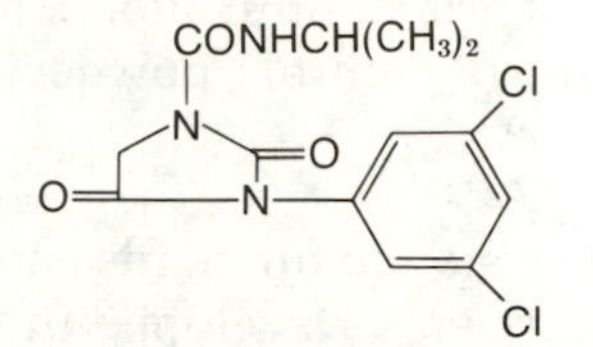

3-(3,5-dichlorophenyl)-*N*-(1-methyl-ethyl)-2,4-dioxo-1-imidazolidine-carboxamide

VINCLOZOLIN (Ronilan®)

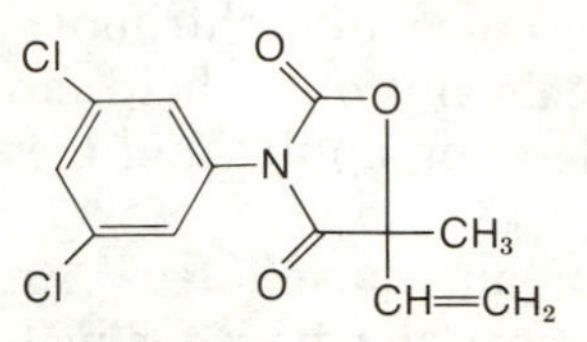

3-(3,5-dichlorophenyl)-5-ethenyl-5-methyl-2,4-oxazolidinedione

Imides. A second group of dicarboximides, the imides, are structurally significantly different from the originals. These systemic chemicals appeared during the 1970s. The group now includes procymidone, iprodione, and vinclozolin. They are particularly effective against *Botrytis*, *Monilinia*, and *Sclerotinia*. Iprodione is also active against *Alternaria*, *Helminthosporium*, *Rhizoctonia*, *Corticium*, *Typhula*, and *Fusarium* species.

Dinitrophenols

We have mentioned the dinitrophenols as insecticides and as herbicides (in Chapters 4 and 11, respectively). Their mode of action as fungicides is the same: the uncoupling of oxidative phosphorylation in cells with an attendant upset of the energy systems within the cells.

Dinocap (Karathane®) has been used since the late 1930s, both as an acaricide and for powdery mildew on a number of fruit and vegetable crops. Dinocap undoubtedly acts in the vapor phase, since it is quite effective against powdery mildews whose spores germinate in the absence of water. This is a popular home fungicide.

DINOCAP

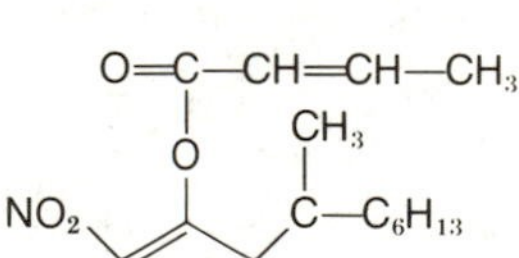

2,4-dinitro-6-(2-octyl)phenyl crotonate

Quinones

The quinones are a fascinating chemical group that offers countless numbers of molecules that are potential fungicides. Chloranil is the first of these to appear (1937). It was used heavily as a seed treatment and for foliar application until the dicarboximides appeared. The most popular of this group, however, is dichlone. It is used on a number of fruit and vegetable crops and for treatment of ponds, to

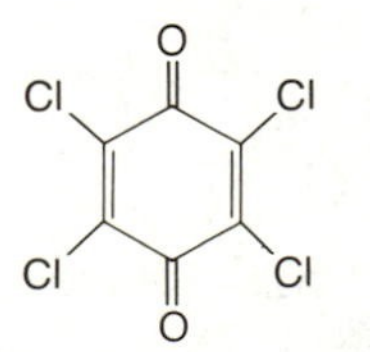

2,3,5,6-tetrachloro-1,4-benzoquinone

DICHLONE

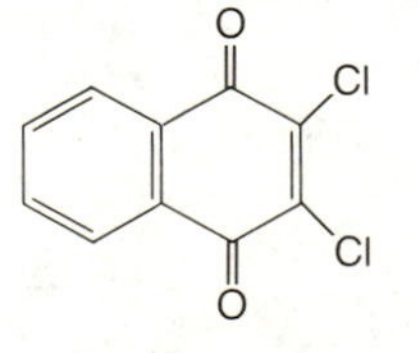

2,3-dichloro-1,4-naphthoquinone

control blue-green algae. Dichlone affects cellular respiration in many fungi and acts by attaching to the —SH groups in enzymes, thus inhibiting their action and indirectly uncoupling oxidative phosphorylation.

Organotins

The organotins were first introduced in the mid 1960s, ten years after their fungicidal properties had been discovered. In general, the trialkyl derivatives are highly fungicidal, but also phytotoxic. The triaryl (three-ring) compounds are suitable for protective use, and also have acaricidal properties. The trisubstituted tin compounds block oxidative phosphorylation.

FENTIN HYDROXIDE (Du-Ter®)

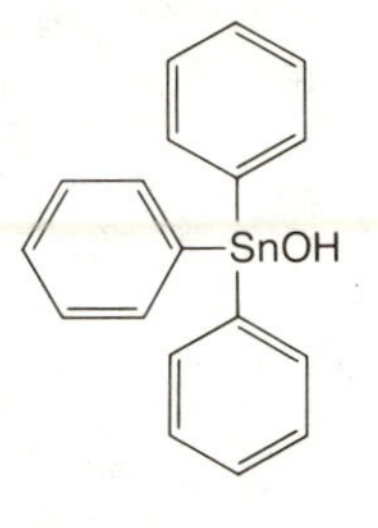

triphenyltinhydroxide

Aliphatic Nitrogen Compounds

Dodine, a fungicide introduced in the mid 1950s, has proved effective in controlling certain diseases such as apple and pear scab and cherry leaf spot. It has disease specificity and slight systemic qualities. Its mode of action is not totally clear, but it is taken up rapidly by fungal cells, causing leakage in these cells, possibly by alteration in membrane permeability. The guanidine nucleus of dodine is also known to inhibit the synthesis of RNA.

DODINE

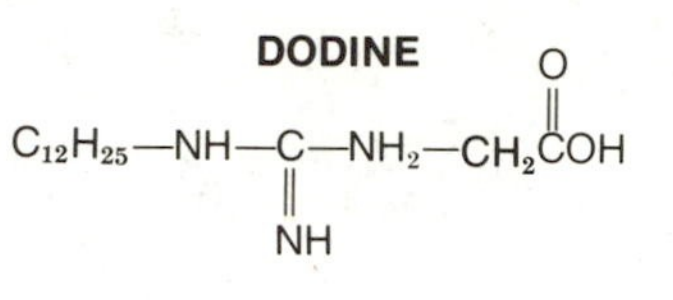

n-dodecylguanidine acetate

FUMIGANTS

As with the insecticides, there are several highly volatile, small-molecule fungicides that have fumigant action. They are unrelated chemically but are handled similarly and are dealt with as a single class in this book. Chloropicrin was mentioned in Chapter 4 as a warning agent in grain fumigants, but it is also an ideal fumigant itself: It controls fungi, insects, nematodes, and weed seeds in the soil. Methyl bromide, also listed as a fumigant insecticide, is equally effective against fungi, nematodes, and weeds. Methylisothiocyanate (MIT) is closely related to the dithiocarbamates and has a similar mode of action against fungi, nematodes, and weeds. SMDC was listed, but not illustrated, in the dithiocarbamate fungicides, where it belongs chemically. SMDC decomposes in the soil to yield methylisothiocyanate. None of these is available for home use.

CHLOROPICRIN

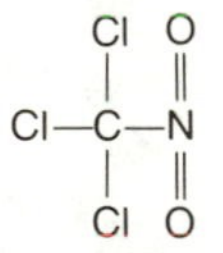

trichloronitromethane

METHYL BROMIDE

CH_3Br

bromomethane

MIT

$CH_3—N{=}C{=}S$

methylisothiocyanate

ANTIBIOTICS

Such antibiotics as penicillin, tetracycline, and chloramphenicol are used medically against human bacterial diseases. These are not involved in our present study. But we note in passing that the oxytetracyclines are therapeutic against some of those mysterious mycoplasmalike diseases. As in the case of the medically important antibiotics, the antibiotic fungicides are substances produced by microorganisms, which in very dilute concentrates inhibit growth and even destroy other microorganisms. To date, several hundred antibiotics have been reported to have fungicidal activity, and the chemical structures of about half of these are already known.

The largest source of antifungal antibiotics is the actinomycetes, a group of the lower plants. Within the actinomycetes is one amazing species, *Streptomyces griseus*, from which we obtain two antibiotics, streptomycin and cycloheximide.

Streptomycin is used as dust, spray, and seed treatment for the control mostly of bacterial diseases such as blight on apples and pears, soft rot on leafy vegetables, and some seedling diseases. It is also effective against a few fungal diseases.

STREPTOMYCIN

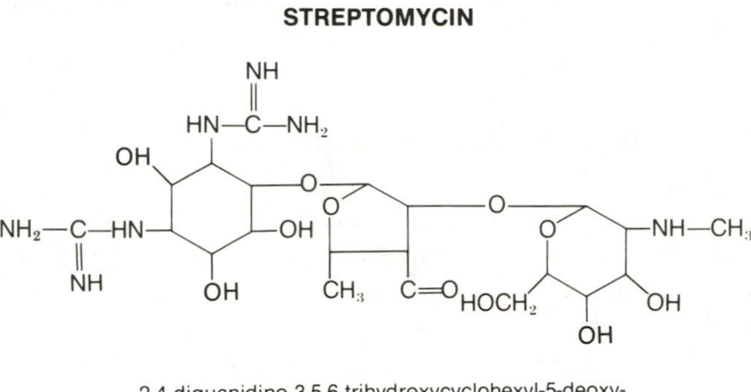

2,4-diguanidino-3,5,6-trihydroxycyclohexyl-5-deoxy-2-*O*-(2-deoxy-2-methylamino-α-glucopyranosyl)-3-formyl pentofuranoside

The mode of action of streptomycin is not clearly understood, but it probably interferes in the synthesis of proteins. Despite the evidence of antibiotic-resistant strains, streptomycin has a place in the control of some bacterial diseases, and the tetracyclines may well play an important part in controlling some mycoplasmalike diseases of plants. Because streptomycin appears to have a single and specific site of action, resistance to it by both bacteria and mycoplasmas is inevitable (see Chapter 20).

Cycloheximide is a smaller, less complicated antibiotic, about which more is understood. First, cycloheximide is toxic to a wide range of organisms, including yeasts, filament-forming fungi, algae, protozoa, higher plants, and especially mammals. Surprisingly, it is

CYCLOHEXIMIDE

CH₃ OH O NH CH₃ O

β[2-(3,5-dimethyl-2-oxocyclohexyl)-2-hydroxyethyl]-glutarimide

BLASTICIDIN-S

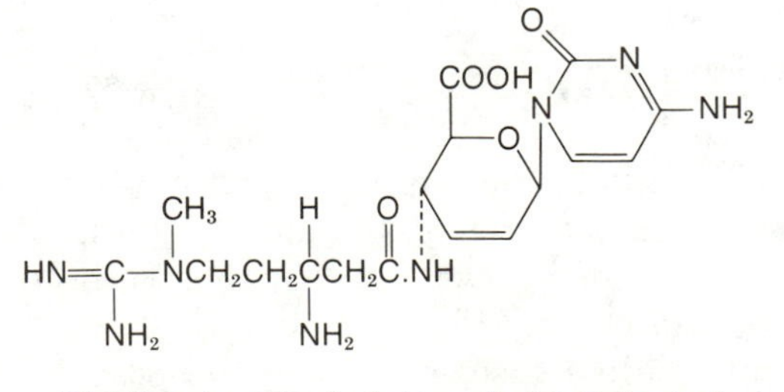

(*S*)-4[[3-amino-5[(aminoiminomethyl)methylamino]-1-oxopentyl]amino]-1-[4-amino-2-oxo-1(2*H*)-pyrimidinyl]-1,2,3,4-tetradeoxy-β-D-erythrohex-2-enopyranuronic acid

KASUGAMYCIN

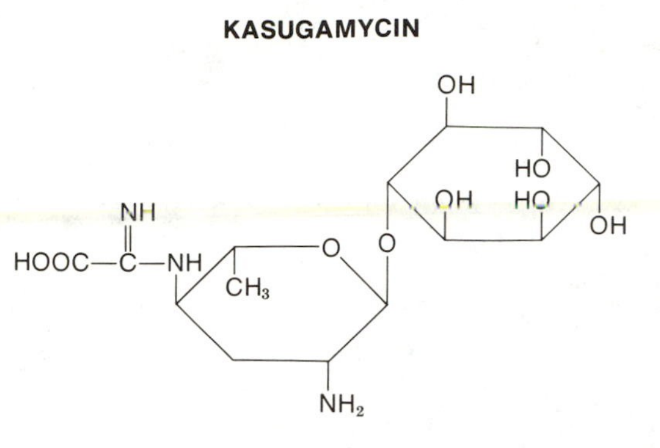

D 3 O [2 amino-4-[(1-carboxyliminomethyl)amino]-2,3,4,6-tetradeoxy-α-D-arabino-hexopyranosyl]-D-chiro-inositol

POLYOXIN B

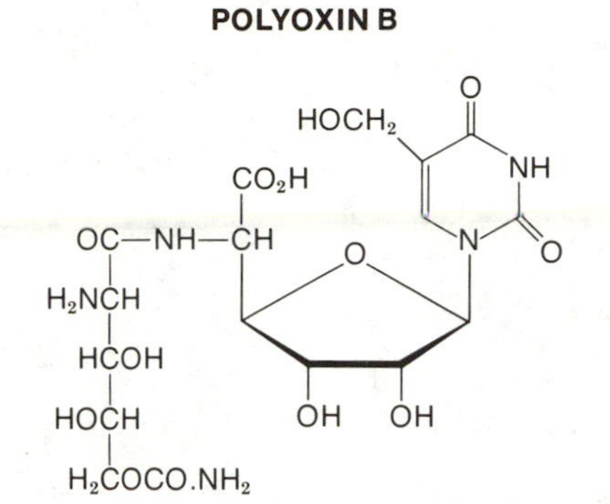

5-[[2-amino-5-O-(aminocarbonyl)-2-deoxy-L-xylonoyl]amino]-pyrimidinyl]-β-D-allofura-1,5-dideoxy-1[3,4-dihydro-5-(hydroxymethyl)-2,4-dioxo-1(2H)-nuronic acid

inactive against bacteria, perhaps because the bacteria fail to absorb it. Cycloheximide causes growth inhibition in yeasts and filament-forming fungi by inhibiting protein and RNA synthesis. Thus, with a fair amount of confidence, we can say that both streptomycin and cycloheximide act by inhibiting the synthesis of nucleic acids.

Cycloheximide was introduced as a fungicide in 1948 and has since become popular in the control of powdery mildew, rusts, turf diseases, and certain blights. It is best known commercially under the name Acti-Dione®. Because of its high acute toxicity, it cannot be purchased for home and garden use.

Blasticidin-S, discovered in 1955, is produced by the fermentation of *Streptomyces griseochromogenes*. Kasugamycin, introduced in 1963, is formed by *Streptomyces kasugaenis*, and polyoxins are extracted from *Streptomyces cacaoi*. All three are Japanese contributions to the systemic antibiotics. The first two are effective against rice blast, and the latter against rice sheath blight.

CHAPTER 15

Somewhere, behind Space and Time,
Is wetter water, slimier slime!

Rupert Brooke

Algicides

Algae are a group of simple freshwater and marine plants, ranging from single-celled organisms to green pond scums and very long seaweeds (kelps). The algicides are chemicals intended for the control of algae, especially in water that is stored or is being used industrially. Both public water systems and swimming pools require algicidal treatment.

INORGANIC CHLORINES

Indeed, the strong smell of chlorine issuing from the YMCA or other public pool is probably released from the ever-popular calcium hypochlorite or chloride of lime, not only a good algicide but an excellent disinfectant as well. It contains 70 percent available chlorine and is the source of most bottled laundry bleaches (7 to 8 percent $CaOCl_2$).

Several chlorine-based inorganic salts are available for pool chlorination. Some of the more popular are: calcium hypochlorite, $Ca(ClO)_2$; sodium hypochlorite, NaClO; and sodium chlorite, $NaClO_2$.

COPPER COMPOUNDS

Any of the copper-containing algicides would be equally effective, and longer lasting, than the chlorine-based materials. However, the copper content may eventually become phytotoxic to grass or other herbaceous plants surrounding the pool that are splashed or drenched occasionally. This principle applies to other algicides of greater potency.

The organic copper complexes are not for the swimming pool, but rather for industrial use, public water systems, and agriculture. Most commonly used among these are the copper-triethanolamine materials. Generally they can be used as surface sprays for filamentous and planktonic forms of algae in potable water reservoirs; irrigation water storage and supply systems; farm, fish, and fire ponds; and lake and fish hatcheries. The treated water can be used immediately for its intended purpose. The same generalities apply also to another organic copper compound, copper ethylenediamine complex.

Copper monoethanolamine was registered in 1980 by the EPA for use as an algicide or aquatic herbicide to control aquatic plants in reservoirs, lakes, ponds, and irrigation ditches and other potential sources of potable water. In establishing this registration, a tolerance of 1 ppm for residues of copper resulting from the use of this compound was set for potential sources of drinking water.

Another of the copper group is the old standby copper sulfate pentahydrate, whose older names are *bluestone* and *blue vitriol*. In all instances, it is the copper ion in solution that is lethal to the simple single-cell and filamentous algae. The control principle is the same as that involved in the copper-based fungicides mentioned in Chapter 14.

QUATERNARY AMMONIUM HALIDES

A host of algicides belong to the quaternary ammonium compounds, which are characterized by a chlorine or bromine ion on one end and a nitrogen atom with four carbon-nitrogen bonds. At least one of the carbons is the extension of an 8- to 18-carbon chain (*R*, in the formula). The straight carbon chain is derived from fatty acids similar to those found in vegetable oil. A generalized formula follows:

$$\begin{array}{c} CH_3 \\ | \\ R{-}N{-}CH_2Cl \\ | \\ CH_3 \end{array}$$

They are general-purpose antiseptics, germicides, and disinfectants, ideal for algae control in the greenhouse—as pot dips and wall, bench, and floor sprays—and in swimming pool and recirculation water systems. Algae control lasts up to several months.

The alkyldimethylbenzylammonium chlorides can be used to give long-lasting control of algae and bacteria in swimming pools, cooling systems, air-conditioning systems, and glass houses. They compose the bulk of the quaternary ammonium halides. As they are toxic to fish, they cannot be used in ponds, lakes, or streams.

ALKYLDIMETHYLBENZYLAMMONIUM CHLORIDE

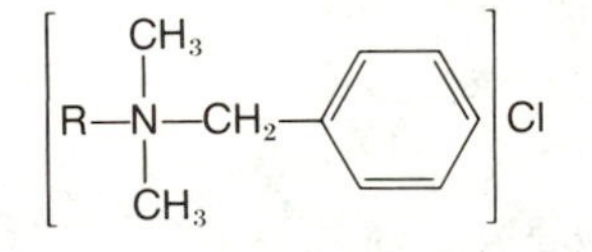

MISCELLANEOUS ORGANIC COMPOUNDS

To be used only by professionals is acrolein, a highly volatile organic molecule with tremendous physiological activity, both against aquatic plants and humans. It is extremely hazardous to work with because of its lachrymatory effect (that is, it causes the eyes to tear) and because skin burns result from contact.

ACROLEIN

$CH_2{=}CH{-}CHO$

2-propenal

Acrolein is an extremely useful aquatic herbicide and algicide but requires application by licensed operators because of its frightening tear-gas effect. Exposed plants disintegrate within a few hours and float downstream. It is a general plant toxicant, destroying plant cell membranes and reacting with various enzyme systems. By the use of spot treatment in lakes, the fish population can be saved, an example of environmental protection by spot application.

Triphenyltin acetate (Brestan®) is another one of the multipurpose pesticides. It can act as a fungicide, algicide, or molluscicide.

Dimanin C (sodium dichloroisocyanurate) also gives long-lasting control of algae in swimming pools and can be used as a disinfectant.

Simazine, mentioned in Chapter 11, is registered by the EPA as an algicide under the name of Aquazine®. It performs well at a concentration of 1.0 ppm against a broad spectrum of algae and may be used in ponds containing fish.

Dichlone, an agricultural fungicide, is also registered as an algicide for use against blue-green algae in lakes and ponds.

ICA and TICA are used as algicides and bactericides in swimming pools. Because they are stable to decomposition, both by other chemicals and by sunlight, they are also added to the water to serve as a "sun screen," thereby extending the life of added chlorine.

TRIPHENYLTIN ACETATE (Brestan®)

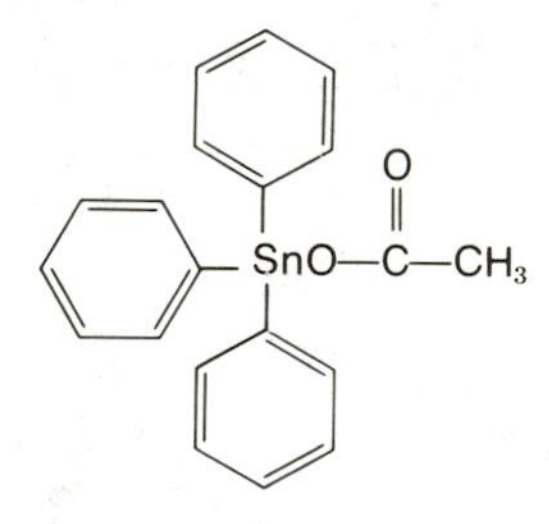

SODIUM DICHLOROISOCYANURATE

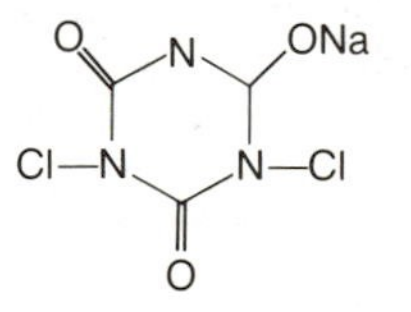

POTASSIUM DICHLOROISOCYANURATE

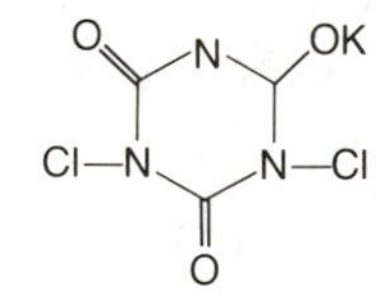

SIMAZINE (Aquazine®)

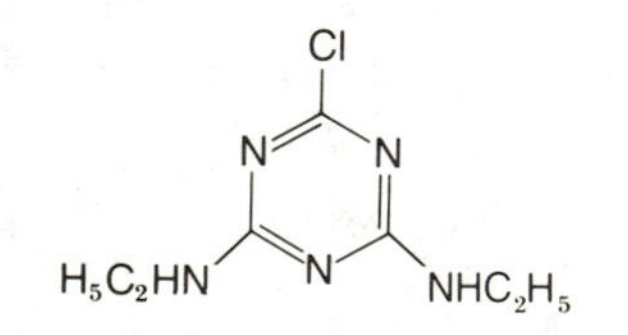

2-chloro-4,6-bis(ethylamino)-*s*-triazine

DICHLONE

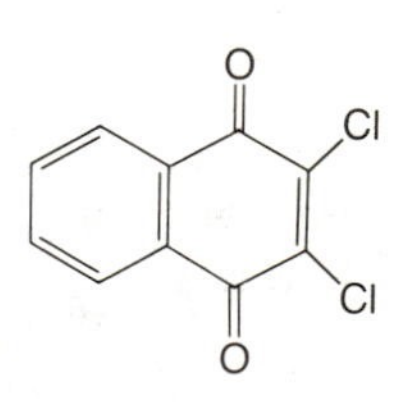

2,3-dichloro-1,4-naphthoquinone

ICA

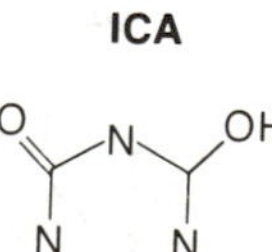
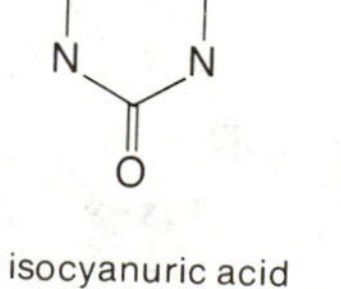

isocyanuric acid

TICA

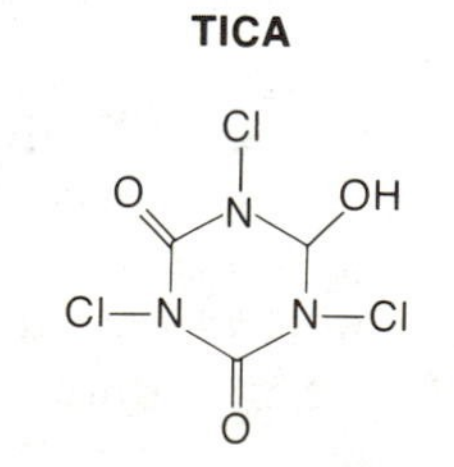

trichloroisocyanuric acid

> Gentlemen, it is the microbes that will have the last word.
>
> Louis Pasteur

Disinfectants

There are an almost limitless number of chemical agents for controlling microorganisms, and new ones appear on the market regularly. These are the disinfectants. A common problem confronting persons who must utilize disinfectants or antiseptics is which one to select and how to use it. There is no single ideal or all-purpose disinfectant, thus the compound to choose is the one that will kill the organisms present in the shortest time, with no damage to the contaminated substrate.

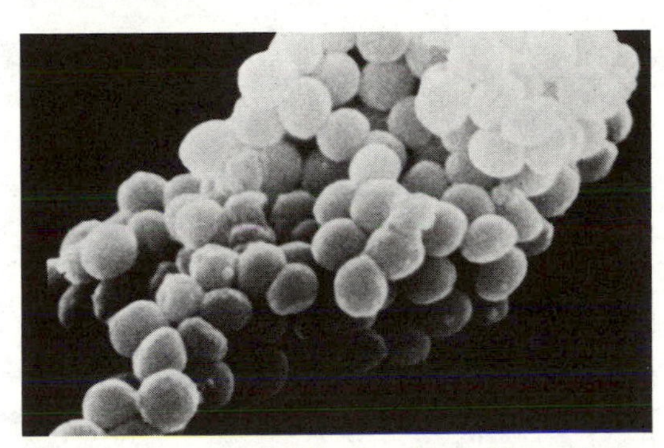

Staphylococcus (×6500). (Courtesy of Stem Laboratories, Inc.)

The EPA *Compendium of Registered Pesticides* devotes an entire volume to disinfectants (all microbiological control producers, e.g., sterilizers, disinfectants, bacteriostats, sanitizers, viricides, and microbiocides). Listed are approximately 600 active ingredients in disinfectant products. The 23 major use categories for disinfectants are further divided into 200 use sites. The major use categories are as follows:

Additives, antimicrobial
Animal (for food use) quarters
Animal (for nonfood use) quarters
Aquatic sites
Barber and beauty shops
Beverage processing plants
Carpets
Chemical lavatory holding tanks
Dairy processing plants
Dry cleaning
Eating establishments (restaurants, bars, and taverns)
Food-processing plants
Funeral homes, mortuaries, and morgues
Hospitals and related institutions
Laundry
Maintenance, commercial and industrial
Maintenance, household
Milk samples

Oil recovery process and oil well drilling
Preservatives, food and feed
Preservatives, industrial
Specialty sites
Toilet bowls

This list should give the reader some idea as to the number of disinfectants available and their range of uses.

The history of the development of disinfectants is a fascinating documentation of the misunderstanding of the importance of microorganisms. During the early 1800s, microorganisms were considered to be biological nonentities. Infections were believed to be caused by magical powers in the air, or by an imbalance of body fluids. Certainly contaminated hands were not involved. It took the dogged tenacity of Louis Pasteur to demonstrate to the world that microbes not only could ferment fruit juice to wine but also could cause the wine to spoil.

The next steps were made in 1867 by an English surgeon, Joseph Lister, who first developed antiseptic surgery with the use of heat-sterilized instruments and the application of carbolic acid (phenol) to wounds by means of soaked dressings. In 1881, the German bacteriologist Robert Koch evaluated 70 different chemicals for use in disinfection and antisepsis. Among these were various phenols and mercuric chloride ($HgCl_2$). From this point on, progress in the science of destroying our microorganismic enemies has been overwhelming.

Before we discuss the disinfectants, we must make a quick distinction between two confusing words, *antisepsis* and *sanitation*. Antisepsis is the disinfection of skin and mucous membranes, while sanitation is the disinfection of inanimate surfaces. Consequently, much more severe treatment can be used for sanitation than for antisepsis. An agent that kills pathogens but not necessarily spores is called a *germicide*. An agent that kills bacterial spores is a *sporicide*. In this chapter, our discussion is directed to the use of disinfectants used in sanitation.

PHENOLS

Dr. Lister used phenol in 1867 as a germicide in the operating room, and phenol can thus probably be considered the oldest recognized disinfectant. Its deadly effect is due to protein precipitation, even when it is greatly diluted. It is used as the standard of comparison for the activities of other disinfectants, expressed in terms of phenol coefficients. Phenol and the cresols (methylated phenol) have very distinct odors, which change little with modification of their chemical structures. The popular household disinfectant sold under the trade name of Lysol® is an example of a cresol. There are 3 cresols, ortho-, meta-, and para-(*o*-, *m*-, and *p*-); only *o*-cresol is illustrated. Cresols are several times more germicidal than phenol, however, they are being replaced by another phenolic compound, *o*-phenylphenol, and are used for disinfection of inanimate objects.

Cresylacetate is used in spray form for antisepsis and as an analge-

Phenol and Several Cresol and Hexachlorophene Structures

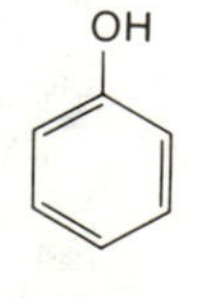

o-CRESOL

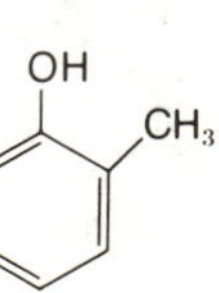

sic on the mucous membranes of the ear, nose, and throat. The addition of chlorine or a short-chain organic compound to phenols increases their activity. For instance, hexachlorophene is one of the most useful of the phenol derivatives. Hexachlorophene, which combines a halogen (Cl) and a phenolic derivative in a soap, is used for surgical scrubs. By order of the Food and Drug Administration (FDA), hexachlorophene is no longer used for infant skin care or over-the-counter products because of its carcinogenic potential.

o-PHENYLPHENOL

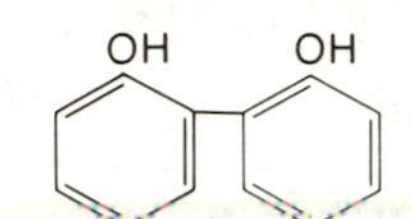

CRESYLACETATE

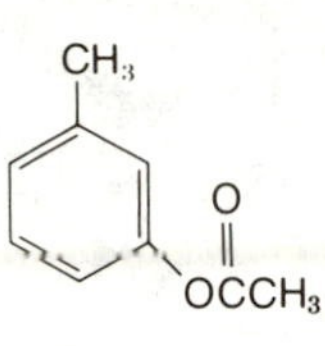

HEXACHLOROPHENE

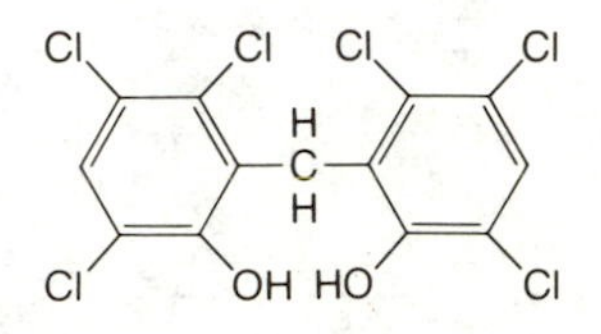

HALOGENS

The halogens are organic and inorganic compounds containing chlorine, iodine, bromine, or fluorine. Generally, the inorganic halogens are deadly to all living cells.

Chlorine

Chlorine, either in the form of gas or in certain chemical combinations, represents one of the most widely used disinfectants. The compressed gas in liquid form is almost universally employed for the purification of municipal water supplies.

Chlorine in its many forms was first used as a deodorant and later as a disinfecting agent. Hypochlorites are those most commonly used in disinfecting and deodorizing procedures because they are relatively safe to handle, colorless, good bleaches, and do not stain. Several organic chlorine derivatives are used for the disinfection of water, particularly for campers, hikers, and the military. The most common of these are halazone and succinchlorimide.

HALAZONE

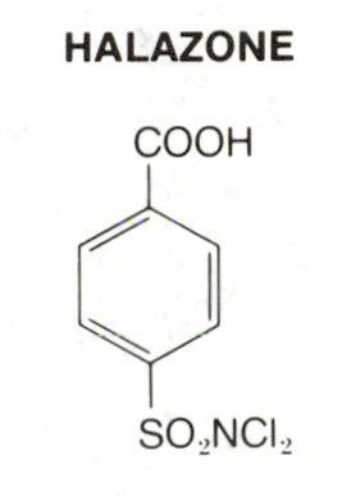

p-sulfone dichloramidobenzoic acid

SUCCINCHLORIMIDE

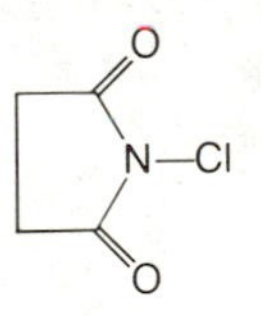

A halazone concentration of 4 to 8 mg/L safely disinfects even fairly hard water containing typhoid bacilli in approximately 30 minutes. A succinchlorimide concentration of 12 mg/L disinfects water in 20 minutes. These organic chlorides are quite stable in tablet form, becoming active when placed in water.

Chlorine is the dominant element in disinfectants, in that roughly 25 percent of those registered with the EPA contain one or more atoms of this important halogen.

Hypochlorites

Ignatz Semmelweis, a Hungarian physician working in Vienna, is credited with having used hypochlorites in 1846 to 1848 in an attempt to reduce the incidence of childbed fever. Medical students were required to wash their hands and soak them in a hypochlorite solution before examining patients. Calcium hypochlorite ($Ca(OCl)_2$) and sodium hypochlorite ($NaOCl$) are popular compounds widely used both domestically and industrially. They are available as powders or liquid solutions and in varying concentrations depending on the use. Products containing 5 to 70 percent calcium hypochlorite are used for sanitizing dairy equipment and eating utensils in restaurants. Solutions of sodium hypochlorite are used as a household disinfectant; higher concentrations of 5 to 12 percent are also used as household bleaches and disinfectants and as sanitizing agents in dairy and food-processing establishments.

Chloramines

Chloramines are another category of chlorine compounds used as disinfectants, sanitizing agents, or antiseptics. Chemically they are characterized by having one or more hydrogen atoms of an amino group of a compound replaced by chlorine. The simplest of these is monochloramine, NH_2Cl. Chloramine-T and azochloramide, two of several germicidal compounds in this general group, have more complex chemical structures. One of the advantages of the chloramines over the hypochlorites is their stability and prolonged chlorine release.

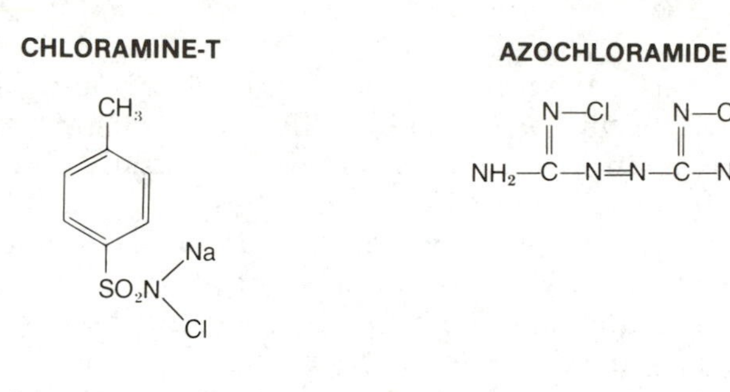

The germicidal action of chlorine and its compounds is produced by the formation of hypochlorous acid when free chlorine reacts with water:

$$Cl_2 + H_2O \rightarrow HCl + HClO \text{ (hypochlorous acid)}$$

Similarly, hypochlorites and chloramines undergo hydrolysis, forming hypochlorous acid. The hypochlorous acid formed in each instance is further decomposed, releasing oxygen:

$$HClO \rightarrow HCl + O \text{ (formed from chlorine, hypochlorites, and chloramines)}$$

The oxygen released in this reaction (nascent oxygen) is a strong oxidizing agent, and microorganisms are destroyed by its action on cellular constituents. The killing of cells by chlorine and its compounds is also in part caused by the direct combination of chlorine with proteins of the cell membranes and enzymes.

Iodine

Iodine is traditionally used as a germicidal agent in a form referred to as tincture of iodine, typically 2 percent iodine plus 2 percent sodium iodide in ethyl alcohol, or 7 percent iodine plus 5 percent potassium iodide in 83 percent ethyl alcohol. Iodine is a highly effective bactericidal agent and is unique in that it is effective against all kinds of bacteria. It also possesses sporicidal activity. Water or alcohol solutions of iodine are highly antiseptic and have been used for decades before surgical procedures. Several metallic salts, such as sodium and potassium iodide, are registered as disinfectants, but far fewer compounds contain iodine than contain chlorine.

Fluorine

Because of its extreme reactivity, and lack of easy-to-handle characteristics, fluorine in combination with other elements is found in only a handful of disinfectants. The same is true for bromine.

PEROXIDES

Hydrogen peroxide (H_2O_2) is an effective and nontoxic antiseptic. The molecule is unstable and when warmed degrades into water and oxygen:

$$2\ H_2O_2 \rightarrow 2\ H_2O + O_2 \uparrow$$

At concentrations of 0.3 to 6.0 percent, H_2O_2 is used in disinfection; and at concentrations of 6.0 to 25.0 percent, in sterilization.

ALCOHOLS

Alcohols denature proteins, possibly by dehydration, and they also act as solvents for lipids. Consequently, membranes are likely to be disrupted and enzymes inactivated in the presence of alcohols. Three alcohols are used: ethanol, CH_3CH_2OH (grain alcohol); methanol, CH_3OH (wood alcohol); and isopropanol, $(CH_3)_2CHOH$ (rubbing alcohol). As a rule of thumb, the bactericidal value increases as the molecular weight increases. Isopropyl alcohol is therefore the most widely used of the three. In practice, a solution of 70 to 80 percent alcohol in water is used. Percentages above 90 and below 50 are usually less effective, except for isopropyl alcohol, which is effective even in 99 percent solutions. A 10-minute exposure is sufficient to kill vegetative cells but not spores.

HEAVY METALS

The heavy metals, either alone or in certain chemical compounds, usually exhibit their deadly effect by precipitating proteins or by reacting with enzymes or other essential cellular components. The same effects were discussed in Chapter 14, which mentioned the use of some of the heavy metals as fungicides. In fact some of the fungicides are disinfectants. The heavy metals in common use are mercury, copper, silver, arsenic, and zinc.

Mercury

Mercury compounds are historic, in that virtually every registration of both the inorganic and organic mercurial fungicides and disinfectants has been canceled by the EPA out of concern about the levels of mercury in the environment. Prominent among the historic compounds is mercuric bichloride, $HgCl_2$, which at one time or another was used in every conceivable situation in which a disinfectant was needed.

Copper

Copper is much more effective against algae and molds than against bacteria and is used in the forms of copper sulfate and copper ethylenediaminetetra-acetate. A concentration of 2 ppm in water is sufficient to prevent algal growth and is used in swimming pools and open-water reservoirs.

Silver

Of the silver compounds, only silver fluoride and silver nitrate (Argyrol®) have retained registrations and are used only as antiseptics.

Arsenic

Arsenic achieved fame as the first known treatment for syphilis. It still finds some use in the treatment of protozoan infections but is not used in sanitation procedures.

Zinc

Fungi are particularly vulnerable to several zinc compounds, two of which are common household garden fungicides, zineb and ziram. A mixture of a long-chained fatty acid and the zinc salt of the acid is commonly used as an antifungal powder or ointment. It is particularly effective for the treatment of athlete's foot. Zinc oxide paste is used for treating diaper rash and superficial bacterial or fungal infections.

Structure of a quaternary ammonium compound shown in relation to the structure of ammonium chloride

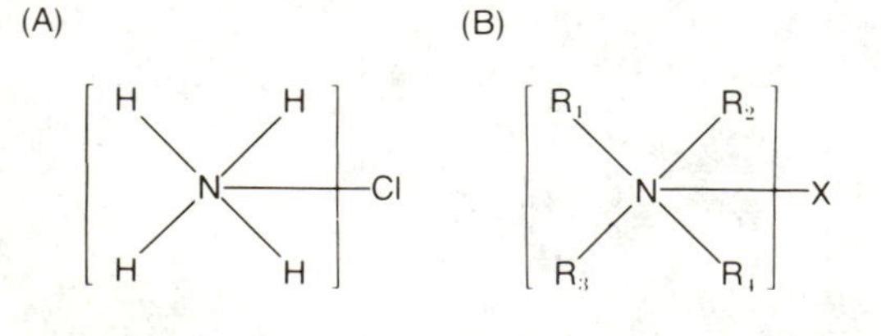

(A) Ammonium chloride; (B) the general structure of a quaternary ammonium compound—R_1, R_2, R_3, and R_4 are carbon-containing groups, and the X^- is a negatively charged ion, such as Cl^- and Br^-.

DETERGENTS

The detergents are organic compounds that have two ends or poles. One is hydrophilic (mixes well with water) and the other is hydrophobic (does not mix well with water). As a result, the compounds orient themselves on the surfaces of objects with their hydrophilic poles toward the water. Basically, these may be classed as *ionic* or *nonionic* detergents. The ionic are either anionic (negatively charged) or cationic (positively charged). The anionic detergents are only mildly bactericidal. The cationic materials, which are the quaternary ammonium compounds, are extremely bactericidal, especially for *Staphylococcus*, but do not affect spores. Hard water, containing calcium or magnesium ions, will interfere with their action, and they also rust metal objects. Even with these disadvantages, the cationic detergents are among the most widely used disinfecting chemicals, since they are easily handled and are not irritating to the skin in concentrations ordinarily used. Three commonly used quaternary ammonium salts of the cationic-detergent class are cetylpyridinium chloride (Ceepryn®, illustrated), Zephiran®, and Phemerol®.

Quaternaries have been shown to be fungicidal as well as destructive to certain of the pathogenic protozoa. Viruses appear to be more resistant than bacteria and fungi. The combined properties of germicidal activity and detergent action, and such other features as relatively low toxicity, solubility, noncorrosiveness, and stability, have resulted in many practical applications of quaternaries for sanitization and disinfection. They are extensively used as skin antiseptics and as sanitizing agents in eating and drinking establishments, dairies, and food-processing plants.

The mode of action of the cationic-detergent disinfectants is not precisely understood. Various modes of antimicrobial action have been proposed, including enzyme inhibition, protein denaturation, and disruption of the cell membrane, causing a leakage of vital constituents. It is likely that a combination of actions is the cause of inhibition or death of affected cells.

Detergent-Disinfectant Molecules

NONIONIC DETERGENT

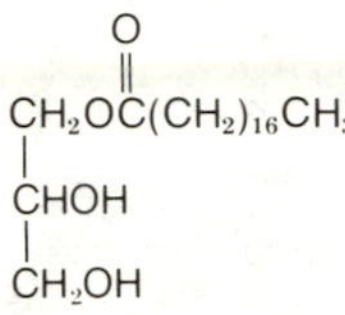

$$\begin{array}{l} \quad\;\; O \\ \quad\;\; \| \\ CH_2OC(CH_2)_{16}CH_3 \\ | \\ CHOH \\ | \\ CH_2OH \end{array}$$

Stearic acid monoglyceride

ANIONIC DETERGENT

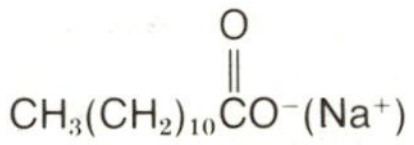

$$\begin{array}{c} O \\ \| \\ CH_3(CH_2)_{10}CO^-(Na^+) \end{array}$$

Sodium laurate

CATIONIC DETERGENT

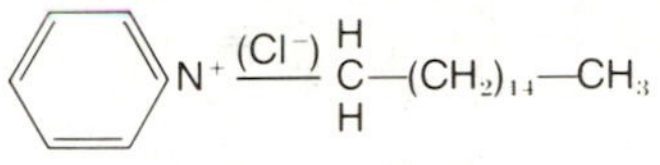

$$C_5H_5N^+ \overset{(Cl^-)}{—} \underset{H}{\overset{H}{C}}—(CH_2)_{14}—CH_3$$

Cetylpyridinium chloride
(Ceepryn®)

ALDEHYDES

Combinations of formaldehyde and alcohol are outstanding sterilizing agents, with the exception of the residue remaining after use. A related compound, glutaraldehyde, in solution is as effective as formaldehyde. Most organisms are killed within a 5-minute exposure to 2% aqueous glutaraldehyde, while bacterial spores succumb in 3 to 12 hours.

Other classes of disinfectants, including a number of dyes, acids and alkalis, alcohols, peroxides, and fumigants (ethylene oxide and methyl bromide), are effective under certain conditions.

FORMALDEHYDE

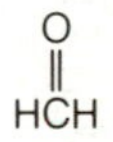

$$\begin{array}{c} O \\ \| \\ HCH \end{array}$$

GLUTARALDEHYDE

$$\overset{O}{\overset{\|}{HC}}CH_2CH_2CH_2\overset{O}{\overset{\|}{CH}}$$

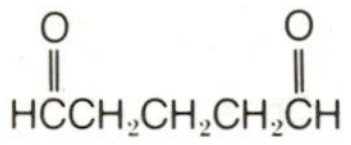

In summary, remember that no single chemical antimicrobial agent is best or ideal for any and all purposes. This is not surprising, in view of the variety of conditions under which agents may be used, the differences in modes of action, and the many types of microbial cells to be destroyed.

PART VI

BIOLOGICAL INTERACTIONS WITH PESTICIDES

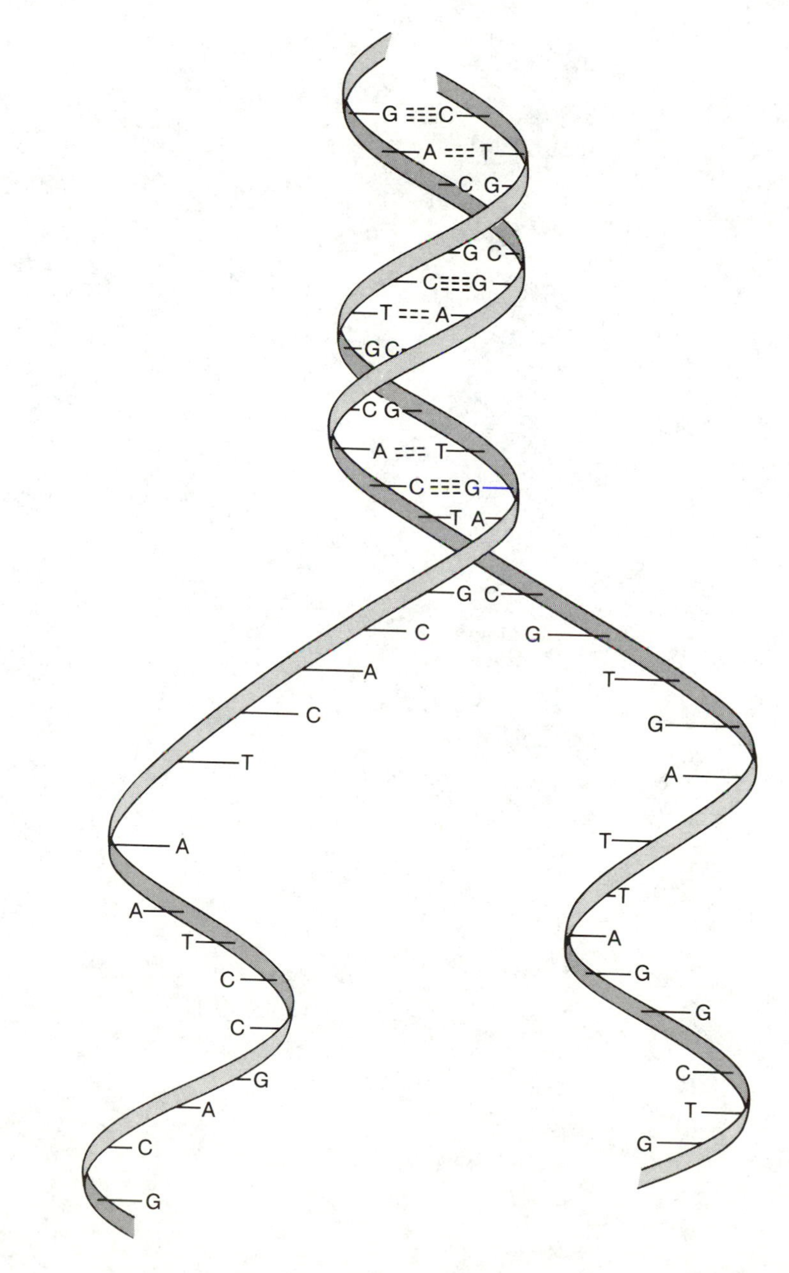

CHAPTER 17

> Nature and the pesticide industry apparently have decided that the best way to poison an animal is through its nervous system.
>
> Dan Shankland (1976)

Modes of Action for Insecticides

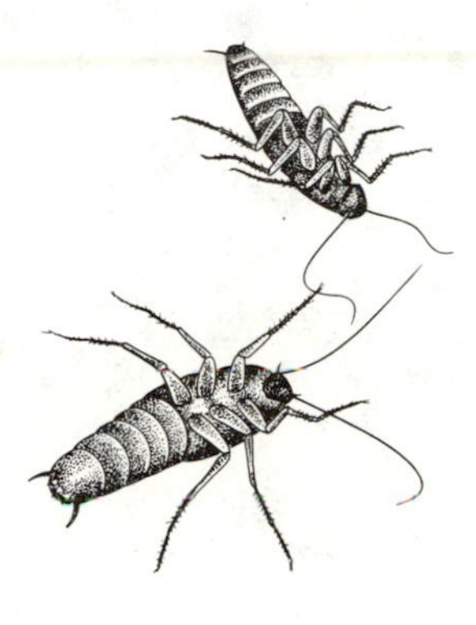

Pesticides, for the most part, are intended to kill selected organisms. All pesticides are alike in blocking some metabolic process; however, how they do this—that is, their mode of action—is sometimes very difficult to determine, and in many instances we still do not know. In rare instances a pesticide may even exhibit more than one mode of action, making classification complex. At present, we have only fragments of knowledge concerning the great array of commonly used pesticides. Even in the well-established field of molecular pharmacology, a similar but somewhat less extreme situation exists. As Zbenden (1968) states, "The exact mode of action of many of our more valuable drugs is not known."

In this chapter, as well as the two following, *mode of action* comprises the sum of anatomical, physiological, and biochemical responses that make up the total toxic action of a chemical, as well as the physical (location) and molecular (degradation) fate of the chemical in the organism. The term *mechanism of action* is restricted to the biochemical and biophysical responses of the organism that appear to be associated with the pesticidal action. These responses may not necessarily involve the primary biochemical lesion, or "site of action," the single enzyme or metabolic reaction that is affected at a dose lower than any other enzyme or metabolic reaction, or the first reaction effected at a given low dose. The primary site of action may be only part of the answer to how a given toxicant is effective as a pesticide.

To retain some form of brevity and simplicity in a subject that can become overwhelming in detail, only selected pesticides and modes of action are covered. Only the insecticides, herbicides, and fungicides are covered in Chapters 17, 18, and 19; together they represent approximately 90 to 95 percent of the number and quantity of pesticides in use today. (The modes of action for other classes of pesticides are given in the chapters dealing specifically with these materials.) Simplified diagrams are included to illustrate the mode-of-action principle.

Detailed mode-of-action studies are required for registration of a compound by the EPA. They must contain supporting data to establish, with very little room for doubt, how a specific chemical affects the target organism. These studies are essential not only for understanding the pesticide's biological behavior but also for understanding how to treat human poisonings, design antidotes, and establish safety criteria for humans and animals. From this creative, imaginative research into chemicals with rather specific modes of action also comes the knowledge needed to design new and better pesticides. Such knowledge can also be used in the exploration of biology, ecology, and even aspects of human health.

In terms of their modes of action, insecticides fall into six classes: physical toxicants, protoplasmic poisons, nerve poisons, metabolic inhibitors, muscle poisons, and alkylating agents.

PHYSICAL TOXICANTS

Physical toxicants are those materials that act to block a metabolic process, not by a biochemical or neurological reaction, but mechanically, such as oils used to control mosquito larvae by blocking or clogging the respiratory openings or the gills. Heavier oils applied to fruit trees during the dormant season control scales by clogging their spiracles. Other physical toxicants are the inert, abrasive dusts such as boric acid, diatomaceous earth, silica gel, and aerosilica gel. These kill insects by adsorbing waxes from the insect cuticle, effecting the continuous loss of water from the insect body. The insects become desiccated and die from dehydration.

PROTOPLASMIC POISONS

Protoplasmic poisons attack all enzymes in the insects' system, seemingly leading to their precipitation. These include mercury and its salts, all strong acids, and several of the heavy metals, including cadmium and lead. Phenomenal quantities are required to kill in this manner as compared to present-day insecticides.

NERVE POISONS

Narcotics

The fumigants, particularly the halogenated (containing chlorine, bromine, or fluorine) fumigants, as a group, are narcotics; that is, their mode of action is more physical than chemical. The fumigants are liposoluble (fat soluble); they have common symptomology; their effects are reversible; and their activity is altered very little by structural changes in their molecular architecture. Pharmacological narcotics induce *narcosis,* sleep, or unconsciousness, which in effect is their action on insects. These narcotics lodge in lipid-containing tissues, including the nerve sheaths and the lipoproteins of the brain. There are a few fumigants whose mode of action goes beyond

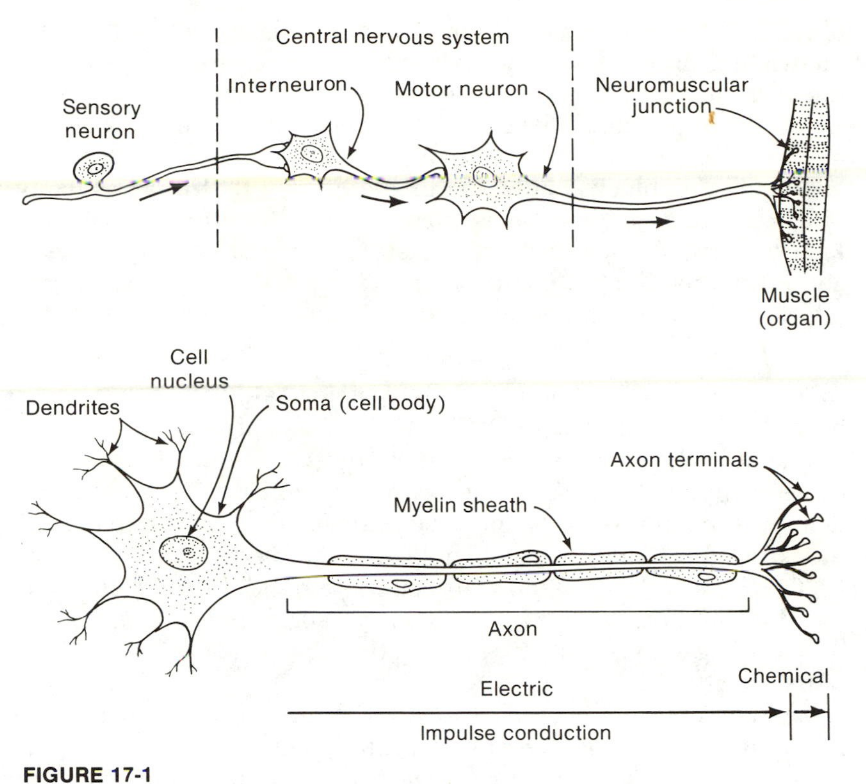

FIGURE 17-1
Typical mammalian nerve cells. (*Source:* After Matsumura, 1975.)

the narcotic; for example, methyl bromide, ethylene dibromide, hydrogen cyanide, and chloropicrin. Many of the fumigants are odorless, and chloropicrin is commonly added in trace quantities to them as an olfactory warning agent to users.

Axonic Poisons

The axon of a nerve cell or neuron is an elongated extension of the cell body and is especially important in the transmission of nerve impulses from the region of the cell body to other cells (Figure 17-1). Virtually all axonic transmission of impulses is electrical. Axonic chemicals are those that in some way affect this impulse transmission in the axon. All the organochlorine or chlorinated insecticides and the pyrethroids are considered axonic poisons.

Pyrethroids. Pyrethroids have a greater insecticidal effect when the temperature is lowered; that is, they have a negative temperature coefficient, as does DDT. They affect both the peripheral and central nervous system of the insect. Pyrethroids initially stimulate nerve cells to produce repetitive discharges and eventually cause paralysis. These effects are produced in insect nerve cord, which contains ganglia and synapses, as well as in giant nerve fiber axons. The stimulating effect of pyrethroids is much more pronounced than that of DDT. The exact sites of action of pyrethroids at synapses are not known. It is probable that the toxic action of pyrethroids is primarily due to its blocking action on the nerve axon since this action shows a negative temperature coefficient. But because the cockroach ganglion is affected by pyrethroid concentrations many fold less than are required to block conduction in giant fibers, it also seems likely that pyrethroids act on some aspect of synaptic function. The fast knockdown of flying insects is the result of rapid muscular paralysis, suggesting that the ganglia of the insect central nervous system are affected. There is some evidence that the neurons are also affected.

DDT. The mode of action, or type of biological activity, has never been clearly worked out for DDT or any of its relatives. It does affect the axons of the neurons in a way that prevents normal transmission of nerve impulses, both in insects and mammals. Eventually the neurons fire impulses spontaneously, causing the muscles to twitch, the "DDT jitters"; this is usually followed by convulsions and death. DDT is a relatively slow-acting insecticide and has the unusual quality of being more toxic to insects as the surrounding temperature drops, a negative temperature correlation that is also characteristic of the pyrethroids.

Because DDT does not seem to react with any particular enzyme, it has been theorized that the physical interference of DDT with membrane permeability is an important factor in its effect. The prevailing theory, developed by Holan (1969), is that DDT acts as a molecular "wedge" held in place by lipophile interactions with membrane lipoproteins (Figure 17-2). In this fashion DDT molecules keep the sodium gates somewhat open or leaking so that the membrane's resting potential cannot be restored. This repeated discharge causes the train of spikes that results in tetanus.

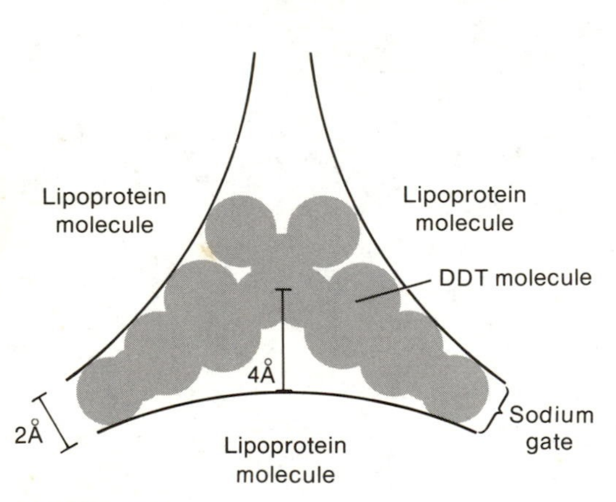

FIGURE 17-2
Intermolecular fit of DDT: Each large arc represents a lipoprotein molecule. (*Source:* L. J. Mullins, *Science* 122: 118–119, July 1955.)

Thus, DDT and its chemical relatives in some complex manner destroy the delicate balance of sodium and potassium around the axon, thereby preventing it from normally conducting electric impulses.

Cyclodienes. Unlike DDT and the pyrethroids, the cyclodienes have a positive temperature correlation; that is, their toxicity increases with increases in the surrounding temperature. Beyond this large difference, the similarity of dieldrin poisoning and that of DDT is striking. Trains of repetitive discharges can be observed for the nerve cords of cockroaches treated with cyclodienes. There are always time lags or latent periods between application and the appearance of symptoms, varying from 2 to 8 hours, depending on the insecticide. It is quite likely that the mechanism of cyclodiene poisoning also involves changes in ion permeability at axonic membrane levels. Because all the cyclodienes have two electron-rich sites positioned opposite each other along the line of symmetry, it is theorized that they should fit into a particular biological site in the nervous system in such a way as to block its normal physiological function. These compounds do become bound with the outer covering or sheath of the nerve; however, this bonding does not completely explain the mode of action.

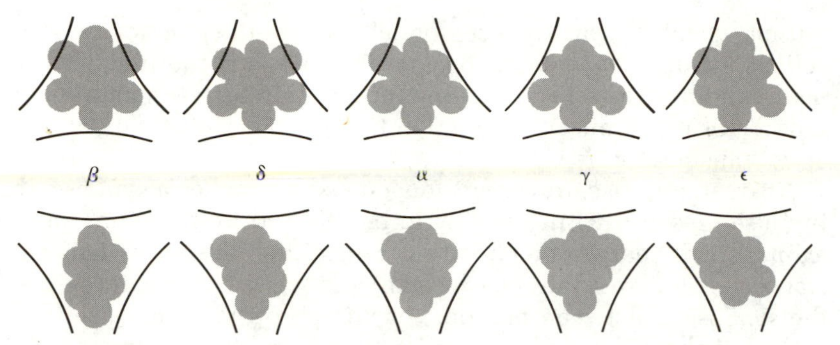

FIGURE 17-3
Stuart models of various isomers of HCH fitting into a membrane interspace for plane (upper row) and one end-on orientation (lower row). The interspace is that employed for DDT (see Figure 17-2). While all end-on orientations are possible, only the γ-isomer (lindane) fits into the interspace in a plane orientation. (*Source:* L. J. Mullins, *Science* 122: 118–119, July 1955.)

HCH (lindane). The active isomer of HCH (hexachlorocyclohexane) is the gamma isomer, known as lindane. Lindane acts much faster than DDT or the cyclodienes and causes a much higher rate of respiration in insects than DDT. Lindane toxicity also exhibits a negative temperature correlation, but not as pronounced as that of DDT. The repetitive spikes seen in treated nerve cord occur in groups of two to four in series, a phenomenon not yet understood but definitely different from the repetitive discharges or trains observed in DDT poisoning. Lindane may be classified as belonging to a group of neurotoxicants similar to DDT, though its mode of action is noticeably different from that of DDT. There are, however, more similarities between lindane and DDT poisonings than differences.

Fascinating in the chemistry of HCH are the tremendous differences in biological activity found among the closely related isomers. These differences suggest that a rigid spatial arrangement of the molecule is necessary to secure its insecticidal activity (see page 38). One widely accepted theory is that when lindane enters the hypothetical interspace (Figure 17-3), the nerve membrane is thrown out of equilibrium by the attractive force of chlorine atoms applied against the membrane components. The membrane then becomes excited due to untimely ion leaks caused by distortion of the lipoprotein molecules.

In summary, it is known that lindane attacks the nervous system and requires a complete reflex arc to be effective, like DDT. Since lindane does not have any special functional group, it appears that physical interaction involving such electronegative centers takes place between lindane and some biological site in the nervous system that normally carries out an important role.

Synaptic Poisons

There are two different methods of nerve impulse transmission in the nervous system: Axonic transmission, discussed earlier, conveys an impulse from its arrival point along the axon to a neuron or a

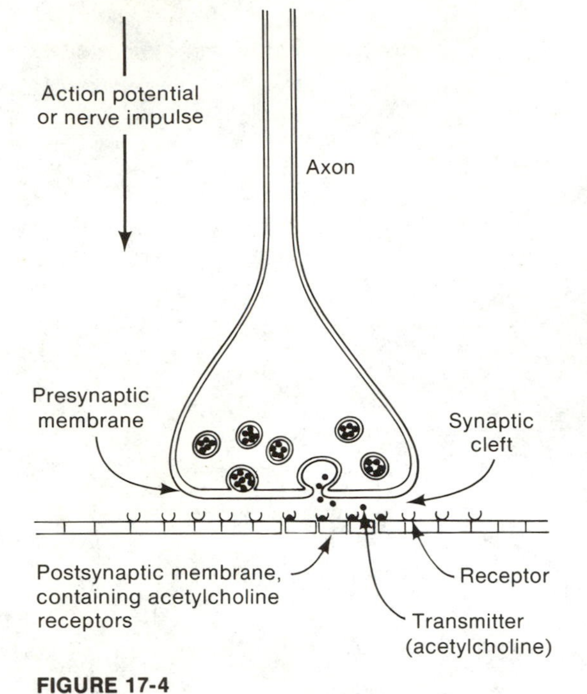

FIGURE 17-4
Diagram of a nerve synapse.

muscle, gland, or sensory receptor cell. Across the junction between cells, *synaptic transmission* occurs. A synapse is the junction of a neuron with other cells, including the junction between neuron and muscle, or neuromuscular junction (Figure 17-4). Virtually all synaptic transmission is chemical.

When an impulse, traveling along an axon, reaches a synapse, the impulse dies out, while causing to be released from the end of the axon a small charge of a chemical transmitter substance. This substance moves across the synapse (gap) and sets off another impulse if the synapse is between neurons, or an appropriate response if the synapse is between a neuron and a muscle or gland. There are two well-known chemical transmitters: acetylcholine and norepinephrine; however, there are probably others. Synapses that utilize acetylcholine are referred to as *cholinergic*. Those that utilize norepinephrine are called *adrenergic* (Figures 17-5 and 17-6).

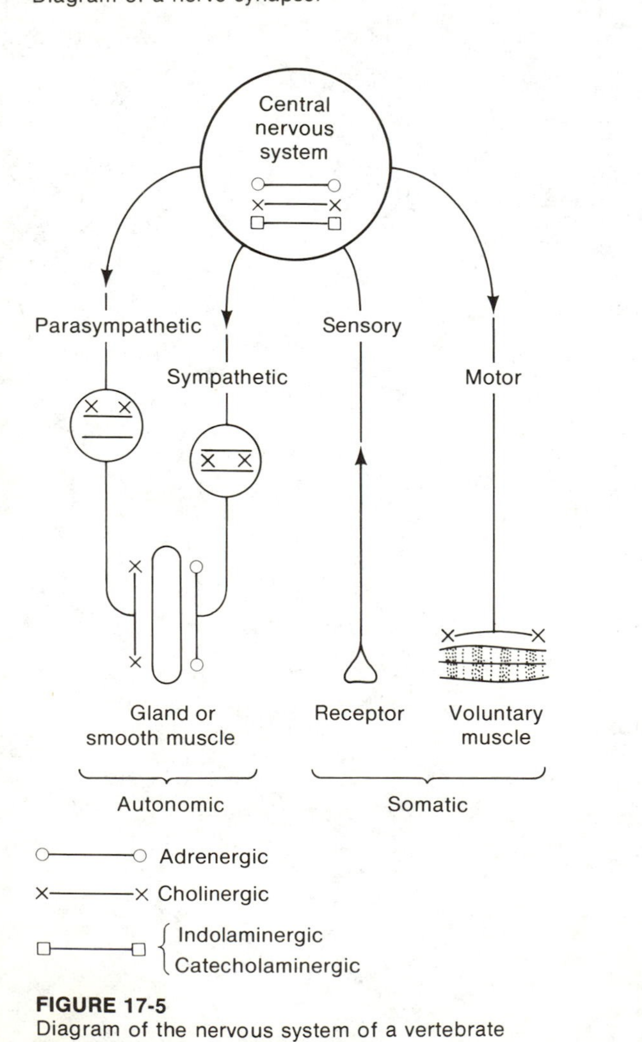

FIGURE 17-5
Diagram of the nervous system of a vertebrate showing locations of adrenergic and cholinergic junctions. (*Source:* O'Brien, 1967.)

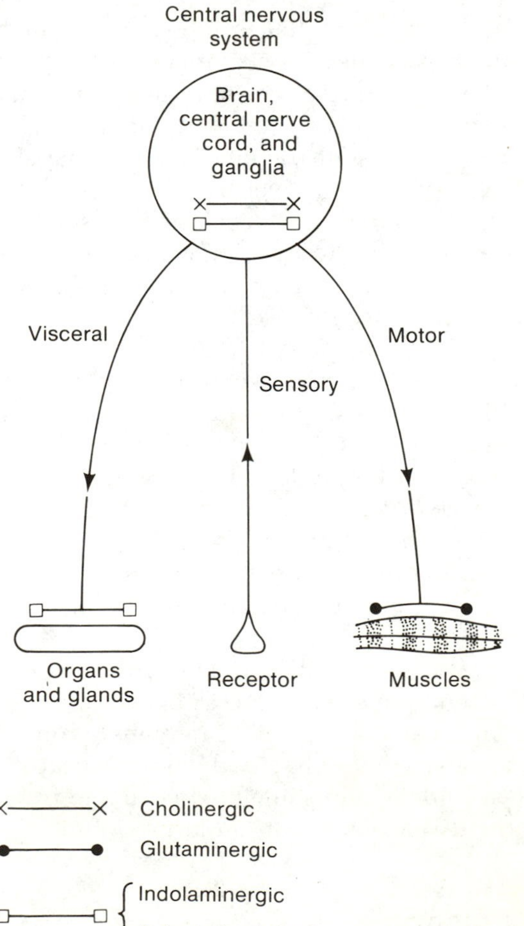

FIGURE 17-6
Diagram of the nervous system of an insect showing the (proposed) locations of cholinergic, glutaminergic, and indolaminergic/catecholaminergic junctions.

Anticholinesterase Insecticides. The organophosphates (OPs) exert their toxic action by tying up or inhibiting certain important enzymes of the nervous system, namely cholinesterases (ChE). At the synapse an impulse is transmitted by acetylcholine (ACh), which is then destroyed by the ChE enzyme so the synapse will be cleared for another transmission. These chemical reactions happen within microseconds and continue constantly, as needed, under normal conditions. However, the OPs attach to the enzyme ChE in a way that prevents it from clearing away the ACh transmitter; in effect, the electric transmission circuits jam because of the accumulation of ACh. In mammals the accumulation of ACh interferes with the neuromuscular junction, producing rapid twitching of voluntary muscles, finally resulting in paralysis and death due to respiratory failure. Symptoms in insects follow the pattern of nerve poisoning: restlessness, hyperexcitability, tremors and convulsions, and paralysis.

A unique feature of all organophosphate (and carbamate) insecticides is their structural "fit" with the ChE enzyme. These insecticides mimic the molecular shape of ACh, the natural substrate of cholinesterase (AChE). AChE has two active sites, the esteratic site and anionic site (Figure 17-7).

The inhibition of AChE by an insecticide is very similar to the early stage of the hydrolysis (splitting) of ACh. The acetyl carbon ($CH_3{-}\overset{\displaystyle O}{\overset{\|}{C}}{-}O{-}$) of ACh, or phosphorus moiety ($RO_2\overset{\displaystyle O}{\overset{\|}{P}}{-}OR'$) of the organophosphate, or the carbamyl moiety ($R_2{-}N{-}\overset{\displaystyle O}{\overset{\|}{C}}{-}O{-}$) of the carbamate binds to the esteratic site on the amino acid serine, while the remainder of the molecule attaches to the anionic site. Through a series of very rapid steps, ACh is hydrolyzed or split into one molecule of acetic acid and one of choline (Figure 17-7). The organophosphate is hydrolyzed so that the alcohol fragment (HOR′) leaves the reaction and the phosphorus remains attached to the serine, phosphorylating the enzyme, thus inhibiting it for an extended period of days.

This phosphorylation prevents the enzyme from acting further in its role of clearing away ACh, and is a strong and difficult bond to break. Phosphorylated ChE is commonly referred to as being irreversibly inhibited, because of this strong bonding. It is not truly irreversible, but rather slowly reversible.

When the carbamate is hydrolyzed, the alcohol fragment (HOR) leaves the reaction and the ($CH_3{-}\overset{\displaystyle H}{\overset{\diagdown}{N}}{-}\overset{\displaystyle O}{\overset{\|}{C}}{-}O{-}$) remains attached to the serine, thus carbamylating the enzyme for a relatively short period of time, usually hours. Carbamylated ChE is reversibly inhibited; the relatively short attachment period is usually several hours.

Carbamates inhibit cholinesterase as organophosphates do, and they behave in an almost identical manner in biological systems, but with two main differences. Some carbamates are potent inhibitors of

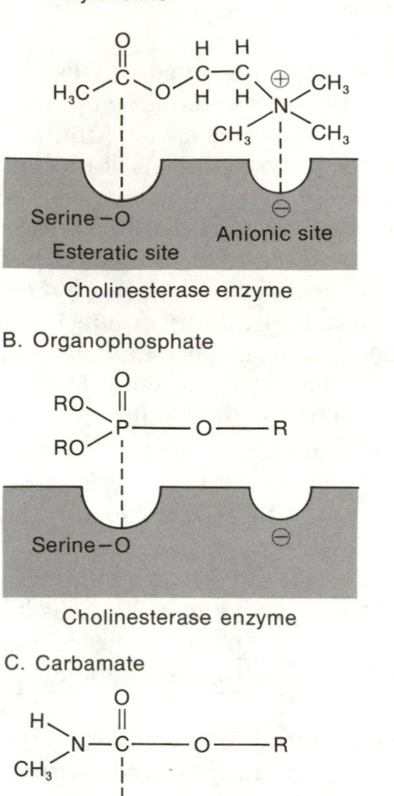

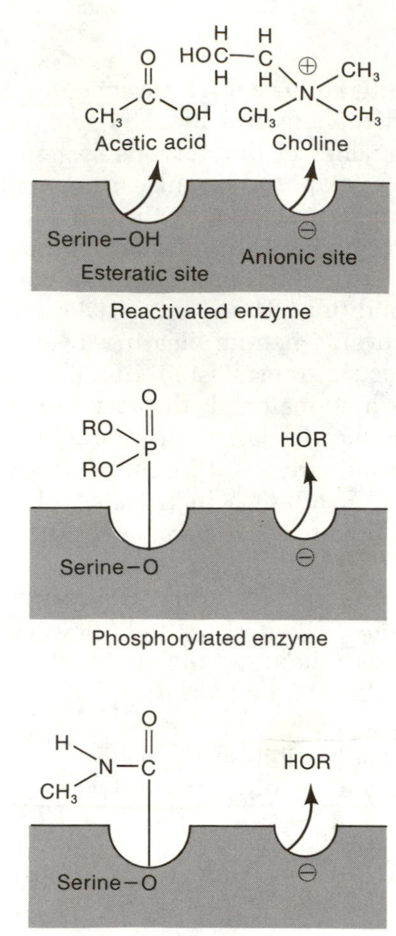

C. Carbamate

Serine–O

Cholinesterase enzyme

FIGURE 17-7
Simplified reactions of acetylcholine, organophosphate, and carbamate with the enzyme cholinesterase. (A) Acetylcholine is hydrolyzed to acetic acid and choline, resulting in reactivated enzyme. (B) Organophosphate is hydrolyzed and phosphorylates the enzyme by attaching to serine at the esteratic site.
(C) Carbamate is hydrolyzed and carbamylates the enzyme by also attaching to serine at the esteratic site.

aliesterase (miscellaneous aliphatic esterases whose exact functions are not known), and their selectivity is sometimes more pronounced against the ChE of different species. Second, ChE inhibition by carbamates is apparently reversible. When ChE is inhibited by a carbamate, it is said to be carbamylated, as when an organophosphate results in the enzyme being phosphorylated.

In insects, the effects of organophosphates and carbamates are primarily those of poisoning of the central nervous system, since the insect neuromuscular junction is not cholinergic, as in mammals (see Figures 17-5 and 17-6). The only cholinergic synapses known in insects are in the central nervous system. The chemical neuromuscular junction transmitter in insects is thought to be glutamic acid, but that has not been proved.

Postsynaptic Insecticides. Nicotine and nicotine sulfate are used very little today. However, many other insecticides produce symp-

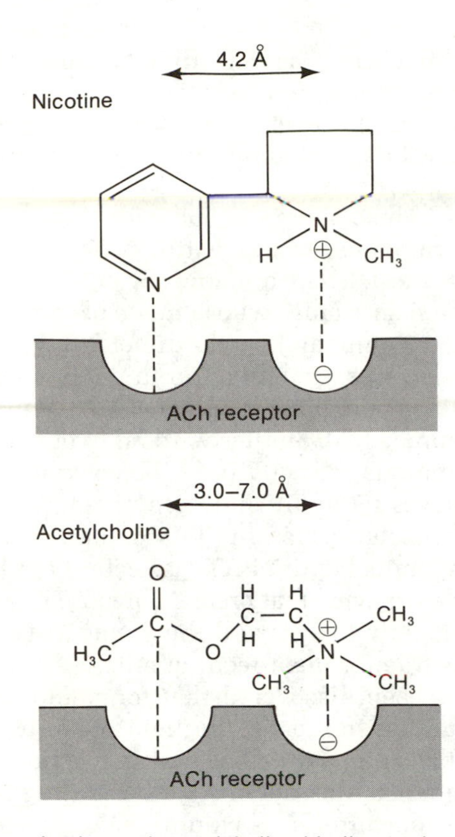

FIGURE 17-8
Similarities between nicotine and acetylcholine binding to the acetylcholine receptor. (*Source:* Matsumura, 1975.)

toms that are described as *nicotinic effects* because they resemble the tremors seen in nicotine poisoning.

Symptoms of nicotine poisoning in insects progress in the following sequence: excitation, convulsions, paralysis, and death. In low concentrations, nicotine stimulates the heartbeat; at high concentrations, it decreases heartbeat. The outstanding feature of nicotine effects in insects is the selective inhibitory action on nerve ganglia and synapses in the voluntary nervous system. When nicotine is applied to the ganglion controlling an isolated leg, the leg muscles tremor violently. When the leg is severed, the tremors stop. Further, nicotine has no effect on the isolated (severed) leg.

In warm-blooded animals, nicotine acts like the transmitter acetylcholine (ACh) and binds to its receptor at the neuromuscular junction and at the ganglion (Figure 17-8). As a result, it causes stimulation of voluntary muscles and of the ganglia, glands, and smooth muscles. Nicotine thus attacks some of the cholinergic junctions and all neuromuscular junctions of mammals by ACh-like action.

Nicotine does not inhibit or bind cholinesterase, but rather acts by mimicking ACh. It is very selective in choosing the binding sites in both insects and mammals: the neuromuscular junction of mammals and the synaptic ganglion of insects. It seems that the receptors cannot distinguish between ACh and nicotine. A comparison of the structures of the two molecules reveals similarities both in dimension and in relative distribution of charge and polarity (Figure 17-8).

Adrenergic Insecticides. The degrading enzyme of the transmitter norepinephrine is monoamine oxidase. Its precise localization in insects is not known; however, it has been identified in the insect nervous system, as has octopamine in the firefly. It is believed that other biogenic amines exist.

Formamidines induce abnormal behavior in pests: reduced feeding, dispersal from plants, erratic mating behavior, and detachment of ticks from their host. There are now several formamidine insecticides, ovicides, and acaricides whose mode of action resembles that of the first of these compounds, chlordimeform. In rats, chlordimeform apparently induces accumulation in the brain of serotonin (5-hydroxytryptamine) and norepinephrine, two important biogenic amines (Matsumura, 1975). Studies with rat liver slices indicate that chlordimeform inhibits the activity of the enzyme monoamine oxidase, which removes serotonin and norepinephrine from their sites of action. Chlordimeform also inhibits oxidative phosphorylation while stimulating mitochondrial ATPase activity in high concentrations; ATPase is an enzyme that breaks down ATP. In insects, chlordimeform is toxic only to eggs and the youngest larvae of caterpillars. The same is true for plant-feeding mites.

Recently it has been discovered that formamidines inhibit octopamine, the neurotransmitter in the light organ of a firefly. It has been suggested that octopamine has multiple actions as a neurohormone and neurotransmitter, actions similar to those of epinephrine in vertebrates. Octopamine is also found in considerable amounts in the central nervous system of several insect species (Hollingworth and Murdock, 1980).

METABOLIC INHIBITORS

Inhibitors of Mitochondrial Electron Transport

The electron transport chain is the series of cytochromes in the mitochondria involved in the production of energy from the oxidation of carbohydrate, lipid, and protein molecules. Pesticides for which there is good evidence that their mode of action is on the electron transport chain are rotenone, fumigants that work through the cyanide ion (CN−), dinitrophenols, organotin acaricides and fungicides, and the oxathiin fungicides.

Rotenone. In insects rotenone poisoning results in slowing of the heartbeat, depression of respiratory movement, reduction in oxygen consumption, and finally limp paralysis (Brown, 1963). Rotenone is a respiratory enzyme inhibitor, acting between NAD^+ (a coenzyme involved in oxidation and reduction in metabolic pathways) and coenzyme Q (a respiratory enzyme responsible for carrying electrons in some electron transport chains), resulting in failure of the respiratory functions. The probable site of inhibition in the electron transport chain is shown in Figure 17-9.

Rotenone is highly toxic to most insects, and its mode of action is interference in energy production. Its extreme toxicity to fish and insects and its low toxicity to mammals are explained by the way each metabolizes rotenone. Insects and fish convert rotenone to highly toxic metabolites in large quantity, while mammals almost exclusively produce nontoxic metabolites.

HCN. HCN is an inhibitor of the electron transport system. It reacts with the terminal cytochrome oxidase, an enzyme in the respiratory chain, by attaching to the heme iron, the same attachment site as for oxygen. Other poisons that also chemically attach to this location and act in the same specific way are CO (carbon monoxide), NO (nitrous oxide), N_3^- (azide ion), H_2S (hydrogen sulfide), and CN^- (cyanide). The site of action for the CN^- ion is shown in Figure 17-9.

Dinitrophenols. The dinitrophenols are "uncouplers" or inhibitors of oxidative phosphorylation. In the biochemical reactions of a cell in contact with a dinitrophenol, the close coupling between the respiratory chain and phosphorylation is broken, respiratory control is lost, and electron transport along the chain occurs at full speed without producing ATP, the high-energy molecule. This type of inhibition is illustrated in Figure 17-9.

Organotins. The organotin compounds used as acaricides inhibit oxidative phosphorylation by blocking rather than uncoupling the formation of ATP, much as in the action of dinitrophenols.

Inhibitors of Mixed-Function Oxidase

Pyrethrins and some of the synthetic pyrethroids are the only insecticides that are used commercially with the addition of a synergist, usually piperonyl butoxide. The synergist, though expensive, is worthwhile because pyrethrins are quite expensive and not persistent; the synergist maximizes the effect of the pyrethrins during their short life.

Several years of mystery surrounded the mode of action of synergists, because, to further confuse matters, these compounds produce no effect with some insecticide classes and even antagonize the action of others. It has since been well established that the synergist mode of action is the inhibition of mixed-function oxidases (MFO), enzymes produced by microsomes, subcellular units found in the liver of mammals and in some insect tissues (e.g., fat bodies). Microsomes were once thought to have no specific function other than to oxidize miscellaneous fat-soluble foreign compounds introduced into the system, such as insecticides. Oxidation in the microsomes leads either to detoxification or activation of an insecticide. If the inhibited enzyme normally detoxifies the insecticide, the insecticide is left intact to exert its effectiveness, and appears to be synergized. If, however, the inhibited enzyme normally activates the insecticide, as with some phosphorothioates, the insecticide is not activated, and appears to be inhibited or antagonized in its effectiveness.

It is now known that most of the biodegradability of current insecticides depends on their successful biotransformation by microsomal oxidases of various types, such as MFO. These enzymes are also responsible for synergism.

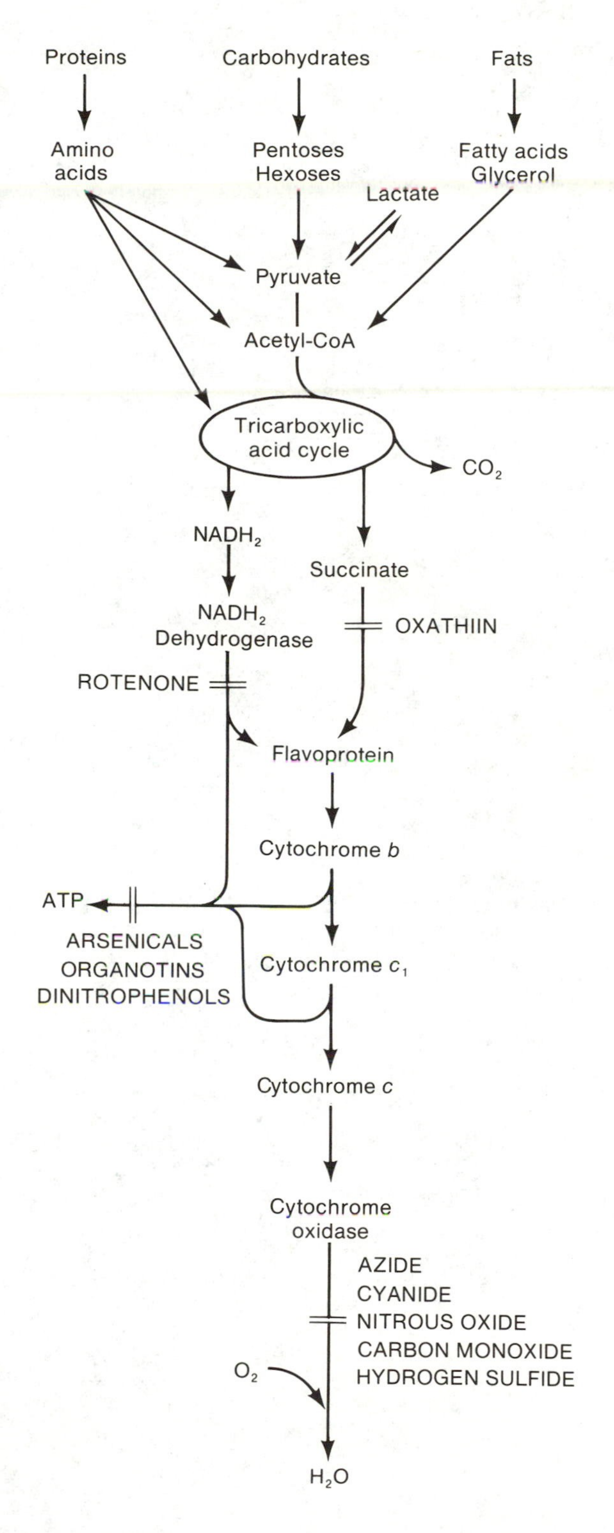

FIGURE 17-9
The respiratory electron transport chain, showing a likely arrangement of electron carriers, the three phosphorylation sites (ATP), and the four points (double lines) where pesticides inhibit this process. Arrows indicate the direction of electron transport. (*Source:* Modified from Corbett, 1974.)

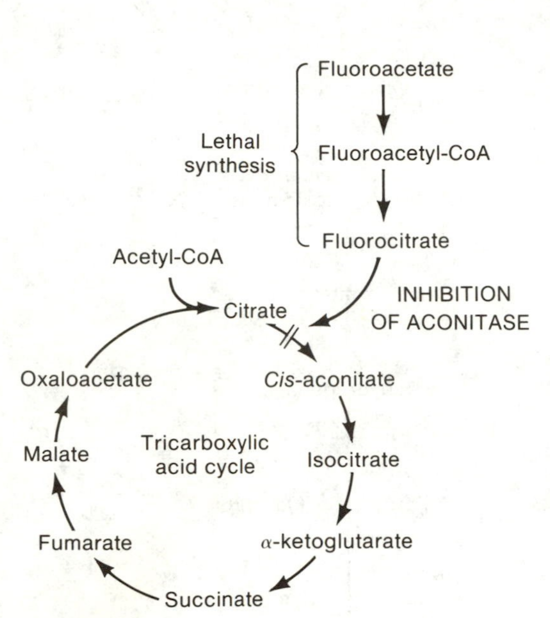

FIGURE 17-10
Inhibition of the tricarboxylic acid cycle enzyme aconitase by lethal synthesis of fluorocitrate from fluoroacetate. (*Source:* Corbett, 1974.)

Glycolysis—Citric Acid Cycle

Fluorine Compounds. The fluorine compounds are discussed briefly because they have a classic mode of action. In mammals the organofluorine compounds (fluoroacetate) cause symptoms only after a 20- to 60-minute delay, when convulsions begin. The heartbeat increases and the body temperature drops. Fluoroacetate is not a direct enzyme inhibitor, but is a latent inhibitor, requiring conversion to a derivative, fluorocitric acid, which is a potent enzyme inhibitor (Figure 17-10). Because of the structural similarity of fluorocitric acid to citric acid, it competes with citric acid for the target enzyme, aconitase, thus blocking the enzyme action. In addition, the result of fluoroacetate poisoning in insects and mammals is the accumulation of citric acid. This probably leads to the complexing of calcium by the citrate, thus reducing free calcium levels at critical sites such as muscles involved in respiration.

Inorganic fluorine compounds—for example, sodium fluoride (NaF)—have rather low toxicity to insects and mammals equally. Fluoride has an unusual and direct effect on the mammalian heart, causing a prolonged increase in the force of contraction. Fluoride has no effect on axonic transmission of nerve impulses. The probable mode of action for inorganic fluoride is the inhibition of a large number of metal-containing enzymes. It forms complexes with those containing iron, calcium, and magnesium, the latter including phosphatases and phosphorylases. Because fluoride may exert its lethal action on any one of a large number of key enzymes, it has not yet been possible to specify the fatal biochemical lesion.

Arsenical Compounds. The arsenicals of interest are lead and calcium arsenate. Calcium arsenate is by far the more toxic to both insects and mammals. Arsenicals kill primarily by inhibiting respiratory enzymes. Three categories of action compose the mode of action for arsenic: (1) uncoupling of oxidative phosphorylation by arsenolysis; (2) combination with various enzymes containing the sulfhydryl group (—SH), especially in pyruvic oxidase; and (3) gross protein precipitation.

Arsenic chemically resembles phosphorus and substitutes partially for phosphorus in some reactions. The best example is arsenolysis, which takes place instead of phosphorylysis (Figure 17-9). Phosphorylysis is essential in forming the high-energy bond of ATP, and oxidative phosphorylation is the major energy-producing step of the cell, but arsenolysis uncouples phosphorylation. In this uncoupling, arsenite is about twice as effective as the arsenate ion.

Numerous enzymes contain the —SH group, and all are susceptible to arsenicals. Pyruvic oxidase is the most arsenic-sensitive enzyme; other enzymes that are equally important but not as sensitive include cytochrome oxidase, lactic acid dehydrogenase, and α-glycerolphosphate dehydrogenase.

Inorganic arsenicals can cause gross coagulation of proteins in high concentrations. This effect, however, is probably the same as that seen on —SH groups; but instead of attacking a particular —SH group at an enzyme's active site, it involves the sulfur bonds that maintain the physical configuration or structure of most proteins.

MUSCLE POISONS

Ryania and sabadilla are classed as muscle poisons because of their direct action on muscle tissue. Ryania contains an alkaloid, ryanodine, the active principle of *Ryania speciosa*, grown in South America. Ryanodine is at least 20 times more toxic to mammals than to most insects. Thus, this natural insecticide is far more hazardous to use than many synthetic compounds.

The mode of action of ryanodine is that of membrane disruption, and its effect is specific for the excitable membrane of muscle. Poisoned insects show tremendous increases in oxygen consumption, as much as tenfold, followed by flaccid paralysis and death. In the frog, complete rigor follows paralysis. In mammals there is progressive rigidity of the muscles with final respiratory failure and death. One theory of ryanodine's effect is that it may interfere with the relaxing system of muscles, which strips calcium ions (Ca^{++}) from contracted muscle and leads to the relaxed state. In addition, however, ryanodine is believed to interfere with metabolism, an effect that may be unrelated to its effects on muscle.

Sabadilla comes from the powdered seeds of the lily *Schoenocaulon officinale* and contains two insecticidal alkaloids, cevadine and veratridine. Especially susceptible are houseflies, household insects, and the Hemiptera and Homoptera, which include the true bugs. Purified cevadine, for example, is approximately 10 times more toxic than DDT to houseflies. In mammals veratridine produces a prolonged rigor in skeletal muscle following the initial twitch, accompanied by repetitive impulse discharge in muscle fibers. Oxygen consumption increases, but not to the extent noted in ryanodine poisoning. Sabadilla appears to have the same general mode of action in insects as ryanodine, resulting in flaccid paralysis and death.

ALKYLATING AGENTS

Alkylating agents are those biologically active substances, found in a broad range of chemical groupings, that replace an active hydrogen in a biologically significant compound by an alkyl group (carbon chain-like structure). Notable among these are the early World War I gases, mustard and nitrogen mustard, and a more recent group of experimental chemosterilants, particularly the aziridines (see Chapter 24). Several of the halogen-containing compounds, mostly fumigants, are also identified as alkylating agents. Methyl bromide is the most prominent of these (see p. 60).

Alkylating agents act in two basic ways. They react directly with cell chromosomes by attacking one or more of the reactive loci on the nucleic acid molecule, and they deactivate essential enzymes, which subsequently cannot fulfill normal functions in the synthesis of nucleic acids. The latter is comparable to the inhibition of —SH enzymes by the arsenicals.

At best we will know that herbicides control weeds. All of their modes of action will probably never be known.

K. C. Hamilton (1982)

CHAPTER 18

Modes of Action for Herbicides

In this chapter, the herbicide mode of action usually pertains to the mechanism of action, that is, the biochemical responses of the plant that appear to be associated with the herbicidal action. Because of the complex interrelations between cellular metabolism and growth and development, responses to chemicals may take place at sites removed from the region of application or the original site of action. In many cases, the mode of action has not been worked out. However, it may be the case that many herbicides are metabolically nonspecific; that is, rather than having a single site of action, there may be several sites and mechanisms involved in inhibition.

In presenting the modes of action for herbicides in a simple and straightforward way, I have drawn heavily on *Mode of Action of Herbicides,* by Floyd M. Ashton and Alden S. Crafts (1973), *Herbicide Handbook,* 4th ed., by the Weed Science Society of America (1979), and *The Biochemical Mode of Action of Pesticides* by J. R. Corbett (1974). Readers desiring more explicit information should consult one of these books.

PHYSICAL TOXICANTS

Petroleum Oils

The action of oils on plants is attributed to disruption of cellular membranes. A plant damaged by oil undergoes darkening and loss of turgor, as if frostbitten. This is presumably due to leakage of the cell contents into the intercellular spaces. The effects of oils are similar to the symptoms produced by the bipyridylium (diquat and paraquat) herbicides, which are also thought to disrupt membranes. The actual molecular mechanism by which oils disrupt membranes is not known, though it is suggested that they act by solubilizing part of the lipid of the membrane.

Bipyridyliums

Paraquat and diquat kill within hours any and all plants onto which they are sprayed, with their effects sometimes visible in minutes. They are highly water soluble and are so tightly bound by soil that they become unavailable to plant roots practically as soon as they reach the ground. Bipyridylium herbicides are almost completely dissociated in solution into positive and negative ions. The herbicidal action is due to the positive ion, which is reduced by photosynthesis to form a relatively stable free radical. The bipyridylium free radical is easily oxidized in the presence of oxygen to re-form the original ion and hydrogen peroxide, which destroys the plant tissue. They cause rapid desiccation of the foliage, with wilting as an early symptom. At the cellular level they cause cell and chloroplast membranes to rupture, giving the tissues their frostbitten appearance. Either or both the hydroxyl radical (OH^-) and hydrogen peroxide (H_2O_2) are the primary toxicants.

AUXINLIKE EFFECTS

Phenoxyaliphatic Acids

The most commonly used herbicides among the phenoxyaliphatic acids are 2,4-D, 2,4,5-T, and MCPA. The effect of these herbicides on broad-leaf plants is auxinlike (auxin is a growth hormone): elongation of the growing terminals, distortion, and in 7 to 10 days collapse, withering, and death. 2,4-D is indeed used as an auxin or growth regulator at extremely low concentrations. The phenoxys increase cellular division, activate phosphate metabolism and usually increase ribonucleic acid (RNA) synthesis, which increases protein synthesis. We can rule out both interference with oxidative phosphorylation in mitochondria and inhibition of the Hill reaction in chloroplasts (see p. 162) as mechanisms to explain the herbicidal action of 2,4-D. Thus the mode of action of the phenoxy herbicides must consist of a large number of structural and biochemical reactions involving prolonged abnormal growth and the failure of those changes characteristic of maturity and senescence.

Arylaliphatic Acids (Substituted Benzoic Acids)

Among the arylaliphatic acids are 2,3,6-TBA, dicamba, fenac, chloramben, and DCPA. These herbicides, related to benzoic acids, act in the same way as the phenoxyaliphatic acids (2,4-D and relatives); that is, they resemble auxins (growth hormones) and are persistent in their actions. For instance, 2,3,6-TBA causes cell elongation and tissue proliferation, induces adventitious roots, modifies the arrangement of leaves and other organs, and causes fruit to develop in the absence of fertilization. Picloram, though not a member of this class but rather a heterocyclic nitrogen, also possesses the typical properties of auxins and herbicidal action similar to that of 2,4-D. The specific mechanism of action for the benzoic acid herbicides is unknown.

METABOLIC INHIBITORS

Inhibitors of the Electron Transport Chain

The phenols include DNOC, dinoseb, DNAP, and pentachlorophenol (PCP). The mode of action for these is twofold. In high concentrations they destroy cell membranes of treated foliage resulting in fluid loss, cell collapse, and desiccation. At lower dosages the phenols block the formation of ATP by uncoupling oxidative phosphorylation. Their actions in plants are the same as in animal tissues, described in Chapter 17.

Nitroanilines, sometimes referred to as dinitroanilines, include trifluralin, benefin, nitralin, and isopropalin. They inhibit growth of the entire plant, apparently by limiting root growth, especially the lateral or secondary roots. They are readily absorbed by both roots and shoots but are only slightly translocated to other plant parts. Generally, the nitroanilines seem to stimulate increased production of nucleic acids in both roots and shoots, the most relevant effect related to the phytotoxic symptoms. With trifluralin, the uncoupling of oxidative phosphorylation and inhibition of the production of several hormone-induced enzymes have also been demonstrated.

Bromoxynil, dichlobenil, and ioxynil (which is not available in the United States) belong to the nitrile herbicides. Ioxynil and bromoxynil are contact herbicides used to control hard-to-kill weeds in cereals. They are strong inhibitors of oxidative phosphorylation and also inhibit CO_2 fixation. These effects do not entirely explain the rapid contact effect of these compounds, but they may be indicators of more drastic effects on cell membrane systems. Dichlobenil is a preemergence herbicide applied by incorporation into the soil, and its site of action is in the meristems of embryos as they begin to germinate. It is strongly adsorbed by the soil, with almost no leaching, making it useful in no-till or minimum tillage vineyard and orchard programs. The precise mode of action of dichlobenil is unknown. It is known that the substituted nitriles release CN^-, which binds to cytochrome *c* oxidase, a respiratory enzyme intermediate in the electron transport chain.

Inhibitors of Phosphorus Metabolism

The arsenicals (cacodylic acid, MMA, MSMA, DSMA, MAMA) act through enzyme systems to inhibit growth. Their actions in plant systems are no different than in animal systems (described in Chapter 17). Arsenicals kill plants relatively slowly, with the first symptom being chlorosis, the appearance of bleached or yellow spots on leaves as a result of the loss of chlorophyll. Arsenic occurs in herbicides in various forms from the inorganic to arsenites, arsenates, and arsonates. Arsenic is known to uncouple phosphorylation, react with —SH groups of enzymes, and act like phosphorus in all the vital roles of phosphorus. Undoubtedly all these effects enter into the broad-spectrum mechanism(s) of the arsenical herbicides.

Enzyme Inhibitors

The substituted amides include the herbicides CDAA, diphenamid, propachlor, and propanil. Apparently their chemical structures are more closely related than their modes of action, for mechanism-of-action studies on the amides have not produced a consistent pattern. The various known actions include inhibition of photosynthesis, respiration, RNA and protein synthesis, amylase, proteinase, dipeptidase, and other enzymes.

Glyphosate is a very broad spectrum herbicide that is relatively nonselective. It appears to inhibit the aromatic amino acid biosynthetic pathway and may inhibit certain enzymes whose role is unclear.

Inhibitors of Nucleic Acid Metabolism and Protein Synthesis

Carbamates. Carbamates include barban, chlorpropham, phenmedipham, propham, swep, and terbucarb. Their chemical structures or derivations are more closely related than their modes of action, for mechanism-of-action studies on the carbamates have not produced a homogeneous pattern. They have been shown to reduce the ATP content of tissue sections, and to inhibit RNA and protein syntheses, oxidative phosphorylation, and the Hill reaction of photosynthesis (see p. 162). Basically, they interfere with cell division, which must relate to the involvement of protein synthesis and associated nucleic acid synthesis, and the energy needs for these reactions in the form of ATP.

Thiocarbamates. All thio- and dithiocarbamates have high vapor pressure, are volatile, and must be soil-incorporated for this reason. They inhibit the growth of shoots of germinating seedlings to a greater degree than they inhibit the roots, and they are absorbed by the shoot more readily than by the roots. Collectively they alter several aspects of plant metabolism: photosynthesis, respiration, oxidative phosphorylation, protein synthesis, and nucleic acid metabolism. The primary site of action may be nucleic acid metabolism and protein synthesis, because CDEC and EPTC inhibit L-leucine incorporation into protein in segments of barley coleoptile or sesbania hemp hypocotyls. Additionally, EPTC inhibits growth and RNA synthesis in soybean tissue.

Aliphatic Acids. Two examples of the chlorinated aliphatic acids are trichloroacetic acid (TCA) and dalapon. The specific modes of action for these herbicides are unknown. It has been suggested that the primary site of action for dalapon is associated with modification of protein structure, including enzymes, not by precipitation but more likely by subtle conformational changes, perhaps by combining with proteins. Dalapon is known to increase ammonia in the cells of treated plants, and susceptible species do not seem to be able to form amides fast enough to prevent toxic levels of ammonia. Even less is known about TCA, but it is assumed that it acts like dalapon.

PHOTOSYNTHESIS INHIBITORS

Among the many modes of actions for herbicides, one of great importance is the inhibition of the Hill reaction. The Hill reaction is a light-initiated reaction that splits water (photolysis), resulting in the production of free oxygen (O_2) by plants. Chlorophyll, the green pigment of plants, is an essential ingredient in the reaction, since it catalyzes the production of oxygen from water and the transfer of the hydrogen to a hydrogen acceptor. A simplified chemical formula for the reaction is written as follows:

$$2H_2O + 2A \xrightarrow[\text{chlorophyll}]{\text{light}} 2AH_2 + O_2$$

where A is some unidentified hydrogen acceptor. The hydrogen and acceptor complex (AH_2) continue on in reactions with CO_2 to form plant sugars and cellulose, while free O_2 is released into the atmosphere.

Triazines

Atrazine and simazine are two of the most prominent of the numerous triazines on the market today. The triazines inhibit the growth of all organs of intact plants, an effect attributed to a deficiency of photosynthate, necessary for growth, caused by inhibition of photosynthesis. The tolerance of certain species of higher plants to the triazines probably results from their ability to rapidly degrade the herbicide to nontoxic metabolites. The mechanism of action of the triazines is blockage of photosynthesis, specifically the photolysis of water. Since this alone cannot account for the great effects of the triazines, it appears that a secondary phytotoxic agent is formed by a reaction that is probably coupled to the water photolysis and carries the burden of triazine activity.

Triazoles

Amitrole is the only member of the unique, 5-member-ring class of triazoles. It stimulates growth at low concentrations and inhibits growth at high concentrations. Treated plants lose their green pigmentation, appearing to become albinos, which is the result of amitrole's interference with the development of chloroplasts. Amitrole is absorbed by leaves and roots and translocates throughout the plant, causing biochemical alterations that affect carbohydrate, lipid, and nitrogen metabolism. Its primary site of action seems to be the biochemistry of chloroplast development. This, however, does not explain the phytotoxic symptoms beyond albinism.

Substituted Ureas

Monuron, diuron, fenuron, linuron, and siduron are only a few of the many urea-derived herbicides. They are not all equally phytotoxic. It is not uncommon to find a four- to sixfold difference in toxicity to some species. Most are readily absorbed by the roots and rapidly translocated to the upper plant parts, producing phytotoxic symptoms that are most visible in the leaves. The primary site of action for the urea herbicides is the inhibition of the Hill reaction of photosynthesis. But since the plants do not merely starve from lack of photosynthate, it is theorized that a secondary toxic substance is formed in the oxygen-releasing pathway of photosynthesis.

Substituted Uracils

Bromacil, isocil, and terbacil are three widely used uracils. All are soil-borne herbicides, readily absorbed by roots and translocated to the leaves. They block the Hill reaction and interfere with a step in the photosynthetic pathway near that of oxygen release. In this way their site and mode of action resemble the substituted ureas.

CHAPTER 19

I have seen some medicate their seed before they sow it. They steep it in nitre and amurca to obtain a fuller produce in the deceitful pods.

Virgil, *Georgics* I

Modes of Action for Fungicides

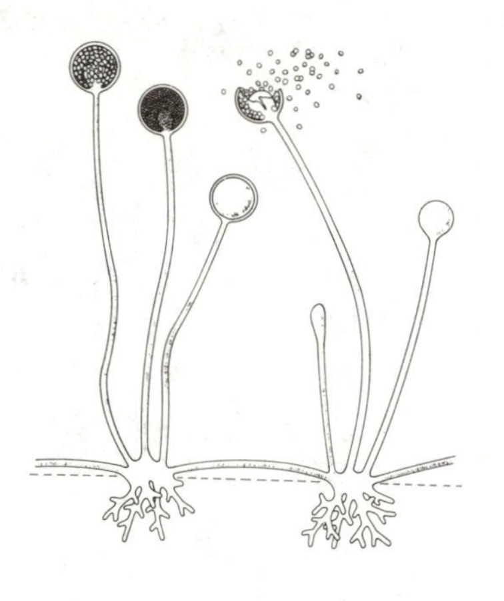

Fungicidal action is usually expressed in one of two physically visible ways: the inhibition of spore germination or the inhibition of fungus growth. Most fungicides prevent spore germination or kill the spore immediately following germination. Some of these chemical inhibitors or toxicants also retard or halt fungus growth when applied after the infectious stage has developed. The newer systemic fungicides have eradicant properties and stop the progress of existing infections.

What happens at the cellular level to cause these readily visible results? As currently viewed, all fungicides are metabolic inhibitors; that is, they block some vital metabolic process. For the sake of organization and simplicity, we can classify the modes of action into three broad groups: inhibitors of the electron transport chain, inhibitors of enzymes, and inhibitors of nucleic acid metabolism and protein synthesis.

INHIBITORS OF THE ELECTRON TRANSPORT CHAIN

Sulfur

Certain pathogenic organisms and most mites are killed by direct contact with sulfur and also by its fumigant action at temperatures above 22°C. The fumigant effect is, however, somewhat secondary at marginal temperatures and under windy conditions. It is quite effective in controlling powdery mildews of plants that are not unduly sensitive to sulfur. Unlike those of any other fungus, spores of powdery mildews will germinate in the absence of a film of water in the penetration court (a spot on the tissue that is "softened" prior to penetration by spore). Sulfur is absorbed by fungi in the vapor state, and its fumigant effect—acting at a distance—is undoubtedly important in killing spores of powdery mildews. Sulfur interferes in electron transport along the cytochromes and is then reduced to hydrogen sulfide (H_2S), a toxic entity to most cellular proteins.

The way fungi respond to sulfur suggests that sulfur can stimulate enzyme activity. The reduction of sulfur to hydrogen sulfide (H_2S) by fungi is a metabolic process that diverts protons (H^+) from steps of normal hydrogenation reactions (see Figure 17-9). This reduction is aerobic (requires oxygen) and involves a cytochrome system, possibly cytochrome *b*, a respiratory enzyme involved in the oxidation of nutritional components and foreign compounds. Sulfur presumably serves as an acceptor of hydrogen atoms from dehydrogenases, and then activates to full capacity the enzymic steps preceding sulfur reduction.

Organotins

Several organotins are now in use, more abroad than in the United States. Fentin hydroxide is the oldest member of those used in this country. They inhibit oxidative phosphorylation by blocking the formation of ATP, like the action of dinitrophenols (Figure 17-9).

Oxathiins

Carboxin and oxycarboxin are the only two oxathiins currently in use. Carboxin inhibits glucose and acetate oxidation by intact fungi, and noncompetitively inhibits succinate, but not $NADH_2$, oxidation by mitochondria. This effect occurs in the electron transport chain of respiration. Succinate accumulates in carboxin-treated cells of fungi, suggesting that inhibition of succinate oxidation is the primary site of action. Oxycarboxin, though less toxic, has the same mode of action (Figure 17-9).

Dinitrophenols

Dinocap is the only dinitrophenol fungicide, and like the others, it uncouples oxidative phosphorylation in the production of energy (see pp. 155, 160).

ENZYME INHIBITORS

Copper

Inorganic copper compounds are protectants, materials applied before the pathogen appears, to protect plants from inoculation. They prevent the spread of an infection but cannot eradicate an existing one. Thus the site of copper fungicidal action must be the fungal spore. Copper ions are concentrated by fungal spores from the surrounding medium, some up to 100-fold over that in the immediate environment. This high concentration of copper inside the spore supports the generally held view that the fungitoxic activity of copper ions is due to a nonspecific attraction for various groups in the cell, such as imidazole, carboxyl, phosphate, or thiol, resulting in nonspecific denaturation of protein and enzymes.

The currently accepted theory for the mode of action of copper's fungistatic action is its nonspecific denaturation of protein. The cupric ion (Cu^{++}) reacts with enzymes having reactive sulfhydryl (—SH) groups, which explains copper's toxicity to all forms of plant life, and especially its toxicity to the vulnerable copper-concentrating spores and cells.

Mercury

Mercury fungicides are purely of historical and classical toxicologic interest, for they are no longer registered except for occasional use. Considerable concern over the widespread use of mercury pesticides focused on the toxic nature of mercury itself and the discovery in 1970 that methyl mercury (CH_3Hg^+)—which is more toxic to life than aryl (aromatic-ring derivatives) mercurials or inorganic salts of mercury—is synthesized in the biosphere. As a result, most registrations for the mercury fungicides were canceled.

The biochemical basis for mercury's toxicity is generally thought to consist of interactions with the thiol groups of proteins, where mercury ions (Hg^{++}) form compounds attaching to one or two sulfurs. Since there are so many functional —SH groups in a living system, it is virtually impossible to select one reaction of mercury as being the most important to its toxic effect.

Organic mercury is different only in that its chemical form is R-Hg^+ rather than Hg^{++}. The organic mercurials are more soluble in lipid than the inorganic Hg^{++}, enabling them to penetrate more readily into cells and tissues. In general, inorganic mercury is a more effective enzyme inhibitor than organic mercury, though the organic forms remain as potent toxicants.

Dithiocarbamates

Included in the dithiocarbamates, the oldest group of the organic fungicides, are maneb, ferbam, zineb, manzate, dithane, polyram, and ziram. The dithiocarbamates may act by being metabolized to the isothiocyanate radical (—N=C=S), which inactivates the —SH groups in amino acids, proteins, and enzymes contained within the individual pathogen. It is difficult to establish exactly how they act since they are unstable, and the chemical nature of the fungitoxic agent(s) is not known for certain. A more recent theory tends to discount the isothiocyanates as toxic products, suggesting instead that ethylene thiuram disulfide

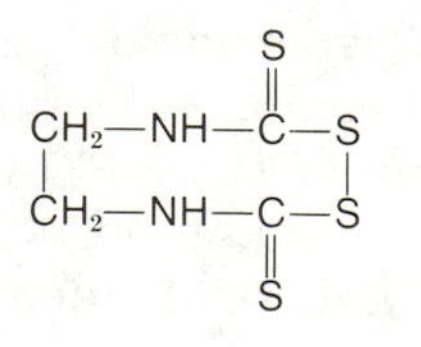

is the toxic moiety. Regardless of the identity of the toxic moiety responsible for the fungitoxicity of these compounds, it is believed that their activity is due to chemical inactivation of important systems in the fungal cell containing —SH or —SR.

An important function in both the lower and higher plants that may help explain the mode of action for not only the dithiocarbamate fungicides but also others, including those derived from the heavy metals, is *chelation*.

A chelate is an organic ring structure composed of a metal atom linked to a ring by nitrogen, oxygen, or sulfur. Particularly as they involve enzymes, chelates are powerful and essential entities in the metabolic processes of plants. Some of the metals required by the higher plants and fungi in trace amounts assist enzymes in conducting their routine duties of metabolism. The metal may be active in this role as a chelate with the biological component.

One of the generally accepted theories to explain the fungicidal activity of copper, mercury, cadmium, and other heavy metals is the formation of chelates within the fungal cells. The chelates, in turn, disrupt protein synthesis and metabolism. And since the most critical proteins of cells are enzymes, the metals required in trace amounts appearing in abundance or excessive quantities are equivalent to the introduction of potent poisons in the cells.

If the chelation theory is correct, it would account for the mode of action for the organic and inorganic heavy metal fungicides, for the formation of isothiocyanates from the dithiocarbamate molecules, and for the potency of the heavy-metal dithiocarbamates. Certain fungicides are in themselves chelating agents. They attach to the scarce metal components, such as Fe, Mg, and Zn, within the cells, literally robbing cells of essential materials.

In summary, chelation of heavy metals plays an important role in both the life and the death of cells.

Thiazoles

Ethazol, tricyclazole, and Busan-72 belong to the thiazole group. The unstable five-member ring of the thiazoles is broken rapidly under soil conditions to form either the fungicidal —N=C=S or a dithiocarbamate, depending on the structure of the parent chemical. The isothiocyanate inactivates —SH or —SR groups in amino acids, proteins, and enzymes contained within the individual pathogen. Thus the site of action is nonspecific.

Substituted Aromatics

The substituted aromatics are diverse in their modes of action. Being generally fungistatic, they reduce growth rates and sporulation of fungi, probably by combining with amino (—NH_2), —SH, or —SR groups in essential amino acids, proteins, or enzymes. Hexachlorobenzene possesses some fumigant qualities, especially with wheat bunt, but its mode of action is unknown. Chlorthalonil is believed to work by inactivation of thiol groups in the fungal cell. Chloroneb may act by inhibiting DNA synthesis. PCNB is effective against many species of fungi having chitin in their cell walls, but is virtually inactive against fungi in which chitin is either absent or present in small quantities, for example, *Pythium* and *Phytophthora*. Thus, there is indirect evidence that PCNB may act by interfering with chitin synthesis. Pentachlorophenol (PCP) blocks the formation of ATP by uncoupling oxidative phosphorylation. (See also Chapter 18, on the phenol derivative herbicides.)

Dicarboximides (Sulfenimides)

Captan, folpet, and captafol are the early members of the dicarboximide class of fungicides. They probably act nonspecifically, because the nature of the reactions between the fungicides and components of the fungal cell are not definitely established. It is generally believed that —SH or —SR groups are likely reaction sites. However, folpet inhibits isolated chymotrypsin, which does not contain thiol groups, so that reactions not involving thiol groups could be involved in fungicidal action.

Quinones

Belonging to the quinones are dichlone and chloranil. Fungal spores take up a large amount of dichlone before expiring, which suggests that dichlone exerts its toxicity by a simultaneous action on a variety of loci in major metabolic pathways. It is not clear what chemical reactions quinones undergo in the fungus. They do react readily with thiols, and they probably enter into a variety of other reactions in the cell, including combination with amino groups. The primary toxic effect of quinones could therefore be due to reaction with —SH or —SR and —NH_2 groups of vital enzymes.

INHIBITORS OF NUCLEIC ACID METABOLISM AND PROTEIN SYNTHESIS

Benzimidazoles

The benzimidazoles include benomyl, thiabendazole, and thiophanate. They are not toxic to fungi in their original state, but must be converted to their ester metabolites, which are known to be the toxic entities. These metabolites cause morphological distortion of germinating spores and probably act by inhibiting DNA synthesis, or by interfering with some closely related aspect of cell or nuclear division.

Antibiotics

Cycloheximide is a protein synthesis inhibitor—it inhibits the incorporation of amino acids into protein. As a consequence of its interference with protein synthesis, it may also inhibit DNA synthesis.

Cycloheximide is toxic not only to fungi but also to plants, and it has an oral LD_{50} in the rat of 2.5 mg/kg, by far the most toxic of the fungicides.

Streptomycin probably inhibits protein synthesis by binding to the ribosome, one molecule of streptomycin per ribosome. It also causes misreading of the genetic code, though this is not likely the primary effect.

Aliphatic Nitrogen Compounds

Dodine has a nonspecific mode of action, but it is generally agreed that it acts by interfering with membrane structure. There is indirect evidence that its primary site of action is the mitochondrial membrane. The quanidine nucleus of dodine is also known to inhibit the synthesis of RNA.

Triazines

Anilazine is the only fungicide belonging to the triazine class of fungicides. The ultimate site and mechanism of action are unknown. However, the fungicide does react readily with $—NH_2$ groups and less readily with —SH or —SR. Consequently, it appears likely that anilazine causes inhibition of a variety of cell processes by nonspecific combination with vital cell components.

CHAPTER **20**

Pesticide Resistance

> To arrest the development of resistance through increased use of insecticides is tantamount to stopping evolution.
>
> Albert S. Perry and Moises Agosin (1974)

Concern among scientists is growing over the increasing resistance of pests to the chemicals that once gave phenomenal control, even with application methods that were much less than ideal. Resistance, as we have encountered it over the past 25 years of chemical control, has been expressed in insects, mites, bacterial and fungal plant pathogens, domestic rodents, weeds, and several disease pathogens of humans and domestic animals. Most *Staphylococcus* or staph infections are resistant to penicillin; houseflies have developed resistance to DDT, malathion, and other insecticides; cockroaches have become resistant to chlordane, and in several areas, to diazinon; some strains of the venereal disease gonorrhea cannot be controlled with any but the very latest antibiotics; some recently developed fungicides with specific modes of action fail to control certain plant diseases; in many parts of the country the Norway rat cannot be controlled with warfarin; and certain weeds, once readily controlled with a particular herbicide, now go virtually unscathed. Certainly other pests are even now developing resistance to various pesticides.

The concept of resistance to chemicals has been understood for approximately 35 years. It is a simple mechanism that can be attributed to the singular mode of action of a particular pesticide, which disrupts only one genetically controlled process in the metabolism of the pest organism. The result is that resistant populations appear suddenly, either by selection of resistant individuals in a population or by a mutation, which appears to be the less common of the two means. Generally, the more specific the site and mode of action, the greater the likelihood that a pest population will develop resistance to that chemical. When a pesticide is lethal and specific in its mode of action, it must be expected to favor the survival of resistant strains of the target population.

INSECT RESISTANCE TO INSECTICIDES

Resistance to insecticides by insects dramatically exemplifies the selection principle of evolution. The susceptible insects are killed, leaving behind only those that are genetically resistant to the toxicant. Resistant individuals make up an increasingly large part of the pest population and pass their resistance on to the next and future generations. Resistance in insect pests simply represents the result of applying a stringent biological selection mechanism, the insecticide, over several generations. The greater the number of generations exposed to an insecticide, the greater the potential for a population to develop resistance as a consequence of this intensive selection mechanism.

Resistance to chemicals by insects is a very serious problem. In 1944, only 44 insect species were known to be resistant to insecticides. Today's estimates place this number well beyond 250, more than half of which are agricultural pests.

As more insecticides become ineffective against certain species, the problems of insect pest control increase proportionally. Populations of several insect pests now have such a high proportion of individuals resistant to all known insecticides that substitute materials are no longer available and insecticides are not recommended as control measures. In these instances chemical control is no longer the method of choice. Nor is resistance to insecticides confined to pests of agriculture. Public health problems resulting from resistance in insects that transmit or vector disease may become far more important than our agricultural difficulties.

One of the less familiar phases of insecticide resistance is cross-resistance. Insect populations that have developed resistance to one organochlorine insecticide are frequently resistant to another organochlorine, even though they have not been previously exposed. In many instances, insect populations that became resistant to the organochlorines quickly became resistant to the organophosphate insecticides. And because the modes of action of the organophosphates and carbamates are similar, populations that became resistant to the former soon became resistant to one or more of the latter.

The seeming ineffectiveness of an insecticide does not indicate with certainty that the insect is resistant. Effectiveness can also be reduced by the destruction of natural controls; for example, of parasites and predators that normally hold the pest species in check. Only a very small fraction of the total number of insects are considered pests. At normal population densities, most insects pose no threat to cultivated crops, and many are important to the health and stability of the environment because they control other potentially damaging species.

Most of the insecticides in use today have broad-spectrum effects, that is, they are lethal to a wide range of insects, including beneficial as well as pest species. If the natural enemies of a pest species are destroyed, a resurgence in the pest population may occur, resulting in increased damage to the crop presumably being protected.

Furthermore, with the use of broad-spectrum insecticides, insects that were once controlled naturally by their enemies and predators sometimes increase in such numbers that they become pests. This is the result of killing the insect's natural enemies. Thus an insect

TABLE 20-1
Susceptibility of some common annual weeds to several selective herbicides at usual selective rate.[a]

	Alachlor	*Atrazine*	*Chloramben*	*Chlorpropham*	*2,4-D*	*Diuron*	*EPTC*	*Linuron*	*Trifluralin*
Barnyardgrass	S	I	S	I	T	S	S	S	S
Chickweed		S		S	I	S	S	S	S
Cocklebur	T	S	I	T	S	I		S	T
Crabgrasses	S	T	S	I	T	S	S	S	S
Fall panicum	S	I	I	I	T	S	S	I	S
Foxtails	S	S	S	I	T	S	S	S	S
Jimsonweed	I	S	I	I	S			S	T
Johnsongrass seedlings	I	T	I	T	T	I	S	I	S
Knotweed	T	S		S	I	S	S	I	S
Lambsquarters	I	S	S	T	S	S	S	S	S
Morningglory annual	T	S	T	I	S	I	S	S	I
Mustards	I	S	I		S	S	T	S	T
Nightshades	I	S	S		I		I	S	I
Nutsedges	I	I	T	T	I		I	T	T
Pigweeds	S	S	S	I	S	S	I	S	S
Purslane	S	S	S	S	I	S	S	S	S
Ragweed	I	S	S	T	S	S	I	S	I
Smartweed	I	S	S	S	I			S	I
Velvetleaf	T	S	I	T	I	I	I	S	T
Wild cucumber	T	S	I	T	I			S	I

[a] S—susceptible; I—intermediate; and T—tolerant.
Source: Klingman, Ashton, and Noordhoff (1975).

species of secondary importance can become a pest if insecticides are applied to control a primary pest.

Resistance is indeed a problem in many insect pest species, and, unfortunately, the exploration and discovery of new insecticides have not kept pace with the loss of once-effective insecticides due to resistance or cancelation for reasons of human and environmental safety. Thus, the long-range answers to insect resistance will have to be found in other control methods or in the combination of methods known as *integrated pest management*. This subject is discussed in more detail under the final section of Chapter 24.

WEED TOLERANCE AND RESISTANCE TO HERBICIDES

Some weeds are remarkably naturally tolerant to the toxic action of herbicides. Such tolerance is based largely on morphological, physiological, and genetic plant characteristics. In contrast, resistance implies a genetic selection over a period of time and several plant generations, in response to a selective herbicide, resulting in a truly resistant weed biotype. Tolerant plants survive chemical applications and flourish in the absence of competition from susceptible

TABLE 20-2
Examples of annual weeds tolerant to specific herbicides.

Herbicides	*Tolerant annual weeds*
Atrazine (AAtrex®)[a]	*Panicum* spp.
Benefin (Balan®)	Groundcherry, mustard, prickly lettuce, spiny sowthistle, sunflower
Bensulide (Prefar®)	Pigweed, groundcherry, horse purslane, mustard, sowthistle, prickly lettuce
Bromacil (Hyvar-X®)	Spurge, horse purslane, plantain
Cycloate (Ro-neet®)	Mustard, little mallow, shepherdspurse, legumes
Difenzoquat (Avenge®)	Canary grass
Diuron (Karmex®)[a]	Plantain, spurge, horse purslane
EPTC (Eptam®)[a]	Legumes, little mallow, mustard
Glyphosate (Roundup®)	Malva, Russian thistle
MSMA (various)	Bermudagrass
Paraquat (Ortho Paraquat CL®)	Mexican sprangletop, plantain
Phenmedipham (Betanal®)	Pigweed, barnyardgrass, knotweed
Prometryn (Caparol®)[a]	Grasses
Propham (Chem-Hoe®)[a]	Summer grasses, most broadleaf weeds
Simazine (Princep®)	*Panicum* spp.
Trifluralin (Treflan®)	Groundcherry, mustard, legumes, sunflower
2,4-D (various)	Most grasses, little mallow, knotweed

[a] Other trade names may be available.
Source: Arizona Agricultural Chemicals Association, *Arizona Study Guide for Agricultural Pest Control Advisors: Control of Weeds* (1980).

species. The rotation of herbicides is useful in these cases if selective chemicals are available. However, most herbicides in common use today are broad spectrum.

Tables 20-1 and 20-2 contain examples of weeds that are difficult to control with specific herbicides because of their inherent or natural tolerance.

The problem of resistance to herbicides by weeds has not yet plagued the weed scientist as seriously as it has the entomologist and the rodent control specialist. However, plant pathologists are already quite familiar with the problem. For more than 20 years, they have seen the antibiotic fungicides fail to control bacterial plant diseases, because antibiotic-resistant strains of plant pathogenic bacteria quickly become prominent in the population of those organisms treated with antibiotics.

Several instances of weed resistance have been observed in the past few years, and with increasing frequency. The most significant is the triazine-resistant biotypes of at least 10 weed species that have been reported since 1970. These are in the plant families Compositae, Chenopodiaceae, and Amaranthaceae.

The species that have received the greatest amount of attention are the common groundsel, common lambsquarters, and redroot pigweed (*Amaranthus retroflexus* L.) (Hensley, 1981). These occur in the corn belt where corn has followed corn year after year without crop rotation, and triazines have been the herbicides of choice, almost to the exclusion of all others. Investigations into the mode of resistance in these resistant species reveal a different mechanism from that found in species tolerant to the triazines, such as corn. Corn's tolerance to the triazine herbicides is attributed to its ability to metabolize the active triazine herbicide to an inactive form. The mechanism of resistance in weeds results from an alteration at the herbicide's site of action. It is suggested that a protein of the photosystem II complex, which normally contains a triazine binding site, is modified in the resistant plants such that triazine affinity is lost (Ahrens et al., 1981).

Another instance is that of Johnsongrass' resistance to dalapon, observed in Arizona (Hamilton, 1981). Johnsongrass is not a native of the area and was brought in as a forage grass; numerous biotypes were introduced, which shortly became perennial weeds. Apparently, resistance to dalapon was observed after five or six years' usage, when the susceptible biotypes had been eliminated, leaving only the resistant strains. Dalapon, once a very effective herbicide against this perennial, is no longer useful.

Why do herbicide-resistant strains of plant pests become so quickly the predominating strains of the population? If a potent herbicide acts specifically, that is, if it kills by blocking a single enzymatically catalyzed metabolic reaction, it exerts a selection pressure on the population of the pest that permits survival of only those strains that are resistant to the poison. The likelihood that such strains exist is high; their resistance is derived from their capacity to bypass the metabolic "roadblock" imposed by the pesticide and continue their metabolic activities using an alternative pathway. Because a single metabolic reaction is governed by a single gene, a one-gene difference between an individual strain and its wild sister strains could make one strain resistant to a herbicide that has a highly specific mode of toxic action.

The potency of a herbicide is also important. A weak herbicide that is effective only in large amounts will be ineffective against a population of pest species that consists largely of strains that tolerate the poison.

Development of herbicide-resistant strains of plant pests is not an effect unique to therapeutic chemicals. As long as the pesticide is potent and specific in its mode of action, it must be expected to favor the survival of resistant strains of the target population.

DISEASE RESISTANCE TO FUNGICIDES

Resistance to chemicals other than the heavy metals occurs commonly in fungal, and on rare occasions in bacterial, plant disease pathogens. Several growing seasons after a new fungicide appears, it becomes noticeably less effective against a particular disease. As our fungicides become more specific for selected diseases, we can expect the pathogens to become resistant. Again, such fungicides exert a selective pressure on the population, and resistant populations appear suddenly, either by selection of resistant individuals or by a single-gene mutation. Generally, the more specific the site and mode of fungicidal action, the greater the likelihood for a pathogen to develop a tolerance to that chemical.

Fifteen years ago, plant pathologists were not concerned with resistance. However, the number of practical field problems caused by resistance in plant pathogens has increased substantially in the last

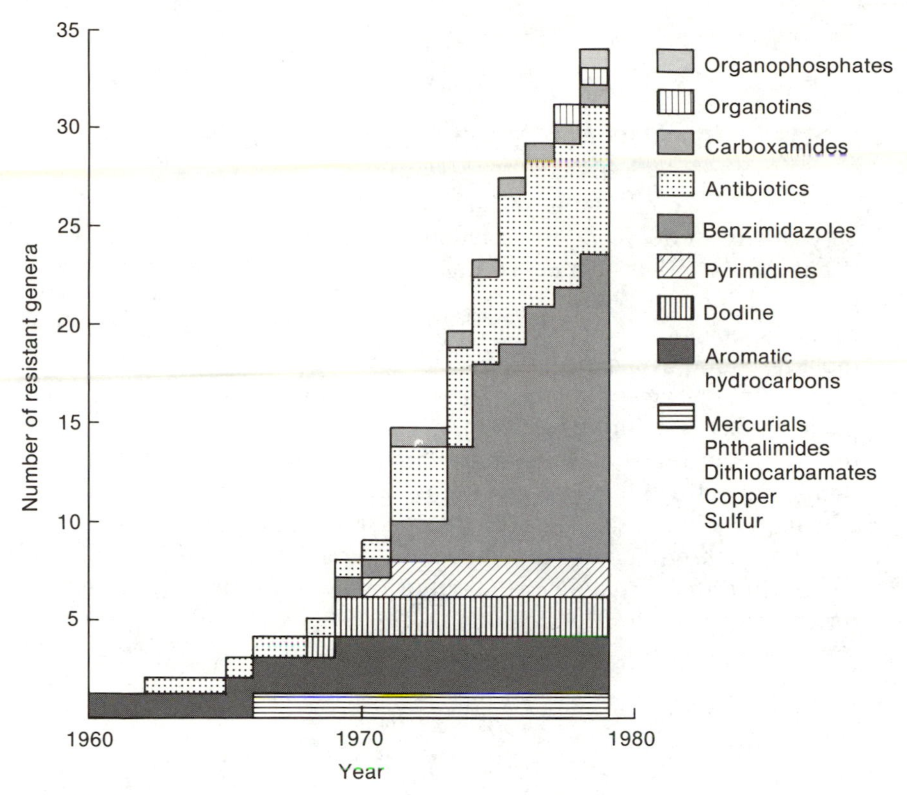

FIGURE 20-1
Increase of resistance by plant pathogens to several classes of fungicides in the past 20 years. (*Source:* Modified from Delp, 1980.)

10 years. This increase is attributed to the exclusive and intensive use of potent new disease control agents. Figure 20-1 illustrates the recent increases in known cases of plant disease resistance. Some resistance has been observed in all major fungicide classifications, and because of single-site modes of action, we can expect resistance to develop in all groups. In contrast, the older, conventional multisite inhibitors interfere with numerous vital metabolic processes of the pathogen. Resistance to such multisite fungicides as the dicarboximides, the dithiocarbamates, the mercurials, sulfur, and the copper compounds is rare because multiple differences or modifications in the pathogen's genetic material are required to circumvent the multisite action.

Cross-resistance (see p. 171) has also been observed in some pathogen populations.

Resistance can be alleviated or prevented by early use of spray programs designed to preclude long-term exposure of the pathogen to a single disease control agent. Once resistance is a significant part of a pathogen population, the only choice is to use disease control agents to which there is no cross-resistance. When a fungicide with a propensity for resistance is to be used, the program must be planned to prevent resistance; for example, by (1) using mixtures of fungicides, (2) rotating fungicides, (3) limiting the use of fungicides, or (4) developing integrated control strategies (Delp, 1980).

TABLE 20-3
Norway rat resistance levels to anticoagulant rodenticides.

Anticoagulant	Resistance level
Brodifacoum	1.3
Difenacoum[a]	1.9
Coumatetralyl[a]	14.2
Chlorophacinone	91.0
Warfarin	167.0
Diphacinone	227.0

[a] Not available in the U.S.
Source: Halder and Shadbolt (1975).

RODENT RESISTANCE TO RODENTICIDES

Because of the continuous and universal use of warfarin for the past 30 years, most domestic rodents have developed some resistance to it and to the other related coumarins. Some have also developed cross-resistance to the indandione rodenticides. Of the several pest rodent species the Norway rat (*Rattus norvegicus*) has become the most resistant, especially in larger cities. Table 20-3 lists resistance levels of the Norway rat to the rodenticides used most frequently by professional pest control operators. These resistance factors are useful in measuring the effectiveness of an anticoagulant against rodents, indicating the relative amounts of a rodenticide needed to kill resistant versus susceptible or nonresistant rats. For example, 1.3 times more brodifacoum is required for a resistant strain than for a susceptible one to achieve the same degree of control, while the resistance factor for diphacinone is 227.

PART VII

LEGALITY AND HAZARDS OF PESTICIDE USE

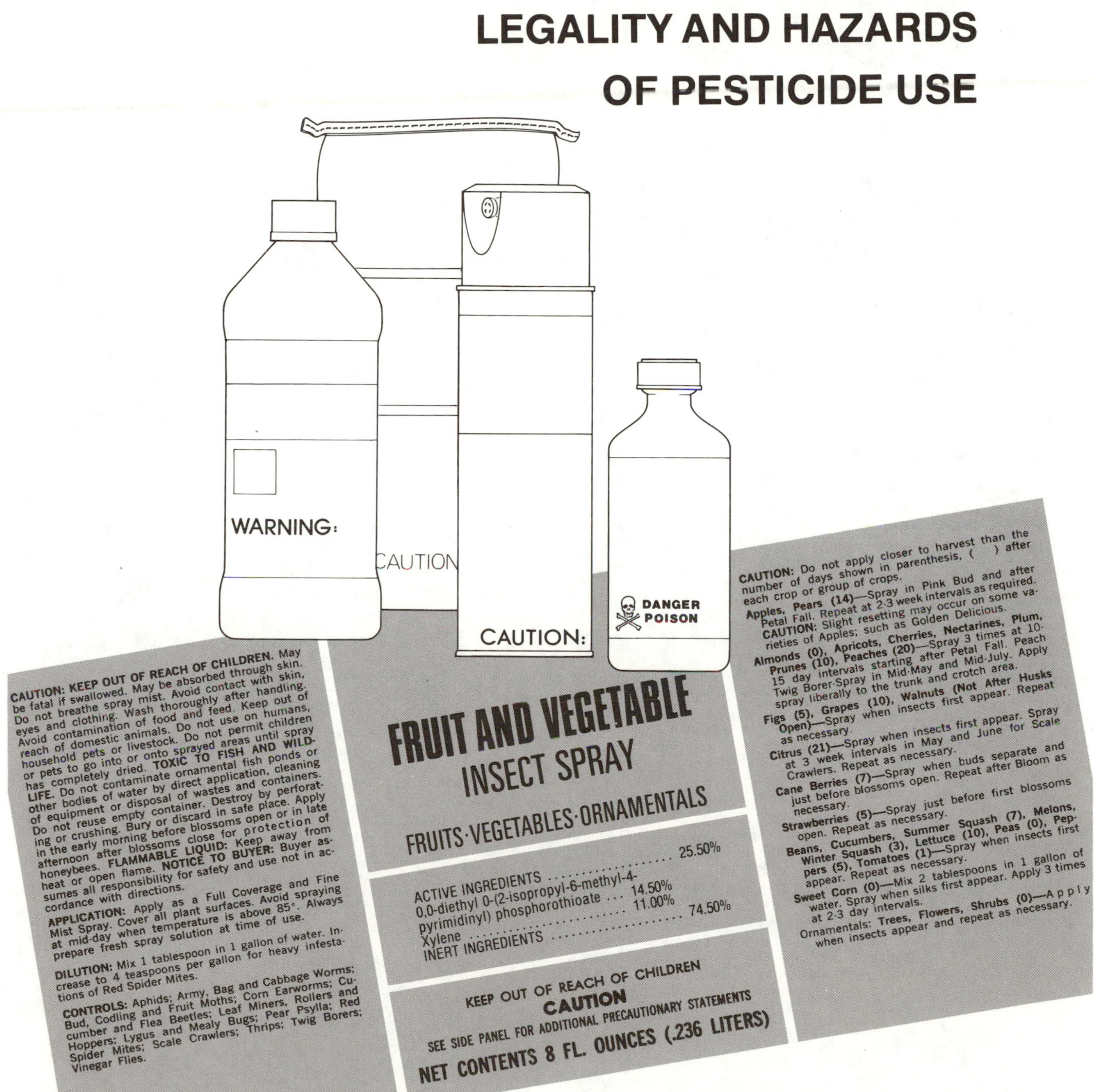

CHAPTER 21

The Toxicity and Hazards of Pesticides

Dosage alone determines poisoning.

Paracelsus (1564)

The terms *toxicity* and *hazard* are not synonymous. *Toxicity* refers to the inherent poisonous potency of a compound under experimental conditions. The word *hazard* refers to the risk or danger of poisoning when a chemical is used or applied, sometimes referred to as *use hazard*. The factor with which the user of a pesticide is really concerned is the use hazard and not the inherent toxicity of the material. Hazard depends not only on toxicity but also on the chance of exposure to toxic amounts of the material.

The dictionary definition of *poison* is "any substance which, introduced into an organism in relatively small amounts, acts chemically upon the tissues to produce serious injury or death." This definition is inadequate, however, for scientific purposes. The "relatively small amount" statement is open to wide interpretation. For instance, many chemical agents to which people are exposed regularly could be termed poisons under this definition. An oral dose of 400 mg/kg of sodium chloride, ordinary table salt, will make a person violently ill. Similarly, a dose of 5 to 15 g of aspirin, or 15 to 45 tablets, is fatal for humans, and approximately 60 deaths occur every year (about one-fifth are children) as a result of overdoses of aspirin. To take a third example, let us consider nicotine. A fatal oral dose of this naturally occurring alkaloid for humans is about 50 mg, approximately the amount of nicotine contained in two unfiltered cigarettes. However, in smoking most of the nicotine is decomposed by burning, and thus it is not absorbed by the smoker.

In each of these cases, humans are not exposed during ordinary use to the amounts of salt, aspirin, and nicotine that cause toxicity problems. Therefore, it is obvious that the *hazard* from normal exposure is very slight even though the compounds themselves are toxic under other circumstances.

A more adequate scientific definition of *poison* is: "A chemical substance that exerts an injurious effect in the majority of cases in which it comes into contact with living organisms during normal use." Salt, aspirin, and nicotine are excluded by such a definition, and so are the majority of pesticides.

Pesticides, by necessity, are poisons, but the toxic hazards of different compounds vary greatly. As far as the possible risks associated with the use of pesticides are concerned, we can distinguish be-

TABLE 21-1
Total deaths from accidental poisoning.

Substance	*1968*	*1974*	*1978*
Medicines	1,692	2,742	1,906
Alcohol	182	370	363
Cleaning and polishing agents	23	13	14
Disinfectants	9	8	3
Paints and varnishes	2	1	3
Petroleum products and other solvents	70	54	40
Pesticides, fertilizers, or plant foods	72 (2.8%)	35 (0.9%)	31 (1.0%)
Heavy metals (and their fumes)	53	23	15
Corrosives and caustics	28	13	17
Noxious foodstuffs and poisonous plants	10	5	0
Other unspecified solid and liquid	442	752	642
Total	2,583	4,016	3,034

Source: National Center for Health Statistics, U.S. Department of Health and Human Services, *Mortality Statistics—Special Reports, Accident Fatalities.*

tween two types: (1) acute poisoning, resulting from the handling and application of toxic materials; and (2) chronic risks from long-term exposure to small quantities of materials or from ingestion of them. The question of acute toxicity is obviously of paramount interest to people engaged in manufacturing and formulating pesticides and to those responsible for their application. Chronic risks, however, are of much greater public interest, because of their potential effect on the consumer of agricultural products.

Fatal human poisoning by pesticides is uncommon in the United States and is due to accident, ignorance, suicide, or crime. Fatalities represent only a small fraction of all recorded cases of poisoning, as demonstrated by recent statistics (Table 21-1). Note that in 1968, 2.8 percent of the deaths from accidental poisoning were from pesticides, while in 1978 this cause dropped to 1.0 percent.

Toxic substances of all kinds pose special hazards for children. In 1968, 11 percent of all accidental poisoning deaths were children under five years of age, by 1977 that figure had dropped to 2.7 percent, a remarkable improvement that is due to educational programs, the development of poison control centers, media publicity, and the services of community groups that provide free information and help. Of the children under five poisoned in 1968, 11 percent ingested pesticides, while in 1978 that percentage had declined to 5.1 percent (11.7 percent were killed by aspirin alone). Most disappointing is the statement that of all deaths attributed to accidental poisoning by pesticides in 1977, 20.6 percent were children under five years of age.

Another set of data for 1979 (Table 21-2) shows that of 85,255 reported accidental ingestions among children under five years of age, pesticides were responsible for less than one-half the number attributed to cosmetics or 30 percent more than those assigned to aspirin. (The child-proof aspirin and other medicine bottles were in effect in 1979.)

Finally, Table 21-3 presents categories of substances most frequently ingested by children under five years of age and reported to Poison Control Centers. In 1978 plants were the most frequently

TABLE 21-2
Accidental ingestions among children under 5 years of age.

Substance	1970 Number	1970 Percent	1973 Number	1973 Percent	1979 Number	1979 Percent
Medicines	35,189	49.6	52,113	44.3	34,710	40.7
Internal	29,923	(42.2)	42,215	(35.9)	25,160	(29.5)
Aspirin	9,610	(13.6)	7,763	(6.6)	3,332	(3.9)
Other	20,313	(28.7)	34,452	(29.3)	21,828	(25.6)
External	5,266	(7.4)	9,898	(8.4)	9,550	(11.2)
Cleaning and polishing agents	10,810	15.2	19,132	16.3	12,692	14.9
Petroleum products	3,254	4.6	4,974	4.2	2,468	2.9
Cosmetics	5,112	7.2	10,362	8.8	9,533	11.2
Pesticides	3,887	5.5	5,591	4.8	4,359	5.1
Gases and vapors	55	0.1	140	0.1	93	0.1
Plants	3,574	5.0	7,032	6.0	10,754	12.6
Turpentine, paints, etc.	4,006	5.7	6,988	5.9	3,520	4.1
Miscellaneous	4,585	6.5	10,517	9.0	7,126	8.4
Not specified	425	0.6	740	0.6	—	—
Total	70,897	100.0	117,589	100.0	85,255	100.0

Source: Individual case reports submitted to the National Clearinghouse for Poison Control centers; 1973, from 517 centers in 45 states; 1970 from 432 centers in 49 states; 1979 unspecified. Bulletin of the National Clearinghouse for Poison Control Centers, U.S. Food and Drug Administration, Bureau of Drugs, U.S. Department of Health and Human Services, May–June 1974, and August 1981.

TABLE 21-3
Categories of substances most frequently ingested by children under 5 years of age.

Substance	1972 Number	1972 Percent	1977 Number	1977 Percent	1978 Number	1978 Percent
Plants	4,759	4.5	11,819	12.4	11,010	11.7
Soaps, detergents, cleaners	5,940	5.7	5,674	6.0	5,836	6.2
Antihistamines, cold medications	4,355	4.1	3,661	3.9	4,003	4.3
Perfume, cologne, toilet water	3,108	3.0	3,681	3.9	3,748	4.0
Vitamins, minerals	5,320	5.1	3,622	3.8	3,677	3.9
Aspirin	8,146	7.8	3,264	3.4	3,577	3.8
Baby	5,305	5.1	2,211	2.3	2,557	2.7
Adult	1,322	1.3	491	0.5	380	0.4
Unspecified	1,519	1.4	562	0.6	640	0.7
Household disinfectants, deodorizers	3,301	3.1	2,813	3.0	2,752	2.9
Miscellaneous analgesics	2,220	2.1	2,699	2.8	2,752	2.9
Insecticides (excluding mothballs)	2,306	2.2	2,723	2.9	2,675	2.9
Miscellaneous internal medicines	3,186	3.0	2,403	2.5	2,303	2.5
Fingernail preparations	1,191	1.1	2,093	2.2	2,270	2.4
Miscellaneous external medicines	1,587	1.5	1,919	2.0	2,151	2.3
Liniments	2,133	2.0	2,060	2.2	2,016	2.2
Household bleach	2,794	2.7	2,029	2.1	1,863	2.0
Miscellaneous products	1,941	1.8	1,475	1.6	1,627	1.7
Cosmetic lotions, creams	1,783	1.7	1,711	1.8	1,625	1.7
Antiseptic medications	1,346	1.3	1,506	1.6	1,603	1.7
Psychopharmacologic agents	2,998	2.9	1,558	1.6	1,463	1.6
Cough medicines	1,279	1.2	1,387	1.5	1,443	1.5
Hormones	1,970	1.9	1,487	1.6	1,386	1.5
Glues, adhesives	1,866	1.8	1,400	1.5	1,384	1.5
Rodenticides	1,508	1.4	1,267	1.3	1,347	1.4
Internal antibiotics	1,187	1.1	1,061	1.1	1,246	1.3
Corrosive acids, alkalies	1,815	1.7	1,264	1.3	1,204	1.3
Paint	1,633	1.6	1,248	1.3	1,204	1.3

Source: Individual poison reports (phone inquiries and treated cases) submitted to the National Clearinghouse for Poison Control Centers by 326 centers in 44 states, the District of Columbia, Panama, and military bases abroad. Bulletin Of the National Clearinghouse for Poison Control Centers, U.S. Food and Drug Administration, Bureau of Drugs, U.S. Department of Health and Human Services, October 1980.

ingested (11.7 percent), while insecticides were ingested at the same frequency as disinfectants-deodorizers and miscellaneous pain pills (2.9 percent).

Thus pesticides have a decent safety record, which is improving each year, mainly through education and labeling of containers.

EFFECTS OF PESTICIDES ON HUMANS

Under certain conditions, pesticides may be toxic to humans, and an understanding of the basic principles of toxicity and the differences between toxicity and hazard is essential. Some pesticides are much more toxic than others, and severe illness may result from ingestion of only a small amount of a certain chemical, while with other compounds no serious effects result even from ingesting large quantities. Some of the factors that influence the effects of ingestion include (1) the toxicity of the chemical, (2) the dose of the chemical, especially concentration, (3) length of exposure, and (4) the route of entry or absorption by the body.

In the early stages of the development of a pesticide for further experiments and exploration, toxicity data are collected on the pure toxicant, as required by the EPA. These tests are conducted on test animals that are easy to work with and whose physiology, in some instances, is like that of humans; for example, white mice, white rats, white rabbits, guinea pigs, and beagle dogs. Intravenous tests are determined usually on mice and rats, whereas dermal tests are conducted on shaved rabbits and guinea pigs. Acute oral toxicity determinations are most commonly made in rats and dogs, with the test substance being introduced directly into the stomach by tube. Chronic studies are conducted on the same two species for extended periods, and the compound is usually incorporated in the animal's daily ration. Inhalation studies may involve any of the test animals, but rats, guinea pigs, and rabbits are most commonly used.

All these procedures are necessary to determine the overall toxic properties of the compound to various animals. From this information, toxicity to humans can generally be extrapolated, and eventually some microlevel portion of the pesticide may be permitted in human food as a residue, which is expressed in parts per million (ppm).

Pesticide toxicologists use rather simple animal toxicity tests to rank pesticides according to their toxicity. Long before a pesticide is registered with the EPA and eventually released for public use, the manufacturer must declare the toxicity of the pesticide to the white rat under laboratory conditions. This toxicity is defined by the LD_{50}, the dose that kills 50 percent of the test animals to which it is administered under experimental conditions, expressed as milligrams of toxicant per kilogram of body weight.

The LD_{50} is measured in terms of oral (fed to, or placed directly in the stomachs of rats), dermal (applied to the skin of rats or rabbits), and respiratory toxicity (inhaled). From two of these tests, oral and dermal, a toxicologic ranking is derived; for example, see Figures 21-1 and 21-2. The materials on the top of the lists are the most

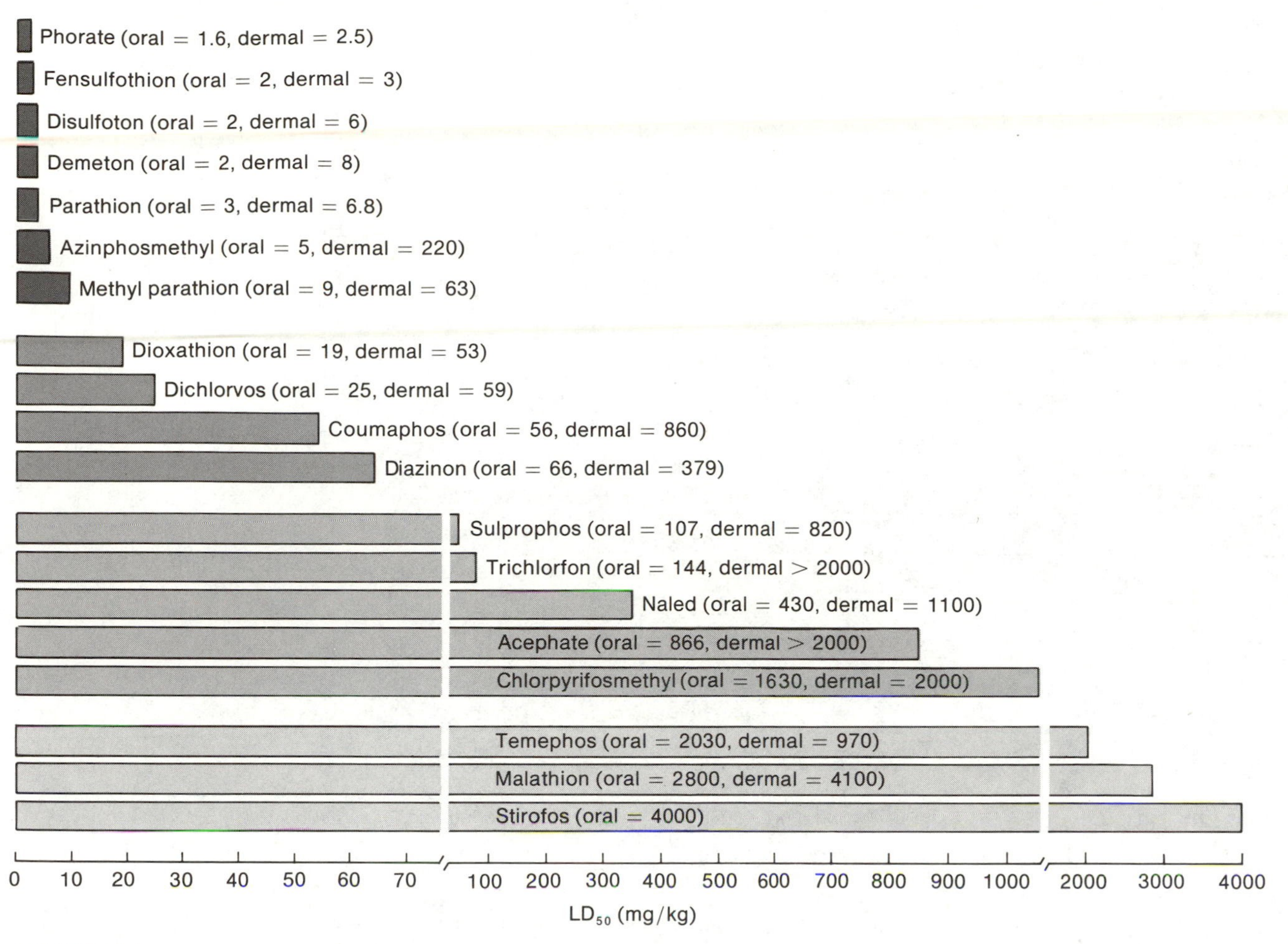

FIGURE 21-1
Acute oral LD_{50}s for rats and dermal LD_{50}s for rabbits for some organophosphate insecticides.

toxic and those at the bottom the least. The size of the dose is the most important single item in determining the safety of a given chemical, and actual statistics of human poisonings correlate reasonably well with these toxicity ratings.

ESTIMATION OF TOXICITY TO HUMANS

The dose, length of exposure, and route of absorption are the other important variables beside toxicity. The amount of pesticide required to kill a human being can be correlated with the LD_{50} of the material to rats in the laboratory. In Table 21-4, for example, the acute oral LD_{50}, expressed as mg/kg dose of the technical material, is extrapolated to estimate the amount needed to kill a 170-lb (77 kg) human. Dermal LD_{50}s are included for a better understanding of the relationship of expressed animal toxicity to human toxicity.

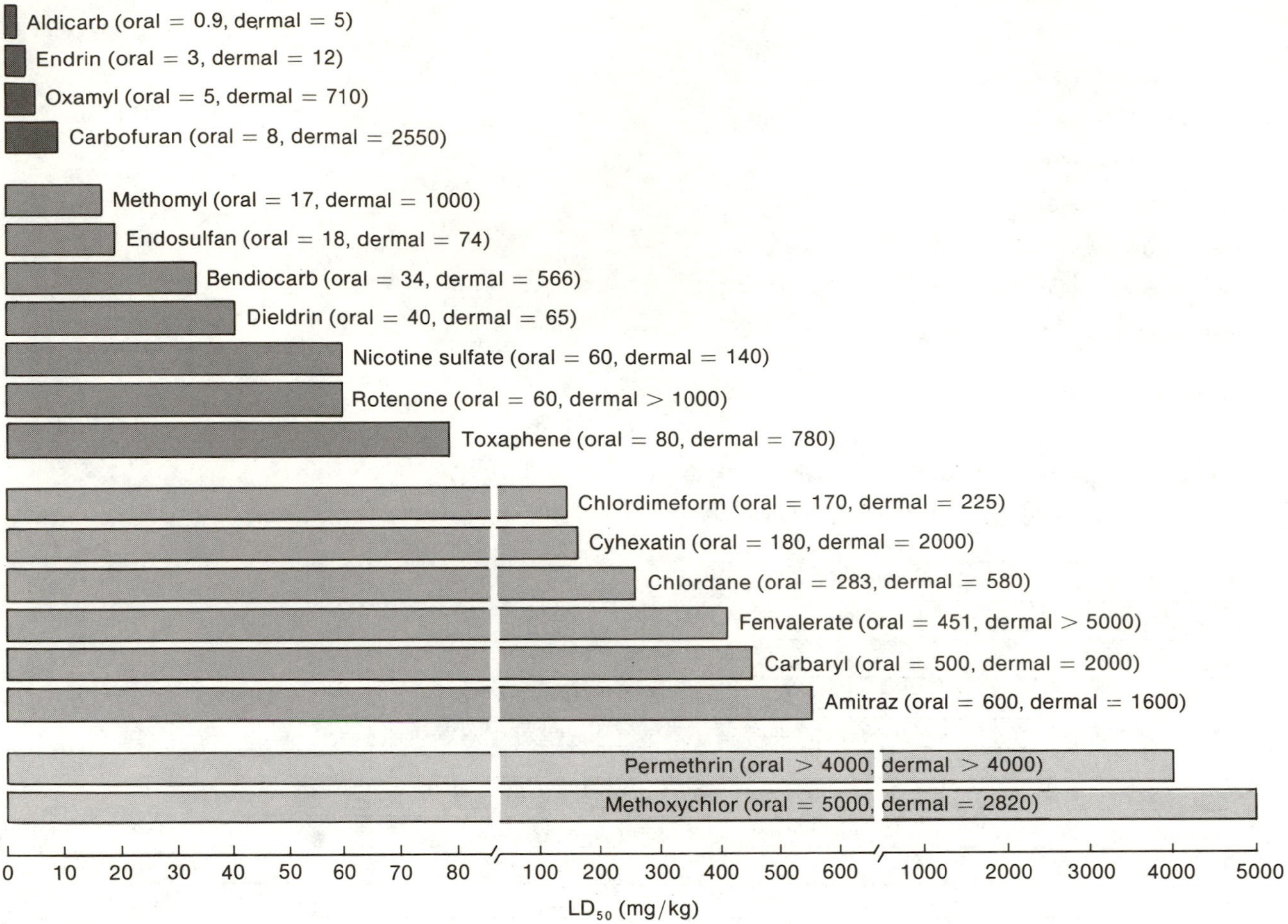

Aldicarb (oral = 0.9, dermal = 5)
Endrin (oral = 3, dermal = 12)
Oxamyl (oral = 5, dermal = 710)
Carbofuran (oral = 8, dermal = 2550)
Methomyl (oral = 17, dermal = 1000)
Endosulfan (oral = 18, dermal = 74)
Bendiocarb (oral = 34, dermal = 566)
Dieldrin (oral = 40, dermal = 65)
Nicotine sulfate (oral = 60, dermal = 140)
Rotenone (oral = 60, dermal > 1000)
Toxaphene (oral = 80, dermal = 780)
Chlordimeform (oral = 170, dermal = 225)
Cyhexatin (oral = 180, dermal = 2000)
Chlordane (oral = 283, dermal = 580)
Fenvalerate (oral = 451, dermal > 5000)
Carbaryl (oral = 500, dermal = 2000)
Amitraz (oral = 600, dermal = 1600)
Permethrin (oral > 4000, dermal > 4000)
Methoxychlor (oral = 5000, dermal = 2820)
0 10 20 30 40 50 60 70 80 100 200 300 400 500 600 1000 2000 3000 4000 5000
LD_{50} (mg/kg)

FIGURE 21-2
Acute oral LD_{50}s for rats and dermal LD_{50}s for rabbits for some organochlorine, carbamate, botanical, pyrethroid, and formamidine insecticides.

TABLE 21-4
Combined tabulation of pesticide toxicity classes.

Toxicity rating	Routes of absorption: LD_{50} Single oral dose for rats (mg/kg)	Routes of absorption: LD_{50} Single dermal dose for rabbits (mg/kg)	Probable lethal oral dose for humans
6—Supertoxic	<5	<20	A taste, a grain
5—Extremely toxic	5–50	20–200	A pinch, 1 teaspoon
4—Very toxic	50–500	200–1,000	1 teaspoon to 2 tablespoons
3—Moderately toxic	500–5,000	1,000–2,000	1 ounce to 1 pint
2—Slightly toxic	5,000–15,000	2,000–20,000	1 pint to 1 quart
1—Practically nontoxic	>15,000	>20,000	>1 quart

Source: Gleason, Gosselin, and Hodge (1976).

TABLE 21-5
EPA labeling toxicity categories by hazard indicator.

Hazard indicators	Toxicity categories I (*Danger—Poison*)	II (*Warning*)	III (*Caution*)	IV (*Caution*)
Oral LD_{50}	Up to and including 50 mg/kg	From 50 to 500 mg/kg	From 500 to 5000 mg/kg	Greater than 5000 mg/kg
Inhalation LD_{50}	Up to and including 0.2 mg/liter	From 0.2 to 2 mg/liter	From 2 to 20 mg/liter	Greater than 20 mg/liter
Dermal LD_{50}	Up to and including 200 mg/kg	From 200 to 2000	From 2000 to 20,000	Greater than 20,000
Eye effects	Corrosive; corneal opacity not reversible within 7 days	Corneal opacity reversible within 7 days; irritation persisting for 7 days	No corneal opacity; irritation reversible within 7 days	No irritation
Skin effects	Corrosive	Severe irritation at 72 hours	Moderate irritation at 72 hours	Mild or slight irritation at 72 hours

Source: "EPA Pesticide Programs, Registration and Classification Procedures, Part II." *Federal Register 40:* 28279.

Generally speaking, oral ingestions are more toxic than respiratory inhalations, which are more toxic than dermal absorption. Additionally, physical and chemical differences between pesticides make them more likely or less likely to produce poisoning. For instance, parathion changes to the more toxic metabolite paraoxon under certain conditions of humidity and temperature. Parathion is more toxic than methyl parathion to field workers; yet there is not a great difference in their oral toxicities. Workers' exposure is usually dermal, which explains why many more illnesses are reported in workers exposed to parathion than those exposed to methyl parathion.

TOXICITY AND LABELING

All pesticide labels must contain "signal words" in bold print, to attract the attention of the buyer/user: *Danger—Poison; Warning;* and *Caution*. These are significant words, since they represent a category of toxicity, and thus give an indication of the potential hazard (Table 21-5).

Category I. The signal words *Danger—Poison* and the skull and crossbones symbol are required on the labels for all *highly toxic* compounds. These pesticides all fall within the acute oral LD_{50} range of 0 to 50 mg/kg.

TABLE 21-6
Estimated relative acute toxic hazards of pesticides to users/applicators.[a]

Most dangerous	*Dangerous*
aldicarb, Temik® (I)[b]	aldrin (I)
demeton, Systox® (I)	carbophenothion, Trithion® (I)
disulfoton, Di-Syston® (I)	cycloheximide (F)
fensulfothion, Dasanit® (I)	dialifor, Torak® (I)
fonofos, Dyfonate® (I)	dichlorvos, Vapona® (I)
mevinphos, Phosdrin® (I)	dicrotophos, Bidrin® (I)
parathion (I)	dieldrin (I)
phorate, Thimet® (I)	dioxathion, Delnav® (I)
TEPP (I)	dinoseb, DNBP (H, D, I)
terbufos, Counter® (I)	DNOC (I, F, H, D)
	endrin (I)
	EPN (I)
	ethion, Nialate® (I)
	fenaminosulf, Lesan® (F)
	isofenfos, Amaze® (I)
	mephosfolan, Cytrolane® (I)
	methamidophos, Monitor® (I)
	methyl parathion (I)
	monocrotophos, Azodrin® (I)
	nicotine sulfate (I)
	paraquat (H)
	pentachlorophenol (D, F)
	phosphamidon (I)
	sodium arsenite (H, I)

[a] Estimates of hazards are based primarily on the observed acute dermal and, to a lesser extent, oral toxicity to experimental animals. When available, use experience is also considered. The classification into toxicity groups is both approximate and relative. These toxicity categories are not related to specific categories for label requirements.

Category II. The word *Warning* is required on the labels for all *moderately toxic* compounds. They all fall within the acute oral LD_{50} range of 50 to 500 mg/kg.

Category III. The word *Caution* is required on labels for *slightly toxic* pesticides that fall within the LD_{50} range of 500 to 5000 mg/kg.

Category IV. The word *Caution* is required on labels for compounds having acute LD_{50}s greater than 5000 mg/kg. However, unqualified claims for safety are not acceptable on any label, and all labels must bear the statement, "Keep Out of Reach of Children."

Table 21-6 shows the relative acute toxic hazards to applicators of many of the commonly used pesticides. Examples of insecticides,

Less dangerous	*Least dangerous*
acifluorfen, Blazer® (H)	alachlor, Lasso® (H)
aminocarb, Matacil® (I)	atrazine (H)
azinphosmethyl, Guthion® (I)	bensulide, Prefar® (H)
bendiocarb, Ficam® (I)	captan (F)
binapacryl, Morocide® (F, I)	carbaryl, Sevin® (I)
bufencarb, Bux® (I)	carbofuran, Furadan® (I)
bupirimate, Nimrod® (F)	chlorpyrifos, Dursban®, Lorsban® (I)
butrizol, Indar® (F)	copper, organic and inorganic (F)
chlordane (I)	cyhexatin, Plictran® (I)
chlordimeform, Fundal®, Galecron® (I)	cypermethrin, Ammo®, Cymbush® (I)
coumaphos, Co-Ral® (I)	2,4-D (H)
crotoxyphos, Ciodrin® (I)	DEF®, merphos (D)
diazinon (I)	dinocap (F)
dimethoate, Cygon® (I)	diquat (H)
endosulfan, Thiodan® (I)	fenvalerate, Pydrin® (I)
endothall (H)	flucythrinate, Pay-Off® (I)
fenthion, Baytex® (I)	folpet, Phaltan (F)
fentin acetate, Brestan® (F)	formetanate, Carzol® (I)
heptachlor (I)	malathion (I)
lindane (I)	maneb (F)
metam-sodium, Vapam® (fumigant F, H)	metalaxyl, Ridomil®, Subdue® (F)
methidathion, Supracide® (I)	oxythioquinox, Morestan® (F)
methomyl, Lannate®, Nudrin® (I)	PCNB, Terraclor® (F)
naled, Dibrom® (I)	permethrin, Ambush®, Pounce® (I)
oxamyl, Vydate® (I)	phosalone, Zolone® (I)
oxydemetonmethyl, Meta-Systox®-R (I)	phosmet, Imidan® (I)
profenofos, Curacron® (I)	simazine, Princep® (H)
propachlor, Bexton® (H)	stirofos, Gardona® (I)
prosulfalin, Sward® (H)	2,4,5-T (H)
sulprophos, Bolstar® (I)	tetradifon, Tedion® (I)
tebuthiuron, Spike® (H)	thiram (F)
temephos, Abate® (I)	thidiazuron, Dropp® (D)
toxaphene (I)	triadimefon, Bayleton® (F)
trichlorfon, Dipterex®, Dylox® (I)	trifluralin, Treflan® (H)
	zineb (F)
	ziram (F)

[b] The pesticide category to which the chemical belongs is designated as follows: D, defoliant; F, fungicide; H, herbicide; and I, insecticide/miticide.
Source: Adapted from Wolfe and Durham (1966).

herbicides, and fungicides in the three label toxicity classifications are presented in Table 21-7.

With regard to the classifications of pesticides, their general toxicity in decreasing order, is insecticides > defoliants > desiccants > herbicides > fungicides. Within the most toxic class, the insecticides, the categories fall in the following general order of their dermal hazards to humans: organophosphates > carbamates > cyclodienes > DDT relatives > botanicals > activators or synergists > inorganics. There are usually exceptions in each category.

The formulations of pesticides, because of their varying kinds of diluents, also pose varying degrees of hazard to humans. Again, we must generalize: liquid pesticide > emulsifiable concentrate > oil solution > water emulsion > water solution > wettable powder/flowable (in suspension) > dust > granular.

TABLE 21-7
Examples of toxicity classes.

Label classification		*Oral LD_{50} (to rats) (mg/kg)*		*Dermal LD_{50} (to rabbits) (mg/kg)*
		INSECTICIDES		
Highly toxic	aldicarb, Temik®	0.65–0.79	parathion	7–21
	fensulfothion, Dasanit®	4.7–10.5	mevinphos, Phosdrin®	4.2–4.7
	monocrotophos, Azodrin®	21	disulfoton, Di-Syston®	2.6–8.6
	phorate, Thimet®	1.1–2.3	demeton, Systox®	2.5–6.0
Moderately toxic	propoxur, Baygon®	95–104	methyl parathion	67
	chlorpyrifos, Dursban®	135–163	dioxathion, Delnav®	63–235
	diazinon	300–850	azinphosmethyl, Guthion®	220
Slightly toxic	malathion	1,000–1,375	toxaphene	780–1,075
	carbaryl, Sevin®	500–850	fenvalerate, Pydrin®	>5,000
	permethrin, Ambush®, Pounce®	450–>4,000[a]	dicofol, Kelthane®	1,000–1,230
	temophos, Abate®	2,030	malathion	>4,444
	stirofos, Gardona®	4,000	carbaryl, Sevin®	>4,000
		HERBICIDES		
Highly toxic	DNOC	25–40	None	
	sodium arsenite	10–50		
Moderately toxic	2,4-D	375	paraquat	236–480
	paraquat	157	acifluorfen, Blazer®	450
Slightly toxic	MSMA	700–1,800	endothall	750
	monuron	2,300–3,700	dichlobenil	500
	simazine, Princep®	5,000	2.4-D acid	1,500
	pendimethalin, Prowl®	1,250	MCPA	>1,000
		FUNGICIDES		
Highly toxic	cycloheximide, Actidione®	1.8–2.5	None	
	fentin chloride, Tinmate®	18		
Moderately toxic	binapacryl, Morocide®	136–225	butrizol, Indar®	315
	tryphenyltin hydroxide, Du-Ter®	108	fentin acetate, Brestan®	500
Slightly toxic	thiram	780	binapacryl, Morocide®	720–810
	anilazine, Dyrene®	2,710	dinoseb	500
	ethazol, Koban®	1,077	maneb	>1,000
	dimethirimol, Milcurb®	2,350	zineb	>1,000
	dicloran, Botran®	5,000	triphenyltin hydroxide, Du-Ter®	5,000

[a] > means LD_{50} is higher than figure shown.

Just a word regarding the toxicity of the petroleum solvents used in formulating pesticides: Currently these are diesel fuel, deodorized kerosene, methanol, petroleum distillates, xylene, and toluene. Of these, only xylene and toluene are aromatics, and they offer by far the greater dermal hazard.

FIELD REENTRY SAFETY INTERVALS

Because insecticides pose the greatest health hazard to the agricultural worker, with the organophosphate insecticides being the most important chemical group in this respect, field reentry safety intervals have been established.

To prevent unnecessary exposure, the EPA now requires safety waiting intervals between application of certain insecticides and worker reentry into all treated fields. Several states (for example, California) have adopted waiting intervals longer than those required by the EPA. The waiting intervals established by EPA are

48 HOURS
- Ethyl parathion
- Methyl parathion
- Demeton (Systox®)
- Monocrotophos (Azodrin®)
- Carbofenothion (Trithion®)
- Oxydemetonmethyl (MetaSystox-R)
- Dicrotophos (Bidrin®)
- Endrin

24 HOURS
- Azinphosmethyl (Guthion®)
- Phosalone (Zolone®)
- EPN
- Ethion

Waiting intervals for other pesticides are being determined by the EPA.

For all other insecticides, it is necessary only that workers wait until sprays have dried or dusts have settled before reentering treated fields. These worker safety intervals are not to be confused with the familiar harvest intervals—the minimum days from last treatment to harvest—indicated on the insecticide label.

If it is necessary for workers to enter fields earlier than the required waiting intervals, they must wear protective clothing: a long-sleeved shirt, long-legged trousers or coveralls, hat, shoes, and socks.

These waiting intervals should not impose any undue hardship on pest management specialists and agricultural pest control advisors, because application of any one of these materials would preclude the necessity for field inspection within the required waiting intervals.

RESTRICTED-USE PESTICIDES

Two classes of pesticides are registered with the EPA: *general-use* and *restricted-use*. General-use pesticides may be purchased and applied by any person. Restricted-use pesticides may be purchased and applied only by certified applicators, that is, by persons having received special training and testing in the use, handling, safety, and application of pesticides. Training and testing are administered in each state by an agency authorized by the EPA to certify applicators, usually the same agency that licenses commercial pesticide applicators.

The criteria for restricted-use classification are usually a factor of human hazard; however, other considerations include effects on aquatic organisms, effects of residues on birds, hazard to nontarget organisms, and accident history.

TABLE 21-8
Restricted-use pesticides and abridged criteria used for classification.

Pesticide	*Restricted uses*	*Criteria for restriction*
acrolein, Aqualin® (H)[a]	All	Human inhalation hazard
acrylonitrile, Acrylon (F)	All	Accident history
aldicarb, Temik® (I)	All	Accident history
allyl alcohol (H)	All	Acute dermal toxicity
aluminum phosphide, Phostoxin® (F)	All	Human inhalation hazard
amitraz, BAAM® (I, M)	All	
azinphosmethyl, Guthion® (I)	Concentrations greater than 13.5%	Human inhalation hazard
calcium cyanide, Cyanogas® (F)	All	Human inhalation hazard
carbofuran, Furadan® (I)	Concentrations 40% and greater	Acute inhalation toxicity and effects on aquatic organisms
chlordimeform, Fundal®, Galecron® (I)	All	Requested by manufacturer
chlorfenvinphos, Supona® (I)	Concentrations 21% and greater	Acute dermal toxicity
chlorophacinone, Rozol® (R)	All	Acute oral toxicity
chloropicrin, Dowfume® (F)	All	
chlorpyrifos, Dursban® (I)	Killmaster® II	Requested by manufacturer
clonitralid, Bayluscide® (M)	All	Acute inhalation toxicity and effects on aquatic organisms
cycloheximide, Acti-Aid® (B)	All	
demeton, Systox® (I)	1% fertilizer and 2% granular	Acute oral and dermal toxicity
dicrotophos, Bidrin® (I)	All	Effects on birds and mammals
diflubenzuron, Dimilin® (I)	All	Effects on aquatic organisms
dioxathion, Delnav® (I)	Concentrations greater than 30%	Acute dermal toxicity
disulfoton, Di-Syston® (I)	Formulations greater than: EC 65%, 95% oil solution, 10% granular	Acute inhalation and dermal toxicity
endrin (I)	All	Acute dermal toxicity and hazard to nontarget organisms
EPN (I)	All	Acute oral and dermal toxicity
ethoprop, Mocap® (I,N)	All	Acute dermal toxicity
ethyl parathion (I)	All	Human inhalation hazard, acute dermal toxicity, effects on nontarget organisms
fenamiphos, Nemacur® (I)	All	Acute dermal toxicity
fensulfothion, Dasanit® (N,I)	All	Acute inhalation and dermal toxicity
fenvalerate, Pydrin® (I)	All	Effects on aquatic organisms

The EPA has prepared a list of 56 pesticides classified as restricted-use (as of July 1981). However, not all formulations of all 56 compounds are restricted. For instance, the systemic insecticide disulfoton can be purchased in the yard-and-garden centers as a 2 or 5 percent granule; formulations with higher percentages of active ingredients are in the restricted-use classification.

No list of restricted-use pesticides is complete or unchanging, for new compounds and formulations dictate additions and occasional deletions. Table 21-8 contains a complete list of the restricted pesticides and an abridged list of formulations, uses, and abbreviated criteria developed by the EPA.

Pesticide	Restricted uses	Criteria for restriction
fluoroacetamide, 1081 (R)	All	Acute oral toxicity
fonofos, Dyfonate® (I)	All	Acute dermal toxicity
heptachlor (I)	All	Hazard to nontarget organisms
hydrocyanic acid, HCN (F)	All	Human inhalation hazard
magnesium phosphide (F)	All	Human inhalation hazard
methamidophos, Monitor® (I)	All	Acute dermal toxicity
methidathion, Supracide® (I)	All	Effects on bees and nontarget organisms
methiocarb, Larvin® (I)	All	Effects on bees and nontarget organisms
methomyl, Lannate®, Nudrin® (I)	All formulations except 1% fly bait	Accident history and effects on nontarget mammals
methyl bromide (F)	Containers greater than 1.5 lb	Accident history
methyl parathion (I)	All	Accident history, acute dermal toxicity, effects on bees and nontarget organisms
mevinphos, Phosdrin® (I)	All	Acute dermal toxicity and effects on nontarget organisms
monocrotophos, Azodrin® (I)	All	Acute dermal toxicity and effects on birds and mammals
nicotine (alkaloid) (I)	All	
nitrofen, TOK® (H)	All	
paraquat (H) (dichloride and bis methyl sulfate)	All formulations except: 0.44% pressurized spray and liquid fertilizer with 0.02% or 0.03%, or 0.04% with atrazine	Use and accident history, human toxicological data
permethrin, Ambush®, Pounce® (I)	All	Effects on aquatic organisms
phorate, Thimet® (I)	All	Acute dermal toxicity, and effects on birds and mammals
phorazetim, Gophacide® (R)	All	Hazard to nontarget species, including aquatic and bird
phosphamidon, Dimecron® (I)	All	Acute dermal toxicity and effects on nontarget organisms
picloram, Tordon® (H)	All formulations except Tordon® 101R	Hazard to nontarget plants, both crop and noncrop
sodium cyanide, Cymag® (R)	All capsules and ball formulations	Human inhalation hazard
sodium fluoroacetate, (R) (P) Compound 1080	All	Accident history, acute oral toxicity, hazard to nontarget organisms
strychnine (P)	All formulations and baits greater than 0.5%	Acute oral toxicity, accident history, hazard to nontarget birds and mammals
sulfotep, Bladafum® (I)	All spray and smoke generator uses	Human inhalation hazard
sulprofos, Bolstar® (I)	All	Effects on nontarget organisms
tepp (I)	All	Human inhalation and dermal hazards, effects on nontarget organisms
zinc phosphide (R)	All	Acute oral toxicity

[a] The pesticide category to which the chemical belongs is designated as follows: F, fumigant; H, herbicide; I, insecticide/acaricide; M, molluscicide; N, nematicide; P, predicide; R, rodenticide; and B, bactericide.

TABLE 21-9
Some of the more significant pesticide cancelations and reduced uses mandated by the EPA, and voluntary cancelations or withdrawals from the market by their basic manufacturers.

Pesticide	*Date and action taken*	*Criteria for action*
aldrin (I)[a]	10/18/74 All uses canceled except: (1) subterranean termiticide (2) nonfood root or top dip (3) mothproofing in manufacturing	Oncogenicity; reduction in nontarget and endangered species
Aramite® (I)	4/12/77 All uses canceled	Oncogenicity
basic copper arsenate (I,F)	4/7/77 Voluntary cancelation by manufacturer	
Benzac® (*see* trichlorobenzoic acid)		
BHC (*see* HCH)		
chloranil (F)	1/19/77 Voluntary cancelation by manufacturer	
chlordane (I)	3/28/78 All uses canceled except: (1) subterranean termiticide (2) nonfood root or top dip	Oncogenicity; reduction in nontarget and endangered species
chlordecone (I) Kepone®	12/13/77 All uses canceled effective 5/1/78	Oncogenicity
DBCP (N)	9/13/78 All uses canceled except for Hawaiian pineapple fields	Oncogenicity; reproductive effects
DDD (TDE) (I)	3/18/71 All uses canceled	Imminent environmental hazard
DDT (I)	7/7/72 Most uses canceled except certain public health applications	Imminent environmental hazard
dieldrin (I)	10/18/74 *See* aldrin	
endrin (I)	7/25/79 Most uses canceled with a few retained	Oncogenicity; teratogenicity; reduction in nontarget and endangered species
erbon (H)	1981 Voluntary cancelation by manufacturer	
ethylan (I) Perthane®	9/4/80 Voluntary cancelation by manufacturer	
HCH (BHC) (I)	10/19/77 All uses canceled	Oncogenicity; fetotoxicity; reproductive effects

[a] The pesticide category to which the chemical belongs is designated as follows: F, fungicide; H, herbicide; I, insecticide/acaricide; A, algicide; and N, nematicide.

CANCELATIONS AND REDUCED-USE PATTERNS FOR PESTICIDES

The EPA was established on December 2, 1970, to develop, coordinate, and monitor federal policy regarding such environmental problems as air and water pollution, pesticide regulation, solid-waste management, and radiation and noise abatement. In the areas of pesticide registration, regulation, and research, the EPA assumed most of its duties from the Department of Agriculture, Department of Health, Education and Welfare, Department of the Interior, Atomic Energy Commission, Federal Radiation Council, and Council of Environmental Quality. By consolidating the broad power for pesticide registration in one agency, the federal government hoped to provide efficient and effective regulation and enforcement policy.

An evaluation of the EPA's first decade shows mixed results. Fewer pesticides are contaminating the environment—but not be-

Pesticide	Date and action taken	Criteria for action
heptachlor (I)	3/28/78 See chlordane, with additional exception: (3) soil insecticide for small grains until 9/1/82, for sorghum until 7/1/83	
isocyanurates (A)	1981 Voluntary cancelation by manufacturer	
Kepone® (*see* chlordecone)		
mercury (F)	8/26/76 Most uses canceled except as fungicide for outdoor textiles, fresh lumber, Dutch elm disease, water-based paints; seed treatment for small grains, cotton, summer and winter turf diseases	Imminent environmental hazard
mirex (I)	12/29/76 Most uses canceled effective 12/1/77	Reduction in nontarget and endangered species
nitrofen (H) TOK®	8/8/80 Voluntary suspension of sales by manufacturer	
OMPA (I) Schradan	5/28/76 Voluntary cancelation by manufacturer	Oncogenicity
Perthane® (*see* ethylan)		
pirimicarb (I) Pirimor®	3/81 Voluntary withdrawal of product by manufacturer	Expense of EPA registration
Safrole repellent	6/10/77 Voluntary cancelation by manufacturer	Oncogenicity; mutagenicity
silvex (H)	12/13/79 Most uses canceled	Oncogenicity; teratogenicity; fetotoxicity
Strobane (I)	6/28/76 Voluntary cancelation by manufacturer	
2,4,5-T (H)	12/13/79 *See* silvex	
trichlorobenzoic acid (H) Benzac®, Trysben®, 2,3,6-TBA	7/7/79 Voluntary cancelation by manufacturer	

cause application methods have improved nor because use has become more selective or judicious. Rather, the EPA has acted by canceling the registration of certain pesticides or mandating highly restrictive use patterns. Furthermore, the tremendous costs of meeting new product registration or old product reregistration requirements have forced many manufacturers to simply withdraw their products from the market. In some instances materials that had been in use 30 years or longer were withdrawn.

In the case of new compounds, the average cost and time for development and research required to meet registration requirements are estimated to be $10 million and 7 years. Since a patent on a pesticide lasts 17 years, manufacturers have 10 years to recover their investment and turn a profit. These two factors, cost and time, tend to discourage innovation in the development of new pesticides.

Table 21-9 presents a partial list of those materials that have been canceled by the EPA or the manufacturer; a large number of pesticides voluntarily removed from the market are not included.

FIRST AID FOR PESTICIDE POISONING

Poisoning symptoms may appear almost immediately after exposure or may be delayed for several hours, depending on the chemical, dose, length of exposure, and the individual. These symptoms may include, but are not restricted to, headache, giddiness, nervousness, blurred vision, cramps, diarrhea, a feeling of general numbness, or abnormal size of eye pupils. In some instances, there is excessive sweating, tearing, or mouth secretions. Severe cases of poisoning may be followed by nausea and vomiting, fluid in the lungs, changes in heart rate, muscle weakness, breathing difficulty, confusion, convulsions, coma, or death. However, pesticide poisoning may mimic brain hemorrhage, heat stroke, heat exhaustion, hypoglycemia (low blood sugar), gastroenteritis (intestinal infection), pneumonia, asthma, or other severe respiratory infections.

Regardless of how trivial the exposure may seem, if poisoning is present or suspected, obtain medical advice at once. If a physician is not immediately available by phone, take the person directly to the emergency ward of the nearest hospital and take along the pesticide label and telephone number of the nearest Poison Control Center. Look it up in the telephone directory under Poison Control Centers, or ask the telephone operator for assistance. Poison Control Centers are usually located in the larger hospitals of most cities and can provide emergency treatment information on all types of human poisoning, including pesticides.

First-aid treatment is extremely important, regardless of the time that may elapse before medical treatment is available. The first-aid treatment received during the first 2 to 3 minutes following a poisoning accident may very well mean the difference between life and death.

FIRST AID IN THE EVENT OF CHEMICAL POISONING

If You Are Alone with the Victim

First—See that the victim is breathing; if not, give artificial respiration.
Second—Decontaminate the victim immediately by washing off any skin residues. Speed is essential!
Third—Call your physician.

Note: Do *not* substitute first aid for professional treatment. First aid is only to relieve the patient before medical help is reached.

In an EMERGENCY
Call toll-free
National Drug and Poison
Information Center:
800-845-7633
(In South Carolina, call 800-922-0193)

If Another Person Is with You and the Victim

Speed is essential; one person should begin first-aid treatment, while the other calls a physician.

The physician will give you instructions. He or she will very likely tell you to get the victim to the emergency room of a hospital. The equipment needed for proper treatment is there. Only if this is impossible should the physician be called to the site of the accident.

General

1. Give the victim mouth-to-mouth artificial respiration if breathing has stopped or is labored.
2. Stop exposure to the poison. If poison is on skin cleanse the victim, including hair and fingernails. If swallowed, induce vomiting as directed (see the following page on swallowed poisons).
3. Save the pesticide container and material in it if any remains; get readable label or name of chemical(s) for the physician. If the poison is not known, save a sample of the vomitus.

Specific

POISON ON SKIN

1. Drench skin and clothing with water (shower, hose, faucet).
2. Remove clothing.
3. Cleanse skin and hair thoroughly with soap and water; rapidity in washing is most important in reducing extent of injury.
4. Dry and wrap in blanket.

POISON IN EYE

1. Hold eyelids open, wash eyes with gentle stream of clean running water immediately. Use copious amounts. Delay of a few seconds greatly increases extent of injury.
2. Continue washing for 15 minutes or more.
3. Do *not* use chemicals or drugs in wash water. They may increase the extent of injury.

INHALED POISONS (DUSTS, VAPORS, GASES)

1. If victim is in enclosed space, do not enter without air-supplied respirator.
2. Carry patient (do not let him walk) to fresh air immediately.
3. Open all doors and windows, if any.
4. Loosen all tight clothing.
5. Apply artificial respiration if breathing has stopped or is irregular.
6. Call a physician.
7. Prevent chilling (wrap victim in blankets but do not overheat him).
8. Keep victim as quiet as possible.
9. If victim is convulsing, watch his breathing and protect him from falling and striking his head on the floor or wall. Keep his chin up so his air passage will remain free for breathing.
10. Do not give alcohol in any form.

SWALLOWED POISONS

1. CALL A PHYSICIAN IMMEDIATELY.
2. *Do not induce vomiting if:*
   a. Victim is in a coma or unconscious.
   b. Victim is in convulsions.
   c. Victim has swallowed petroleum products (that is, kerosene, gasoline, lighter fluid).
   d. Victim has swallowed a corrosive poison (strong acid or alkaline products)—symptoms: severe pain, burning sensation in mouth and throat.
3. If the victim can swallow after ingesting a corrosive poison (any material that in contact with living tissue will cause destruction of tissue by chemical action, such as lye, acids, Lysol) give the following substances by mouth. For acids or alkali: milk or water; for patients one to five years old, 2 to 4 ounces; for patients five-years and older, up to 8 ounces.
4. If a noncorrosive substance has been swallowed, induce vomiting, if possible, EXCEPT if patient is unconscious, in convulsions, or has swallowed a petroleum product.
   a. Induce vomiting with *syrup of Ipecac*. For persons over one year of age, including adults, give 15 cc (1 tablespoon) of syrup of Ipecac followed by water. For children: one to five years old, give up to 6 ounces of water, and up to 8 ounces for persons over five years of age. Ipecac requires about 20 minutes to produce vomiting.
   b. When retching and vomiting begin, place victim face down with head lowered, thus preventing vomitus from entering the lungs and causing further damage. Do not let him lie on his back.
   c. Do not waste excessive time in inducing vomiting if the hospital is a long distance away. It is better to spend the time getting the victim to the hospital where drugs can be administered to induce vomiting and/or stomach pumps are available.
   d. Clean vomitus from victim. Collect some in case physician needs it for chemical tests.

CHEMICAL BURNS OF SKIN

1. Wash with large quantities of running water.
2. Remove contaminated clothing.
3. Immediately cover with loosely applied clean cloth (any kind will do), depending on the size of the area burned.
4. Avoid use of ointments, greases, powders, and other drugs in first-aid treatment of burns.
5. Treat shock by keeping victim flat, keeping him warm, and reassuring him until arrival of physician.

> Safety is a state of mind.
>
> World War II production safety slogan

Safe Handling and Storage of Pesticides

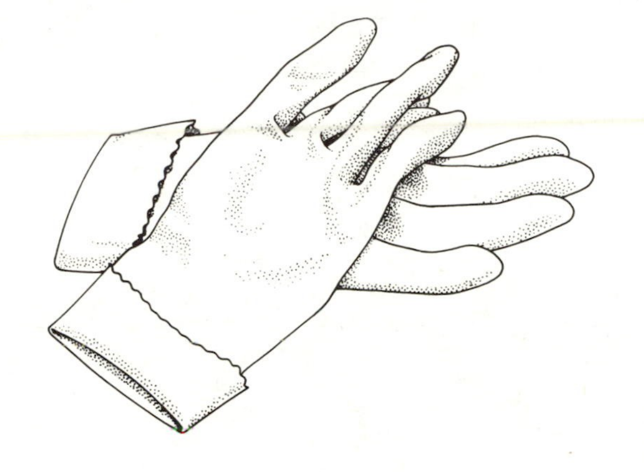

Most householders use the pesticide label as their primary source of information regarding the usage, storage, and disposal of pesticides. The first rule of safety in using any pesticide is to read the label and follow the directions and precautions printed on it. Pesticides are safe to use, provided common-sense safety precautions are practiced and provided they are used according to the label instructions. Safety requires that pesticides be kept away from children and illiterate or mentally incompetent persons.

"Safety is a state of mind" is an old adage in industrial safety engineering. But pesticide safety is more than a state of mind. It must become a habit with those who handle, apply, and sell pesticides—and certainly with those who supervise those who do. Householders can safely control pests in their home and garden if they use pesticides properly.

SELECTING PESTICIDES

Before buying a pesticide, check the label. Make sure it lists the name of the pest you want to control. If in doubt, consult your county agent or other authority. Select the pesticide that is recommended by competent authority and consider the effects it may have on nearby plants and animals. Make certain that the label on the container is intact and up to date; it should include directions and precautions. And finally, purchase only the quantity needed for the current season.

MIXING AND HANDLING PESTICIDES

If the pesticide is to be mixed before applying, carefully read the label directions and current official recommendations of your state's Cooperative Extension Service. This information can be obtained from your local county agent. It is always a good idea to wear rubber gloves when mixing pesticides and to stand upwind of the mixing container. Handle the pesticides in a well-ventilated area. Avoid dusts and splashes when opening containers or pouring liquids into the spray apparatus. Do not use or mix pesticides on windy days. Measure accurately the quantity of pesticide required using the proper equipment. Overdosage is wasteful; it will not kill more pests, it may be injurious to plants, and it may leave an excess residue on fruits and vegetables. Do not mix pesticides in areas where there is a chance that spills or overflows could get into any water supply. Clean up spills immediately. Wash pesticides off skin promptly with plenty of soap and water and change clothes immediately if they become contaminated.

APPLYING PESTICIDES

If the label calls for it, wear the appropriate protective clothing and equipment. Make certain that equipment is calibrated correctly and is in satisfactory working condition. Apply only at the recommended rate, and, to minimize drift, apply only on a calm day. Do not contaminate feed, food, or water supplies, especially pet food and water bowls. Avoid damage to beneficial and pollinating insects by not spraying during periods when such insects are actively working on flowering plants. Honeybees as a rule are inactive at dawn and dusk, which is a good time for outdoor applications. Keep pesticides out of mouth, eyes, and nose. Do not use your mouth to blow out clogged hoses or nozzles. Observe precisely the waiting periods specified on the label between pesticide application and harvest of fruit and vegetables. Clean all equipment used in mixing and applying pesticides according to recommendations. Do not use the same sprayer for insecticide and herbicide applications.

After handling pesticides, wash the sprayer, protective equipment, and hands thoroughly. And, if you should become ill after using pesticides and believe you have the symptoms of pesticide poisoning, call your physician and take the pesticide label with you. This situation is highly unlikely, but safety requires users to be prepared for an unexpected emergency.

STORING PESTICIDES

Around the home, the rule of thumb is to lock up all pesticides. Lock the room, cabinet, or shed where they are stored, to discourage children. Do not store pesticides where food, feed, seed, or water can be contaminated. Store in a dry, well-ventilated place, away from sunlight, and at temperatures above freezing. If your operation is larger than a typical homeowner's, mark all entrances to your storage area with signs bearing this caution: "PESTICIDES STORED HERE—KEEP OUT."

Keep pesticides only in original containers, closed tightly and labeled. To keep the label intact and legible, cover it with transparent tape or lacquer. Examine pesticide containers occasionally for leaks and tears. Dispose of leaking and torn containers and clean up spilled or leaked material immediately. Eliminate all outdated materials. Because many pesticide spray formulations are flammable, take precautions against potential fire hazards.

Small amounts of pesticides can be stored in a cupboard or storage cabinet that is locked and out of the reach of small children. Larger amounts require a locked shed or room in a building. The shed or room should be well lit, well ventilated, and constructed of fire-resistant materials. It should have a smooth (no cracks or crevices) cement floor painted with a hard sealer to simplify cleanup of pesticide leaks and spills.

Store granular pesticides on shelves if there is any possibility of dampness on the floor. Separate volatile herbicides and other pesticides as a precaution against cross-contamination. Keep all corrosive chemicals in proper containers to prevent leaks that might result in serious damage. Even the simple step of tightly closing lids and bungs on containers provides a measure of safety and can help extend the shelf life of pesticides.

When you buy pesticides, date them and keep a current inventory of your supplies. Don't tear open the tops of new bags or boxes of pesticides. Keep a sharp knife handy for this purpose, and clean it each time a container is opened. Partly empty paper containers should be sealed with tape or staples. Avoid stockpiling; buy what you need, but not to excess. This eliminates waste and the problem of what to do with old materials.

PESTICIDE SHELF LIFE

Pesticides are manufactured, formulated, and packaged to exacting standards. However, they can deteriorate in storage, especially under conditions of high temperature and humidity. Some pesticides lose active ingredients due to chemical decomposition or volatilization. Dry formulations become caked and compacted and the emulsifier in emulsifiable concentrates may become inactivated. Some pesticides are converted into more toxic, flammable, or explosive substances as they decompose.

Pesticide formulations that contain low concentrations of active ingredients generally lose effectiveness faster than more concentrated forms. Sometimes a liquid pesticide develops gas as it deteriorates, and this can make opening and handling containers quite hazardous. In time, the gas pressure may cause explosive rupture of the containers.

Certain pesticide chemicals have a characteristic odor. If this odor grows stronger in the storage area, it could indicate a leak or spill, a defective closure or an improperly sealed container. It may also be a clue that the pesticide is deteriorating, since the smell of some materials intensifies as they break down in storage. If none of these problems is found, chemical odors can be reduced by installing an exhaust fan or lowering the temperature of the storage area.

Containers have an important effect on the shelf life of pesticides. To maximize shelf life, follow the storage recommendations in the preceding section. Similarly, most pesticides will have a longer shelf life if the storage area is cool, dry, and out of direct sunlight. At below-freezing temperatures some liquid formulations separate into various components and lose their effectiveness. High temperatures cause many pesticides to volatilize or break down more rapidly. Extreme heat may also cause glass bottles to break or explode.

Other characteristics of a pesticide product that affect its shelf life include the formulation, the types of stabilizers and emulsifiers used, and the chemical nature and stability of the material.

Even with careful planning it is sometimes necessary to carry pesticide stocks over from one year to the next. Check dates of purchase at the beginning of each season, and use the older materials first.

If given proper storage, some pesticides may remain active for several years. However, storage conditions vary so widely that it is difficult to predict long-term shelf life for a pesticide product. This is one reason that most pesticides are not backed by the manufacturer if stored longer than 2 years; so plan every purchase of pesticides to be completely used within this 2-year period.

Table 22-1 lists some common pesticides with comments on their stability and storage requirements. Always read the label for specific information on this subject. If you have a question about the shelf life or storage of a product, a local pesticide distributor may be able to help.

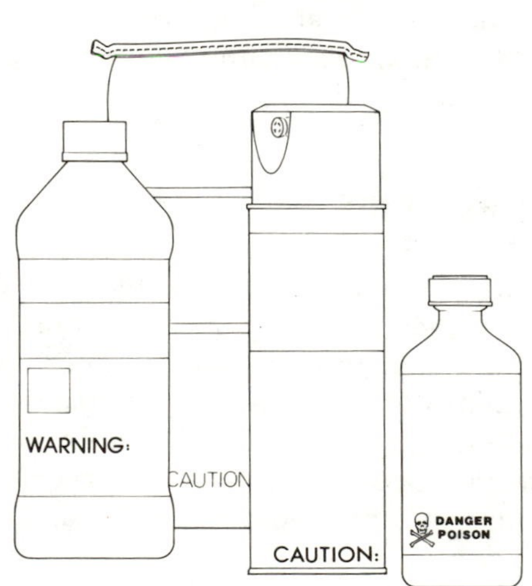

DISPOSING OF PESTICIDE CONTAINERS

The proper way to discard empty pesticide containers is a matter of concern to agricultural applicators, structural pest control operators, and growers. Each state has developed its own specific disposal regulations. The state lead agency for pesticides, as designated by the EPA, should be consulted for EPA-approved toxic waste disposal sites and landfills. When this information is not readily available, the following procedures are appropriate and should be used.

TABLE 22-1
Pesticides, storage, and stability.

Pesticide	*Storage and Stability Comments*
	INSECTICIDES
Carbaryl	Repeated freezing-thawing cycles may decrease effectiveness of flowable formulations. Wettable powders quite stable under normal storage conditions.
Diazinon	Use 4E within 6 months after opening container. Do not store near heat source. Keep lids tightly closed. Keep granular materials and dusts dry.
Dimethoate	Liquid formulations should be stored above freezing temperatures. Flammable. Keep away from heat and open flame. Flash point range 23°C to 38°C.
Malathion	Wettable powders are stable at least 2 years when stored properly. Do not store liquid formulations below 18°C. Keep away from heat source.
	MITICIDES
Chlorobenzilate	Keep emulsifiable solution away from heat or open flame. Store at temperatures above 0°C.
Kelthane	Wettable powders stable under normal storage conditions.
	FUNGICIDES
Benomyl	Stable at least 2 years with proper storage. Not flammable. Decomposes if exposed to moisture. Keep dry and tightly sealed.
Captan	Stable at least 2 years under normal storage conditions. Protect from extreme heat.
Folpet	Store in a cool dry place. Protect from excessive heat.
Karathane	Wettable dust stable under normal storage conditions. Do not store liquid formulations near heat or open flame.
Zineb	Decomposes when exposed to moisture, heat, or air. Flammable derivatives may be formed on decomposition. Shelf life is limited.
	HERBICIDES
Bensulide	Granules are stable. Emulsifiable liquid may crystallize below 5°C, but crystals redissolve if stored or warmed at high temperatures.
Fluometuron	Granules stable at least 2 years if tightly sealed and stored in a cool dry place.
DCPA (Dactal)®	Store in a dry place. Wettable powder stable for at least 2 years under proper storage conditions.
Glyphosate	Store above −12°C to keep from freezing, which results in crystals that settle to the bottom of the container. Do not store, mix, or apply in galvanized steel or unlined steel containers.
Simazine	Wettable powders and granules stable at least 2 years under normal conditions. Nonflammable.
Trifluralin	If stored long periods below 4°C, emulsifiable concentrate formulations may give poor weed control. Flash point is 48°C. Do not store near heat source. Stable at least 2 years with cool, dry storage.
2,4-D	Esters, amines, and salts and their formulations vary in volatility, flammability, and other properties. Follow label directions carefully.

Pesticide containers that have been triple-rinsed and punctured (making them unfit for further use) may be disposed of as nonhazardous wastes, in the normal manner. They are not subject to any special regulations, other than those imposed by the local sanitation agency or trash disposal firm. Large amounts should be taken to a sanitary landfill, while smaller quantities may be placed in regular trash containers.

It is very important to remember, however, that the water that is used to rinse the containers must either be used again to dilute or mix with pesticides, or disposed of as a hazardous waste.

Glass containers that are triple-rinsed and crushed may be disposed of in a sanitary landfill. Empty paper containers in amounts of less than 50 pounds may be burned in an open area if local regulations permit.

Instructions for homeowners: Because empty containers are never completely empty, do not reuse them for any purpose. Instead, break glass containers, triple-rinse metal containers with water, punch holes in top and bottom, and leave in your trash barrels for removal to the municipal landfill. Empty paper bags and cardboard boxes should be torn or smashed to make unusable, placed in a larger paper bag, rolled, and relegated to the trash barrel. All such materials should be packaged so that children and pets cannot accidentally come in contact with them.

UNWANTED PESTICIDES AND HAZARDOUS CONTAINERS

The most practical method for disposing of an unwanted pesticide is to use it according to label directions. Next best is to offer it to a responsible grower or neighbor in need of the materials. If this is not practical, *homeowners* may dispose of small quantities of pesticides by leaving them in their original containers, wrapping in several layers of newspapers, and placing them in the trash. Do not bury them; do not take them to an incinerator; and do not incinerate them yourself.

Sacks that held inorganic pesticides or organic mercury, lead, cadmium, or arsenic compounds may not be burned but must be delivered to an EPA- or state-approved Class I treatment, storage, or disposal (TSD) facility. In addition, empty unrinsed containers may not be stored more than 90 days. This also applies to contaminated or useless pesticides, whose disposal must be handled as a hazardous waste, according to toxicity. They are to be disposed of in an EPA- or state-approved Class I TSD facility.

PESTICIDE LABELS

Most household users consult the label to obtain information regarding application procedures and preventive measures, while relatively few read the label to learn about pesticide ingredients or antidotes. This is unfortunate because the single most important tool to the layman in the safe use of pesticides is the label on the container. The Federal Environmental Pesticide Control Act (FEPCA), which is discussed in Chapter 23, contains three very important points concerning the pesticide label that should be further emphasized. They pertain to reading the label, understanding the label directions, and following these instructions carefully.

Two of the provisions of the FEPCA are that (1) the use of any pesticide inconsistent with the label is prohibited, and (2) deliberate violations by growers, applicators, or dealers can result in heavy fines or imprisonment or both. The third provision is found in the general standards for certification of commercial applicators that in essence licenses them to use *restricted-use* pesticides. For certification, applicators are to be tested on (1) the general format and terminology of pesticide labels and labeling; (2) the understanding of instructions, warning, terms, symbols, and other information commonly appearing on pesticide labels; (3) classification of the product (general or restricted use); and (4) the necessity for use consistent with the label.

Figures 22-1 and 22-2 show the format labels for general-use and restricted-use pesticides as required by the EPA. These labels are keyed as follows:

1. Product name
2. Company name and address
3. Net contents
4. EPA pesticide registration number
5. EPA formulator manufacturer establishment number

6A. Ingredients statement
6B. Pounds/gallon statement (if liquid)
7. Front-panel precautionary statements
7A. Child hazard warning, "Keep Out of Reach of Children"
7B. Signal word—*DANGER*, *WARNING*, or *CAUTION*
7C. Skull and crossbones and word *Poison* in red
7D. Statement of practical treatment
7E. Referral statement
8. Side- or back-panel precautionary statements
8A. Hazards to humans and domestic animals
8B. Environmental hazards
8C. Physical or chemical hazards
9A. "Restricted Use Pesticide" block
9B. Statement of pesticide classification
9C. Misuse statement
10A. Reentry statement
10B. Category of applicator
10C. "Storage and Disposal" block
10D. Directions for use

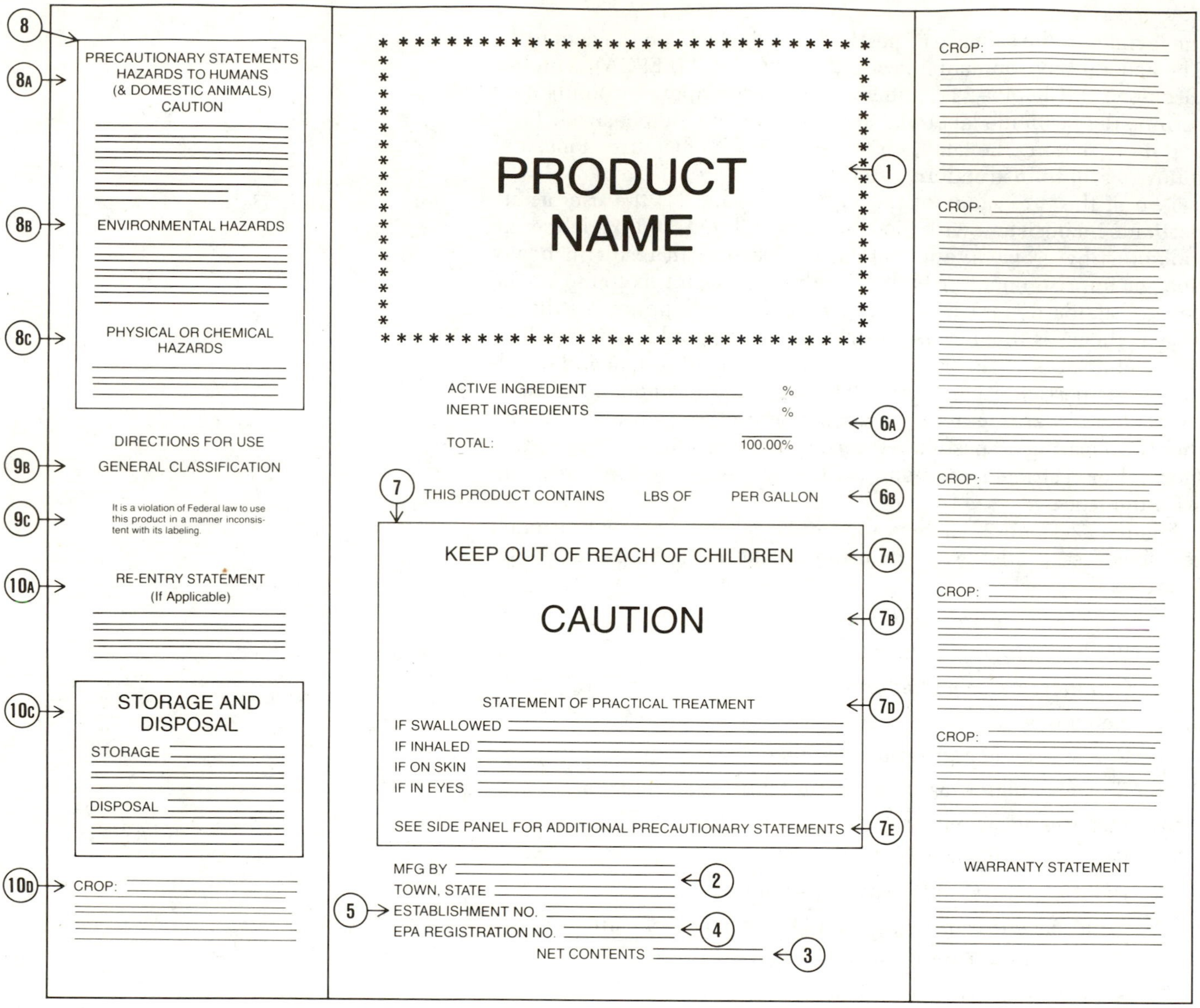

8

8A PRECAUTIONARY STATEMENTS
HAZARDS TO HUMANS
(& DOMESTIC ANIMALS)
CAUTION

8B ENVIRONMENTAL HAZARDS

8C PHYSICAL OR CHEMICAL
HAZARDS

DIRECTIONS FOR USE

9B GENERAL CLASSIFICATION

9C It is a violation of Federal law to use this product in a manner inconsistent with its labeling.

10A RE-ENTRY STATEMENT
(If Applicable)

10C STORAGE AND
DISPOSAL
STORAGE
DISPOSAL

10D CROP:

PRODUCT
NAME
1

ACTIVE INGREDIENT ______ %
INERT INGREDIENTS ______ %
TOTAL: 100.00%
6A

7 THIS PRODUCT CONTAINS LBS OF PER GALLON 6B

KEEP OUT OF REACH OF CHILDREN 7A

CAUTION 7B

STATEMENT OF PRACTICAL TREATMENT 7D
IF SWALLOWED
IF INHALED
IF ON SKIN
IF IN EYES

SEE SIDE PANEL FOR ADDITIONAL PRECAUTIONARY STATEMENTS 7E

MFG BY
TOWN, STATE 2
5 ESTABLISHMENT NO.
EPA REGISTRATION NO. 4
NET CONTENTS 3

CROP:

CROP:

CROP:

CROP:

CROP:

WARRANTY STATEMENT

FIGURE 22-1
EPA format for general-use pesticide label. (*Source:* U.S. EPA, Pesticide Registration Guidelines, 1975.)

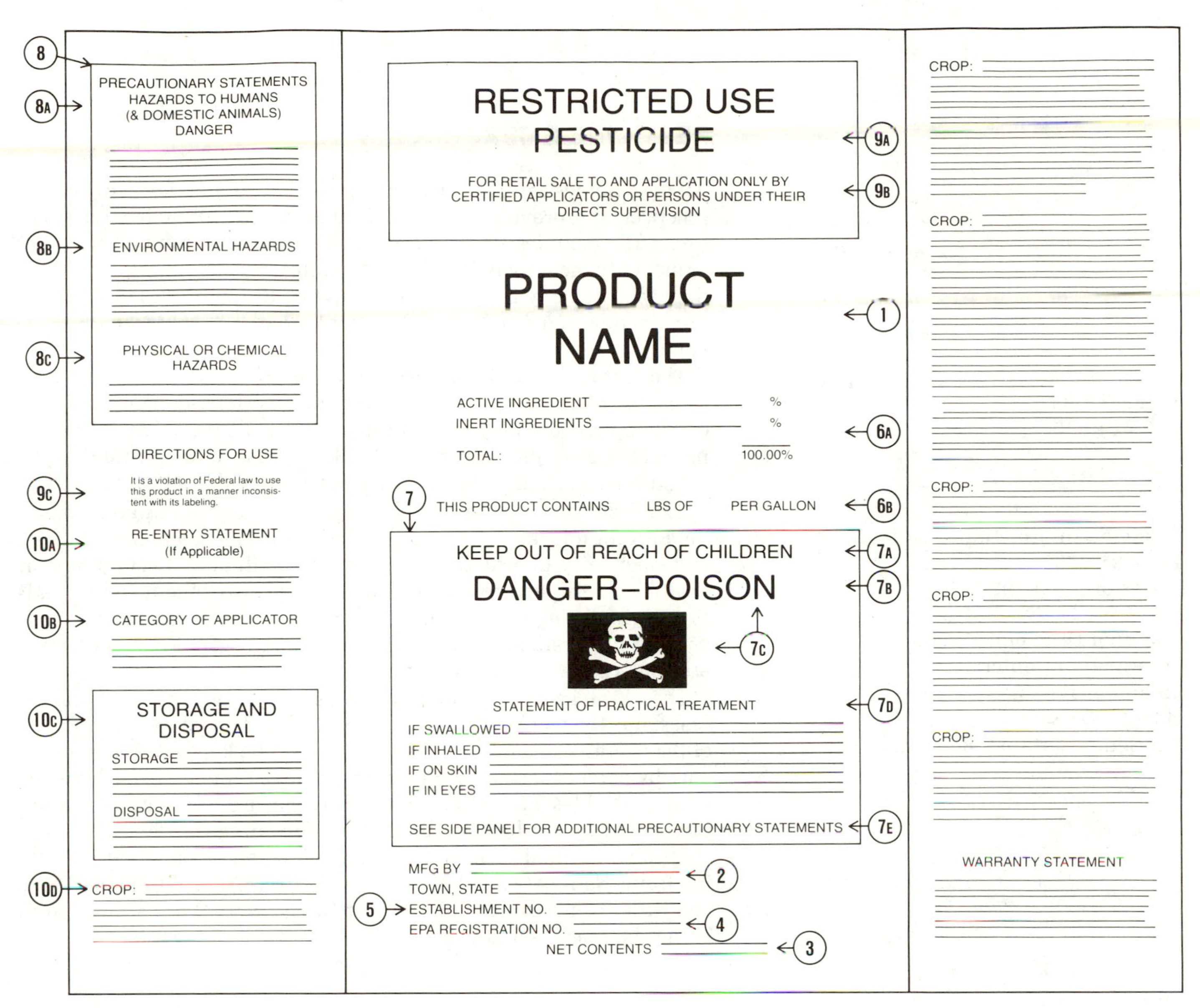

FIGURE 22-2
EPA format for restricted-use pesticide label. Products bearing this kind of label are not available to the layperson. (*Source:* U.S. EPA, Pesticide Registration Guidelines, 1975.)

PESTICIDE EMERGENCIES

All insecticides can be used safely, provided common-sense safety is practiced and provided they are used according to the label instructions; this includes keeping them away from children and illiterate or mentally incompetent persons. Despite the most thorough precautions, accidents will occur. The following three sources of information are important in the event of any kind of serious pesticide accident.

CHEMTREC
800-424-9300

The first and most important source of information is the CHEMTREC telephone number. From this toll-free long-distance number can be obtained emergency information on all pesticide accidents, pesticide-poisoning cases, pesticide spills, and pesticide-spill cleanup teams. This telephone service is available 24 hours a day.

Pesticide Information Clearinghouse
800-531-7790
(in Texas, 800-292-7664)

The Pesticide Information Clearinghouse has been established by the EPA for toll-free telephone service to provide general, technical, and *emergency information* on pesticides and their effect on human health and the environment. Callers may also obtain names, addresses, and telephone numbers of persons or organizations concerned with pesticides, including government agencies, manufacturers, and special interest groups.

National Drug and Poison
Information Center:
in an emergency
800-845-7633
(in South Carolina, 800-922-0193)

The third source of information is only for human-poisoning cases: It is the nearest Poison Control Center. The telephone number of the nearest Poison Control Center should be kept as a ready reference by parents of small children or employees of persons who work with pesticides and other potentially hazardous materials. (See Chapter 19 for further information on poisoning.)

Specific pesticide poisoning information can be obtained in writing from: National Clearinghouse for Poison Control Centers, Food and Drug Administration, Bureau of Drugs, 5401 Westbard Avenue, Bethesda, Maryland 20016.

CHAPTER 23

Those who strive to eliminate all risk are aiming at the end of progress.

Dick Beeler

The Law and Pesticides

Federal laws protect users of pesticides, consumers of treated products, pets and domestic animals, and the environment. The first federal law—the Federal Food, Drug, and Cosmetic Act of 1906, known as the Pure Food Law—required that food (fresh, canned, and frozen) shipped in interstate commerce be pure and wholesome. No provision of that law pertained to pesticides.

The Federal Insecticide Act of 1910, which covered only insecticides and fungicides, was signed into law by President Taft. The act was the first to control pesticides and was designed mainly to protect the farmer from substandard or fraudulent products, which were plentiful at the turn of the century. This was probably one of our earliest consumer protection laws.

The Pure Food Law of 1906 was amended in 1938 to include pesticides on foods, primarily the arsenicals, such as lead arsenate and Paris green. It also required the adding of color to white insecticides, including sodium fluoride and lead arsenate, to prevent their accidental use as flour or other look-alike cooking materials. This was the first federal effort toward protecting the consumer from pesticide-contaminated food, by providing tolerances for pesticide residues, namely arsenic and lead, in foods where these materials were necessary for the production of a food supply.

An Act

To amend the Federal Insecticide, Fungicide, and Rodenticide Act, and for other purposes.

Be it enacted by the Senate and House of Representatives of the United States of America in Congress assembled, That this Act may be cited as the "Federal Environmental Pesticide Control Act of 1972".

AMENDMENTS TO FEDERAL INSECTICIDE, FUNGICIDE, AND RODENTICIDE ACT

SEC. 2. The Federal Insecticide, Fungicide, and Rodenticide Act (7 U.S.C. 135 et seq.) is amended to read as follows:

"SECTION 1. SHORT TITLE AND TABLE OF CONTENTS.

"(a) SHORT TITLE.—This Act may be cited as the 'Federal Insecticide, Fungicide, and Rodenticide Act'.

"(b) TABLE OF CONTENTS.—

FEDERAL INSECTICIDE, FUNGICIDE, AND RODENTICIDE ACT (FIFRA)

The Federal Insecticide, Fungicide, and Rodenticide Act (FIFRA) became law in 1947. It superseded the 1910 Federal Insecticide Act, extended the coverage to include herbicides and rodenticides, and required that these products be registered with the U.S. Department of Agriculture before they could be marketed in interstate commerce. Basically, the law was one requiring good and useful labeling, making the product safe to use if label instructions were followed. The label was required to contain the manufacturer's name and address; name, brand, and trademark of the product; its net contents; an ingredient statement; an appropriate warning statement to prevent injury to humans, animals, plants, and useful invertebrates; and directions for use adequate to protect the user and the public.

The Miller Amendment to the Food, Drug, and Cosmetic Act (1906, 1938) was passed in 1954. It provided that any raw agricultural commodity may be condemned as adulterated if it contains any pesticide chemical whose safety has not been formally cleared or that is present in excessive amounts (above tolerances). In essence, this amendment clearly set tolerances on all pesticides in food products; for example, 10.0 ppm carbaryl in lettuce, or 1.0 ppm ethyl parathion on string beans.

FIFRA and the Miller Amendment supplement each other and are interrelated by law in practical operation. Today they serve as the basic elements of protection for the applicator, the consumer of treated products, and the environment, as modified by the following amendments.

In 1958, the Food Additives Amendment to the Food, Drug, and Cosmetic Act (1906, 1938, 1954) was passed. It extended to all types of food additives the same philosophy that had been applied to pesticide residues on raw agricultural commodities by the 1954 Miller Amendment. Of great importance, however, was the inclusion of the Delaney clause, which states that any chemical found to cause cancer (a carcinogen) in humans or animals when administered in appropriate tests may not appear in foods consumed by humans. This has become a most controversial segment of the entire spectrum of federal laws applying to foods, mainly with regard to the dosage found to produce cancer in experimental animals.

The various statutes mentioned so far apply only to commodities shipped in interstate commerce. In 1959, FIFRA (1947) was amended to include nematicides, plant regulators, defoliants, and desiccants as economic poisons (pesticides). (Poisons and repellents used against amphibians, reptiles, birds, fish, mammals, and invertebrates have since been included as economic poisons.) Because FIFRA and the Food, Drug, and Cosmetics Act are allied, these additional economic poisons were also controlled as they pertain to residues in raw agricultural commodities.

In 1964, FIFRA (1947, 1959) was again amended to require that all pesticide labels contain the federal registration number. It also required caution words, such as *Warning, Danger, Caution,* and *Keep Out of Reach of Children,* to be included on the front label of all poisonous pesticides. Manufacturers also had to remove safety claims from all labels.

The administration of FIFRA was the responsibility of the Pesticides Regulation Division of the U.S. Department of Agriculture until December 1970. At that time, the responsibility was transferred to the newly designated Environmental Protection Agency (EPA). Simultaneously, the authority to establish pesticide tolerances was transferred from the Food and Drug Administration (FDA) to the EPA. The enforcement of tolerances remains the responsibility of the FDA.

FEDERAL ENVIRONMENTAL PESTICIDE CONTROL ACT (FEPCA)

In 1972, FIFRA (1947, 1959, 1964) was revised by the most important pesticide legislation of this century: the Federal Environmental Pesticide Control Act (FEPCA), most commonly referred to as FIFRA Amended, 1972. Some of the provisions of this amendment are as follows:

1. Use of any pesticide inconsistent with the label is prohibited.
2. Deliberate violations of FEPCA by growers, applicators, or dealers can result in heavy fines and/or imprisonment.
3. All pesticides will be classified into (a) general-use or (b) restricted-use categories.
4. Anyone applying restricted-use pesticides must be certified by the state in which he or she lives. This provision includes both farmers and commercial applicators.
5. Pesticide manufacturing plants must be registered and inspected by the EPA.
6. States may register pesticides on a limited basis when intended for special local needs.
7. All pesticide products must be registered by the EPA, whether shipped in interstate or intrastate commerce.
8. For a product to be registered, the manufacturer is required to provide scientific evidence that the product, when used as directed, will (a) effectively control the pests listed on label, (b) not injure humans, crops, livestock, wildlife, or damage the total environment, and (c) not result in illegal residues in food or feed.

It is the responsibility of the EPA to interpret the law and implement its provisions; the EPA does this by preparing regulations. Once a regulation is duly processed, it has the force of law. Usually regulations are developed in consultation with those who will be most affected. Such was the case when the applicator training and certification programs were developed.

By regulation, the EPA established ten categories of certification for commercial applicators: (1) agricultural pest control (plant and animal); (2) forest pest control; (3) ornamental and turf pest control; (4) seed treatment; (5) aquatic pest control; (6) right-of-way pest control; (7) industrial, institutional, structural, and health-related pest control; (8) public health pest control; (9) regulatory pest control; and (10) demonstration and research pest control.

General standards of knowledge were also set for all categories of certified commercial applicators. Testing is based, among other things, on the following areas of competency: (1) label and labeling comprehension, (2) safety, (3) environment, (4) pests, (5) pesticides, (6) equipment, (7) application techniques, and (8) laws and regulations. In each state, certification is carried out by an appropriate regulatory agency (lead agency), often the State Department of Agriculture. Training of pesticide applicators is the function of the Cooperative Extension Service (FIFRA, Section 23 (c)).

FIFRA was further amended in 1975, 1978, 1980, and 1981. These provisions clarified the intent of the law and greatly influence the way pesticides are registered and used.

Many of the changes were designed to improve the registration process, which was slowed significantly by regulations resulting from the 1972 act. The more important provisions of the new amendments are as follows:

1. Generic standards are to be set for the active ingredients rather than for each product. This change permits the EPA to make safety and health decisions for the active ingredient in a pesticide, instead of treating each product on an individual basis. (This provision will speed registration, since there are only about 1000 active ingredients in the 35,000 formulations now on the market.)
2. Reregistration of all older products is required. Under reregistration, all compounds are to be reexamined to make certain that the supporting data for a registered pesticide satisfy current requirements for registration, in light of new knowledge concerning human health and environmental safety. Pesticides registered under the old data requirements (prior to August 1975) must be submitted to the new requirements in order to be reregistered.
3. The EPA may grant a conditional registration for a pesticide even though certain supporting data have not been completed. The information is still required, but its submission may be deferred. Conditional registration can be granted by the EPA if:
   a. The uses are identical or similar to those of already registered products with the same active ingredient;
   b. New uses are being added, providing a notice of Rebuttable Presumption Against Registration (RPAR) has not been issued on the product, or in the case of food or feed use, there is no other available or effective alternative;
   c. New pesticides have had additional data requirements imposed since the date of the original submission.
4. Efficacy data may be waived. The EPA may choose to set aside requirements for providing the efficacy of a pesticide before registration. The manufacturer must decide whether a pesticide is effective enough to market, and final proof will depend on product performance.
5. The use of data from one registrant can be used by other manufacturers or formulators if paid for. All data provided from 1970 on can be used for a 15-year period by other registrants, if they offer to pay "reasonable compensation" for this use. In the future, registrants will have 10 years of exclusive use of data submitted for a new pesticide active ingredient. During that time, other applicants may request and be granted permission to use the information but must obtain approval.
6. Trade secrets will be protected. The EPA may reveal data on most pesticide effects (including human, animal, and plant hazard evaluation), efficacy, and environmental chemistry. Four categories of data are generally to be kept confidential but may be released under certain circumstances:

   a. Manufacturing and quality control processes;
   b. Methods of testing, detecting, or measuring deliberately added inert ingredients;
   c. Identity or quality of deliberately added inerts;
   d. Production, distribution, sale, and inventories of pesticides.
7. The states now have primary enforcement responsibility, referred to as "state primacy." Any suspected misuse of pesticides is to be investigated by and acted upon by state regulatory boards. The state must indicate that its regulatory methods will meet or exceed the federal requirements. If a state does not take appropriate action within 30 days of alleged misuse, the EPA can act. Enforcement authority can be taken away from any state that consistently fails to take proper action.
8. States can register pesticides for Special Local Needs (SLN), that is, to meet unusual situations. Registrations can be made for new products, using already registered active ingredients. Existing product labels can be amended for new uses, including chemicals that have been canceled or suspended. Only those specific uses that were canceled or suspended may not be registered by the state for SLNs.
9. The phrase, "to use any registered pesticide in a manner inconsistent with its labeling" is defined. Users and applicators may now:
   a. Use a pesticide at less than labeled dosage, providing the total amount applied does not exceed that currently allowed on the labeling.
   b. Use a pesticide for control of a target pest not named on the label, providing the site or host is indicated.
   c. Apply the pesticide using any method not specifically prohibited on the label.
   d. Mix one or more pesticides with other pesticides or fertilizers, provided the current labeling does not actually prohibit this practice.

Remember! Other than these allowances, users must strictly adhere to the label!

REBUTTABLE PRESUMPTION AGAINST REGISTRATION (RPAR)

The Rebuttable Presumption Against Registration (RPAR) process by the EPA is designed to ensure a full gathering of scientific information on pesticide safety and a thorough assessment of risks and benefits of pesticide products. This process allows the EPA to study chemicals in detail before determining whether lengthy administrative hearings are necessary to cancel registrations or place restrictions on the uses of pesticides suspected or known to possess one or more of the "triggers" for RPAR.

Definitions of the risk criteria that, when met or exceeded, can trigger RPAR analyses, are found in the *Federal Register* (July 3, 1975; 40 CFR 162.11). These risk criteria are concerned with the following areas:

Note added in proof: The name RPAR has been changed to Special Review as of July 1982. New procedures will soon be implemented to simplify and speed up the RPAR process.

1. Acute toxicity
2. Chronic toxicity
   a. Oncogenic
   b. Mutagenic
3. Other chronic effects
   a. Reproductive
      (1) Fetotoxicity
      (2) Teratogenicity
   b. Spermatogenicity
   c. Testicular effects
4. Significant reduction in wildlife, reduction in endangered species, and reduction in nontarget species
5. Lack of emergency treatment or antidote

Pre-RPAR

The initial investigation of risk involves an intensive review of the scientific studies that suggest the RPAR criteria have been exceeded by the chemical in question. A validation of these studies is conducted by EPA scientists and their contractors. During this period, the registrants of the pesticide are informed of the review and are often requested to submit additional information. An extensive literature search is initiated in an attempt to identify all possible risk triggers. If trigger studies are found to be valid, an effort is made to gather and preliminarily assess available information on exposure to the pesticide in question. If such data are not available, the EPA will extrapolate from models or analogs and develops worst-case exposure assumptions to hypothesize the potential risk of the pesticide under review.

The validation of the trigger studies, the literature search, and the exposure analysis inform the EPA's preliminary position on the potential risk of the pesticide. The document describing this position on risk is referred to as position document one (PD 1). Upon approval by the EPA, the PD 1 is published in the *Federal Register* along with a formal Notice of Presumption Against Registration.

Issuance of an RPAR

After a PD 1 is published, the public process begins. A 45-day period is allowed for comment on the presumptions against registration presented in the PD 1. A 60-day extension is granted, if justified. Therefore, there is typically a 105-day period within which the EPA's presumptions may be rebutted. The period for submitting benefits information closes 180 days after issuance of the RPAR.

Most risk rebuttals are conducted by the pesticide's registrant, however, rebuttals may also be submitted by anyone (e.g., the U.S. Department of Agriculture (USDA), individual states, grower or commodity groups, or private parties). In fact, the EPA may rebut its own presumptions when appropriate.

Benefits assessment and determination of exposure under use conditions are determined as a standard policy by the National Agricultural Pesticide Impact Assessment Program's (NAPIAP) assessment teams. The NAPIAP rebuttal, which involves every state, is at least as important as the EPA's Presumption Against Registration, for it provides a way for the people to be heard in the regulatory process.

The NAPIAP team is usually assembled prior to the RPAR review, in response to the EPA's determination that a specific RPAR will be issued. A technical advisory group, composed of representatives from several major USDA groups, recommends, concurs upon, and then selects various persons to serve on specific assessment teams. Those persons are often associated with State Agricultural Experiment Stations and Cooperative Extension Services within the Land-grant Universities. The Environmental Quality Activities Office (EQAO) of the EPA also obtains its participants and the joint USDA–EPA team commences its review. This team is primarily concerned with collecting biological and economic benefits information, and locating exposure information that argues against or confirms acute and chronic toxicity that may result from use of the pesticide under consideration. The assessment team is also charged with identifying short-term, researchable gaps in the available data.

Thus, in the RPAR process, risks may be rebutted by any interested party submitting to the EPA proof that the study or studies upon which the presumption is based are not scientifically valid or proof that actual exposure to the compound will not cause the effects described. The EPA solicits rebuttal comments, additional information on risk, and responses from other agencies, primarily the USDA (for information on benefits and exposure) and the Department of the Interior, Office of Endangered Species (for information on the implications of an environmental trigger).

If all triggers are successfully rebutted, the pesticide is returned to the registration process and the RPAR is terminated for all or some uses. A second position document (PD 2) is drafted to state the EPA's regulatory action for the pesticide. Any new information obtained during the rebuttal period is, of course, incorporated. If the rebuttal is successful for all or some uses, PD 2 is published and becomes the terminal document of the RPAR process.

The EPA produces and publishes a separate PD 2 only in those cases where the rebuttal is successful. When the rebuttal is not successful, the rebuttal assessment, the risk analysis, the benefits analysis, the risk-benefit synthesis and the proposed regulatory position are presented in a document termed PD 2/3.

A PD 2/3 contains (1) an introductory chapter that reviews the RPAR criteria that gave rise to the RPAR on the pesticide in question; (2) a risk analysis that represents the risks of the currently registered uses of the pesticide; (3) a benefit analysis that represents the impact of total cancelation of the pesticide; (4) a risk-benefit analysis of each of the regulatory options under consideration for resolution of the RPAR; and (5) the recommended regulatory position.

The EPA examines the full range of available regulatory options, from full registration to full cancelation. Intermediate options include labeling changes, classification changes, and use restrictions. The effect of each of these options on the benefits and risks of each use of the pesticide are considered. The potential risks of alternative pesticides are considered. The recommended decision presents the best balance of risks and benefits in the interest of public health and the environment.

At the conclusion of the risk-benefit evaluation, the EPA reviews the completed PD 2/3. If the Assistant Administrator of the Office of Pesticides and Toxic Substances approves this document, a notice of determination is published in the *Federal Register*.

The FIFRA requires that the PD 2/3 be submitted to a scientific advisory panel for review of the scientific basis of the proposed decision and to the Secretary of Agriculture for comment. These comments, as well as any industry or public comments on the proposed decision, are evaluated. The EPA's assessment of these comments, including the final decision on the regulatory action, are announced in the *Federal Register* and in publication PD 4.

TOLERANCES, ADI, AND NOEL

A pesticide *tolerance* is the maximum amount of a pesticide residue that can legally be present on a food or feed. The tolerance is expressed in ppm, or parts of the pesticide per million parts of the food or feed by weight, and usually applies to the raw agricultural commodity. Pesticide tolerances are set by the EPA and enforced by the FDA or, in the case of meat, poultry, and eggs, by USDA agencies.

The tolerance on each food is set sufficiently low that daily consumption of the particular food or of all foods treated with the particular pesticide will not result in an exposure that exceeds the acceptable daily intake (ADI) for the pesticide. The tolerance is set still lower if the effective use of the pesticide results in lower residues.

The acceptable daily intake (ADI) is the level of a residue to which daily exposure over the course of the average human life span appears to be without appreciable risk on the basis of all facts known at the time. The ADI is usually set one hundred times lower than the no observable effect level (NOEL). A much greater safety factor is required if there is evidence that the pesticide causes cancer in test animals. Although the Delaney Amendment to the Federal Food, Drug, and Cosmetic Act prevents the addition of an animal carcinogen to foods, it does not apply to pesticide residues that occur inadvertently in the production of the crop.

The no observable effect level (NOEL) is the dosage of a pesticide that results in no distinguishable harm to experimental animals in chronic toxicity studies that include the minute examination of all body organs for abnormality.

It is extremely unlikely that any American will ever be exposed to anything near the ADI in his or her food. Numerous, detailed, continuing, nationwide studies show the actual pesticide residues in food to be far below the ADI and the established tolerances.

The pesticide residues on crops at the time of harvest are usually less than the tolerances. Residues decrease during storage and transit. They are reduced further by operations such as washing, peeling, and cooking when the food is prepared for eating. Every hectare of a food crop will not have been treated with the same pesticide, if any. And no one eats every food for which there is a tolerance for a particular pesticide every day. Further, the ADI refers to lifetime exposure. A minor excess of the ADI for a short period should be inconsequential.

STATE REGULATION

Beyond the federal laws providing rather strict control over the use of pesticides, most states have two or three similar laws controlling the application of pesticides and the sales and use of pesticides. Some of these laws concern the licensing of aerial and ground applicators as one group, and the licensing of structural applicators or pest control applicators as another. The latter is of concern to householders, who should contract for pest control services only from applicators who are both licensed within that state and certified to apply restricted-use pesticides.

EPILOG

We have to find in the next 25 years, food for as many people again as we have been able to develop in the whole history of man.

Jean Mayer (1975)

Pesticides for the Future

The pesticides that we will use in the future are now in their early stages of development. In this chapter, we examine some of the exciting work that research scientists are currently pursuing in experimental field studies and in the laboratory.

INSECT CONTROL

Not all insect-control chemicals are insecticides. Our discussion of the following topics represents the fruition of thinking that began after the Civil War in the United States, with the introduction of arsenicals as stomach poisons for insects—the first true insecticides. This was the beginning of manipulating the environment of the insect to human advantage.

Traditional Insecticides

An average of fewer than two new synthetic insecticides have been registered with the EPA each year for the past 3 years, a much smaller number than a decade ago. As we noted in Chapter 21, to register a new active ingredient requires, on the average, 7 years and $10 million. Obviously, such expenses are passed on to the consumer when the product finally reaches the market.

Table 24-1 presents detailed information and chemical structures, where appropriate, on 16 prospective insecticides, microbials, and miticides that are close to receiving full approval and registration by the EPA. Of these, there are four pyrethroids, three carbamates, three microbials, two organophosphates, one specific for fire ant control, and one unique acaricidal moiety. These compounds represent only about 60 percent of those nearing registration throughout the industry.

TABLE 24-1
Insecticides for the future.

Common name, trade name, and manufacturer	*Chemical structure, chemical name, and insecticide classification*	*Proposed uses*	*Oral LD_{50} to rats*	*Dermal LD_{50} to rabbits*
Amdro® AC 217, 300 American Cyanamid	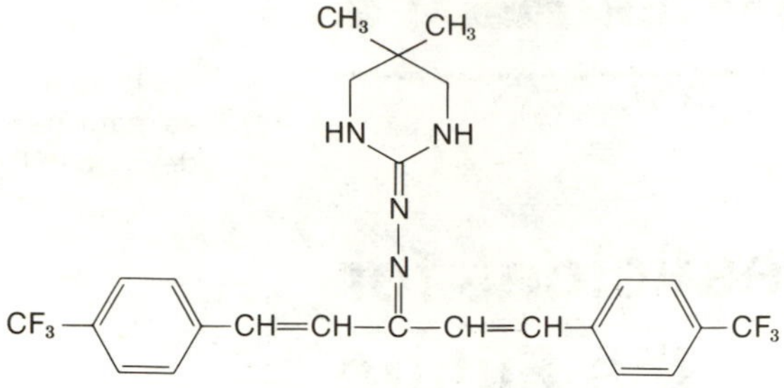 tetrahydro-5,5-dimethyl-2(1H)-pyrimidinoine (3-[4-(trifluoromethyl)-phenyl]-1-(2-[4-(trifluoromethyl)phenyl]-ethenyl)-2-propenylidene) hydrazone Amidinohydrazone insecticide	Fire ant control	1131	>5000
CARBOSULFAN Advantage® FMC 35001 FMC	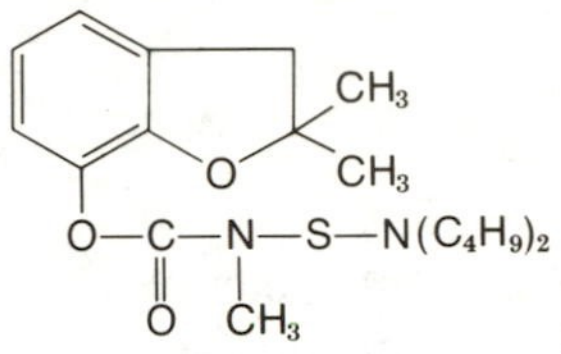 2,3-dihydro-2,2-dimethyl-7-benzofuranyl-[(dibutylamino)thio] methyl carbamate Carbamate insecticide	Soil and foliar insects on alfalfa, citrus, corn, deciduous fruit, and certain nematodes	209	>2000
CLOETHOCARB BAS 263-I BASF Wyandotte	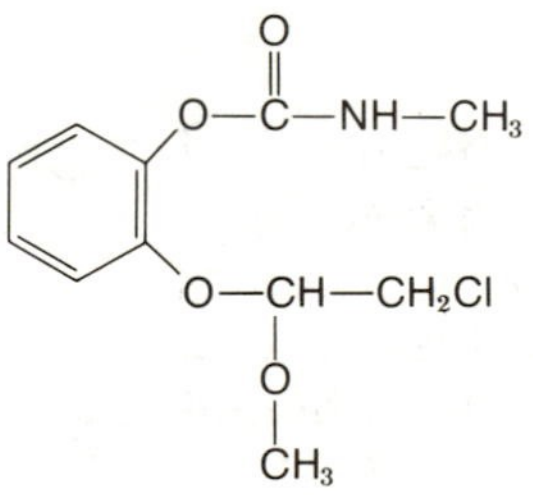 2-(1-methoxy-2-chloro)-ethoxy-phenyl-*N*-methylcarbamate Carbamate insecticide	Insecticide-nematicide against soil insects and nematodes and certain foliar insects	35	>2500
CYPERMETHRIN Ammo® (FMC) Cymbush® (ICI) FMC ICI Americas (Ripcord® and Barricade® are Shell International's, not marketed in the U.S.)	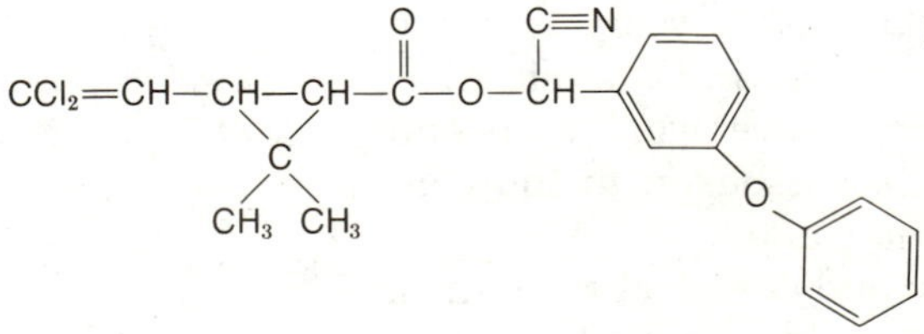 (±)-α-cyano (-3-phenoxyphenyl)-methyl (±)*cis, trans* 3-(2,3-dichloroethenyl)-2,2-dimethylcyclopropanecarboxylate Synthetic pyrethroid insecticide	High level of activity against a broad spectrum of insects	247	>2000

Common name, trade name, and manufacturer	Chemical structure, chemical name, and insecticide classification	Proposed uses	Oral LD_{50} to rats	Dermal LD_{50} to rabbits
FENPROPATHRIN S-3206 Sumitomo	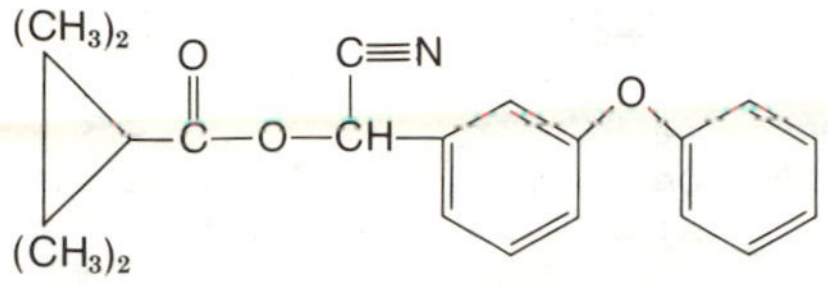 $(CH_3)_2$ O C≡N ‖ | —C—O—CH— O $(CH_3)_2$ α-cyano-3-phenoxybenzyl 2,2,3,3-tetramethyl cyclopropanecarboxylate Synthetic pyrethroid insecticide-acaricide	High level of activity against a broad spectrum of insects and mites	49	—
FLUCYTHRINATE Pay-Off® AC 222, 705 American Cyanamid	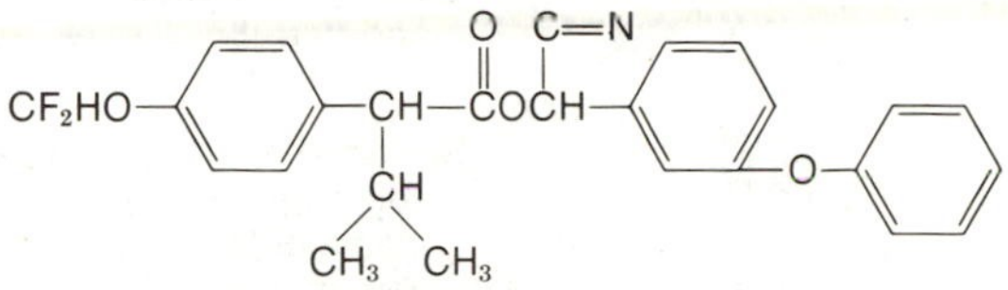 O C≡N ‖ | CF_2HO— —CH—COCH— —O— | CH / \ CH_3 CH_3 (RS)-α-cyano-3-phenoxybenzyl (S)-2-(4-difluro-methoxyphenyl)-3-methylbutyrate Synthetic pyrethroid insecticide-acaricide	High level of activity against a broad spectrum of insects, mites, and ticks	67	>1000
FLUVALINATE Mavrik® ZR-3210 Zoecon	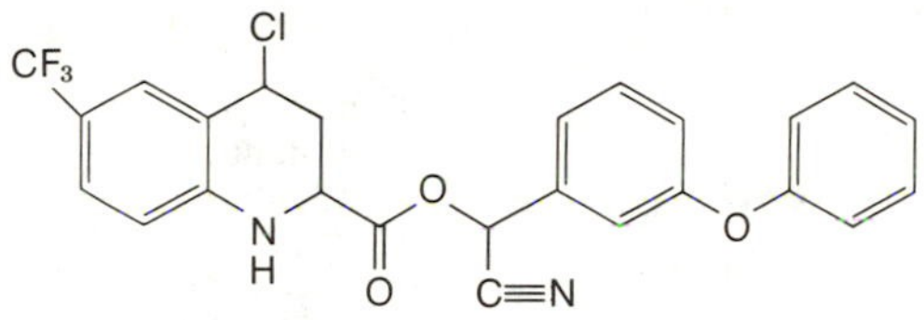 Cl CF_3 N H O O O C≡N N-[2-chloro-4-(trifluoromethyl) phenyl]-DL-valine (±)-cyano (3-phenoxyphenyl) methyl ester Synthetic pyrethroid insecticide	High level of activity against a broad spectrum of insects	>6299	>20,000
PROPETAMPHOS Safrotin® Sandoz	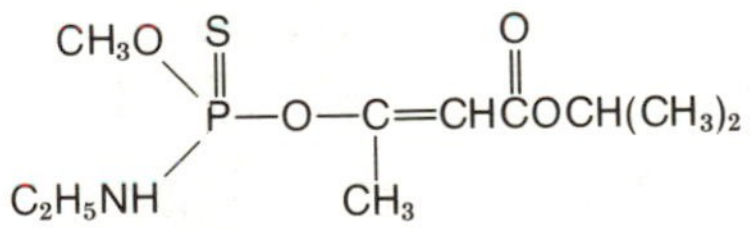 CH_3O S O ‖ ‖ P—O—C=CHCOCH$(CH_3)_2$ C_2H_5NH CH_3 (E)-O-2-isopropoxycarbonyl-1-methylvinyl O-methyl ethyl-phosphoramidothioate Organophosphate insecticide	Household pests: cockroaches, mosquitoes, flies, fleas, carpet beetles, etc.	75	2300
Certan® Sandoz	(No chemical structure) *Bacillus thuringiensis* var. *aizawai* (serotype 7)	Waxmoth in beehives	Nontoxic	
Teknar® Sandoz	(No chemical structure) *Bacillus thuringiensis* var. *israelensis*	Larvae of mosquitoes and blackflies	Nontoxic	
UBI-T930 Micromite® Uniroyal (acaricide)	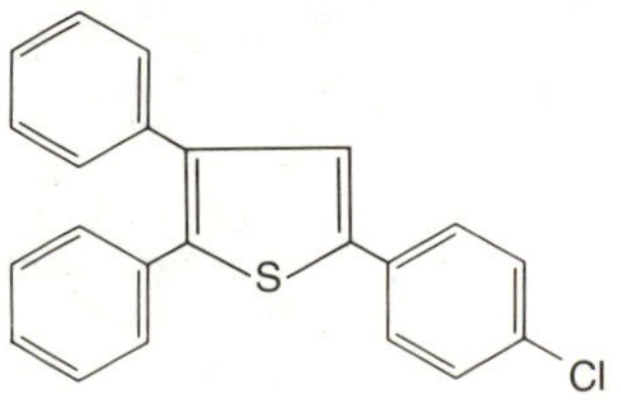 S Cl 5-(4-chlorophenyl)-2,3-diphenylthiophene	Citrus rust mites	>3000	>2000

(continued)

TABLE 24-1 (*continued*)

Common name, trade name, and manufacturer	*Chemical structure, chemical name, and insecticide classification*	*Proposed uses*	*Oral LD_{50} to rats*	*Dermal LD_{50} to rabbits*
Mycar® Abbott	(No chemical structure) *Hirsutella thompsonii*	Biological (fungal) acaricide or myco-acaricide for citrus rust mite and certain other mites	Nontoxic	
ISAZOPHOS Miral® CGA 12223 Ciba-Geigy	C_3H_7; Cl; N; N; N; S; $O-P(OC_2H_5)_2$ O-(5-chloro-1-(methylethyl)-1H-1,2, 4-triazol-3-yl)O,O-diethyl phosphorothioate Organophosphate insecticide	Systemic soil-applied insecticide-nematicide, and certain foliar insects	40–60	290–700
THIODICARB, Larvin® (old name: dicarbasulf) UC-51762 Union Carbide	$CH_3-C(S-CH_3)=N-O-C(=O)-N(CH_3)-S-N(CH_3)-C(=O)-O-N=C(S-CH_3)-CH_3$ dimethyl *N,N′*(thiobis((methylimino) carbonoyloxy)) bis(ethanimidothioate)	Experimental contact and stomach insecticide for caterpillars and other foliage-eating insects; provides long residual activity	1600	6400
SERTAN U.S. Forest Service	Virus polyhedral inclusion bodies of *Neodiprion sertifer* (European pine sawfly)	Experimental control of several sawfly pests of conifers	Nontoxic	
UC-55248 Tranid® Union Carbide Registration efforts for Tranid® were discontinued by the manufacturer in April 1982.	O; C_2H_5; $O-C-CHC_4H_9$; CH_3; CH_3; CH_3; O 3-(2-ethylhexanoyloxy)-5,5-dimethyl-2-(2′-methylphenyl) -2-cyclohexen-1-one	Experimental miticide with fumigant activity; controls a broad range of mites including strains resistant to organochlorine, organophosphate, and carbamate miticides; lacks insecticidal activity	>2000	>4000
BAY SIR 8514 Mobay	Cl; $C(=O)-NH-C(=O)-NH$; OCF_3 2-chloro-N-[[4-(trifluoromethoxy)phenyl]-amino]carbonyl) benzamide	Insect growth regulator; chitin synthesis inhibitor; not useful against sucking insects and spider mites	>5000	>5000

Insect Pheromones

Most insects appear to communicate by releasing molecular quantities of highly specific compounds that vaporize readily and are detected by insects of the same species. These delicate molecules are known as *pheromones*. The word *pheromone* comes from the Greek *pherein*, "to carry," and *hormon*, "to excite or stimulate."

Probably the most potent physiologically active molecules known today are insect pheromones. Pheromones are excreted outside the insect's body, where they cause specific reactions from other insects of the same species; they are also referred to in older literature as *social hormones*.

Pheromones can be classified into the following behavioral categories, based on the behavioral response of the receiving insect: sexual behavior, aggregation (including trail following), dispersion, oviposition, alarm, and specialized colonial behavior.

Of the different types of pheromones, the sex pheromones presently offer the greatest potential for insect control. For example, sex pheromones were recently used in eastern Arizona, in the long-staple cotton production area of Graham County. Pheromone traps containing microquantities of the synthetic sex lure of the pink bollworm, gossyplure, captured a sufficient number of early emerging male moths to prevent mating and reproduction from the first generation of the season. The population was suppressed sufficiently to avoid the use of insecticidal control for this pest for most of the remaining growing season.

The sex pheromones of pest moths have received the most detailed chemical study to date. For instance, after 30 years of trial and error, the gypsy-moth sex pheromone was isolated, identified, and synthesized in the laboratory in 1960. Since then great quantities of disparlure, the synthetic female gypsy-moth sex pheromone, have been used in male-trapping programs for this forest pest. We have no examples of complete insect control using synthetic pheromones, but many species are under investigation.

There are four principal uses for sex pheromones in current insect control programs: (1) male trapping, to reduce the reproductive potential of an insect population; (2) movement studies, to determine how far and where insects move from a given point; (3) population monitoring, to determine when peak emergence or appearance occurs; and (4) detection programs, to determine if a pest occurs in a limited trapping area, such as around international airports or quarantined areas. A fifth use for sex pheromones is the "confusion" or mating disruption technique. Gossyplure, the pink bollworm pheromone, has been incorporated into small, hollow, polyvinyl fibers (Nomate®) that permit slow release of the pheromone. Distributed heavily and uniformly over cotton fields, the concentration of pheromone is intended to confuse or disrupt the mating search of the male moths. This principle is now being applied to other insect species once their synthetic pheromones are available in large and economic quantities.

A list of available synthetic sex pheromones is presented in Table 24-2. This list contains 90 pheromones, whereas a similar list in *The*

TABLE 24-2
Commercially available insect sex pheromones and related compounds, and the species attracted.

Species attracted	*Pheromone common or trade name*	*Species attracted*	*Pheromone common or trade name*
alfalfa looper	looperlure	maggot fruit fly	MAGO®
almond moth	—[a]	Mediterranean flourmoth	—
ambrosia beetle	sulcatol	Mediterranean fruit fly	trimedlure
angoumois grain beetle	angoulure	Mediterranean melon fly	cue-lure
apple maggot fruit fly	AM[b]	Mountain pine beetle	pondelure
artichoke plume moth	—	naval orange worm	—
beet armyworm	—	oak leaf roller	—
black carpet beetle	megatomic acid	oblique-banded leaf roller	riblure
black cherry fruit fly	AM®	omnivorous leafroller	OLR®
black cutworm	BCW®	orange tortrix	OT®
boll weevil	grandlure	oriental fruit moth	orfralure, molestalure
cabbage looper	looplure	Pales weevil	eugenol[c]
California red scale	CRS®	peachtree borer	Nomate® Borer-Gard
carpenterworm	—	peach twig borer	PTB®
cherry fruit fly	AM®	pine bark beetle	Ipsdienol®
citrus leafroller	CACO®	pine beetles	Frontalin®
codling moth	codlelure	pink bollworm	gossyplure
corn earworm	zealure	potato tuberworm moth	PTM®
dermestid beetle	—	raisin moth	—
diamondback moth	DBM®	red-banded leafroller	Redlamone®, RBLR®
Douglas fir beetle	douglure	red plum maggot fly	Funamone®, GFUN®
Douglas fir tussock moth	tussolure	San Jose scale	SJS®
Egyptian cotton leafworm	SPOD®	smaller tea tortrix	ADOX-F®
elm bark beetle	multilure	Southern armyworm	—
European corn borer	nubilure	Southern pine beetle	frontalure
European grape vine moth	Grapamone®, EGVM®	soybean looper	looperlure
European pine shoot moth	EPSM®	spruce budworm	soolure
face fly	—	summer fruit tortrix	ADOX-O®
fall armyworm	frugilure	tentiform leafminer	TLM®
false codling moth	FCM®	Texas leaf-cutting ant	attalure
filbert leafroller	AROS®	three-lined leafroller	PL®
fruit-tree leafroller	FTLR®	tiger moth	—
fruit-tree tortrix	ARPO®	tobacco budworm	virelure
furniture carpet beetle	—	tobacco hornworm	—[c]
gelechiid moths	—	tobacco moth	—
grape berry moth	vitelure	tomato pinworm	TPW®, Nomate® Wormgard
greater peachtree borer	GPTB®	tufted apple budmoth	TBM®
greater wax moth	undecanal	variegated leafroller	VLR®
gypsy moth	disparlure	walnut husk fly	AM®
house fly	muscalure	wasp	citronellal[c]
Indian meal moth	PLOP®	Western pine beetle	brevilure
Japanese beetle	japonilure	Western pine shoot borer	Nomate® Shootgard
khapra beetle	—	yellow jacket	—
leaf cutter ants	—		
lesser appleworm	LAW®		
lesser peachtree borer	—		
lone star tick	—		

[a] Known only by its chemical name; no common or trade name assigned to pheromone.
[b] All 2-, 3-, and 4-letter trade names are those of Zoecon, the currently available source.
[c] Not proven to be the natural pheromone.

Pesticide Book in 1978 contained only 9. Thus we see a 1000 percent increase in 5 years of research and development.

The naming of pheromones may appear to be a bit inconsistent. At this time three kinds of names are used: common, trade or proprietary, and chemical. Common names are applied to materials that have been in use for a substantial period, and perhaps manufactured by more than one company, for example, disparlure, the gypsy-moth lure. Trade names are those given to the newer pheromones synthesized by only one manufacturer, for example, Redlamone®, a pheromone for the red-banded leafroller. Chemical names are usually the first designation given a compound; for example, (Z)-13-nonacosene is one of three constituents in the face-fly pheromone.

Currently, only two companies are considered basic manufacturers of pheromones. ChemSampCo, of Columbus, Ohio, manufactures pheromones, selling them in quantities for research and to formulators. The Zoecon Corporation, of Palo Alto, California, synthesizes pheromones and sells them in traps under the name Pherocon®, pheromone kits, replacement caps, and bait refills for traps. Among the formulators, Bend Research, Inc., of Bend, Oregon, formulates and sells controlled-release pheromones and attractant baits. Herculite Products, Inc., of New York, New York, formulates controlled-release pheromones under the product name of Hercon® Disrupt——(insect name). Additionally, they make pheromone traps for insects under the name Hercon® Luretrap———(insect name). Albany International, Controlled Release Division, of Needham, Massachusetts, also markets several disruptants and lures for use in pest management programs under the name of Nomate® (insect lure).

Despite the exuberance with which the potential of sex pheromones has been praised, pheromones are most practically used in survey traps to provide information about population levels, to delineate infestations, to monitor control or eradication programs, and to warn of new pest introductions. It is likely that future insect control programs will rely heavily on the uses of pheromones as a survey tool and to suppress early-emerging populations through trapping and confusion.

Chemosterilants

Chemicals used to sterilize insects, thus preventing reproduction, are known as *chemosterilants*. More than 1000 compounds that affect reproduction in insects have been described under the very broad classification of chemosterilants. The massive research effort to uncover these sterilizing chemicals is a direct spin-off of the successful programs for the eradication of the screwworm, a severe pest of beef cattle, by the release of males sterilized by gamma radiation. This technique involves the use of insects for the destruction of their own species through the induction of sterility in a large proportion of the males. It utilizes their mating behavior to affect female insects that would not be controlled by the usual insecticidal techniques.

The advantages of using chemicals instead of rearing, irradiating, and releasing massive numbers of males to induce sterility in a population are obvious. In some cases, both males and females can be

TABLE 24-3
Common and chemical names, mammalian oral LD_{50}s, and chemical structures of the more common chemosterilants.

Common and chemical names	Oral LD_{50} (mg/kg)	Chemical structure
Apholate 2,2,4,4,6,6-hexakis(1-aziridinyl)-2,2,4,4,6,6-hexahydro-1,3,5,2,4,6-triazatriphosphorine	98	$(H_2C-CH_2>N)_2$ on each P of a P=N ring
Tepa tris(1-aziridinyl)phosphine oxide	37	$O=P-(N<CH_2-CH_2)_3$
Metepa tris(2-methyl-1-aziridinyl)phosphine oxide	136	$O=P-(N<CHCH_3-CHCH_3)_3$
Thiotepa tris(1-aziridinyl)phosphine sulfide	9	$S=P-(N<CH_2-CH_2)_3$
Tretamine 2,4,6,-tris(1-aziridinyl)*s*-triazine	1	s-triazine ring with three $N<CH_2-CH_2$ groups
Busulfan 1,4-butanediol dimethanesulfonate	18	$CH_3-SO_2-O-C_4H_8-O-SO_2-CH_3$
Hempa hexamethylphosphoric triamide	2,650 ♂ 3,360 ♀	$O=P-(N(CH_3)_2)_3$
Thiohempa hexamethylphosphorothioic triamide	20	$S=P-(N(CH_3)_2)_3$
Hemel hexamethylmelamine	350	s-triazine ring with three $N(CH_3)_2$ groups

sterilized simultaneously for even more effective control. The development of chemosterilants is relatively new, and no definitive statements can be made about all types of compounds with sterilizing activity.

Chemosterilants are divided into four classes: alkylating agents, phosphorus amides, triazines, and antimetabolites.

The alkylating agents constitute the largest and most active class. They are moderately to highly reactive compounds with proteins and nucleic acids. They are sometimes referred to as *radiomimetics* (radiation-mimicking materials) in that their effects are similar to those of x rays or gamma rays. These agents replace hydrogen in fundamental genetic material with an alkyl group (—CH_3 or —C_2H_5, and so on), which results in an effect similar to irradiation. They are highly effective in producing mitotic disturbances or nucleotoxic conditions, particularly in tissues where cell division and multiplication take place at a high rate. This results in the production of multiple dominant lethal mutations or severely injured genetic material in the sperm or the egg. Although fully alive, zygotes (fertilized eggs), if formed, do not complete development into mature progeny.

The two most widely investigated types of alkylating chemosterilants are aziridines and alkanesulfonates. The chemical and physical properties of these compounds are quite variable, but their cytotoxic and mutagenic effects are closely related. Although the alkylating agents are relatively unstable and degrade rapidly, the possible contamination of large areas, even with small residues, makes their use as crop sprays or dusts hazardous and undesirable. Safe applications are possible, however, when these chemosterilants are used to sterilize reared or collected insects under controlled conditions and when personnel are adequately protected. Alkylating agents shown in Table 24-3 are apholate, tepa, metepa, thiotepa, tretamine, and busulfan.

The alkylating chemosterilants have been used in the field with moderate to good success for experimental housefly control around garbage and trash dumps. Busulfan fed to boll weevils that were later released in the field resulted in only moderate success. Penfluron, which is not an alkylating agent, but closely resembles another material, diflubenzuron (Dimilin®, the chitin synthesis inhibitor), has been more successful than busulfan when fed to female boll weevils.

Some of the newer experimental chemosterilants are listed here with their chemical names:

Aphamide	N,N-ethylene bis (P,P-bis(1-aziridinyl)-N-methylphosphinic amide
Morzid, OPSPA	bis(1-aziridinyl)morpholinophosphine sulfide
Methiotepa, MAPS Metapside®	tris(1-aziridinyl)phosphine sulfide tris(methylethylenimido)-triophosphate
Methotrexate, Amethopterin®	N-(p-(2,4-diamino-6-pteridyl)methyl-methylaminobenzoyl) glutamic acid
Methyl Apholate	2,2,4,4,6,6-hexahydro-2,2,4,4,6,6-hexakis (2-methyl-1-aziridinyl)-1,3,5,2,4,6-triazatriphosphorine
AI 3-63220	N-(((4-bromophenyl)amino)carbonyl)-2,6-difluorobenzamide
Penfluron	2,6-difluoro-N-((4-(trifluoromethyl) phenyl)amino)benzamide

Numerous laboratory studies indicate great potential for the chemosterilant principle, but the obvious problem is the hazardous nature of the residues on food and feed crops, especially with the alkylating agents. Again, the chemosterilants represent a new form of chemical control, whose modes of action differ from those of conventional insecticides.

Microbials

Microbials are disease organisms of insects (see Chapter 4). Microbial control is the control of insect pest infestations using these disease organisms. Although microbial control is currently in limited use, its full potential remains to be developed.

Microbial control as a potential tool in insect control programs owes allegiance to two currently used methods, chemical and biological controls. Some of the advantages and disadvantages of both methods apply also to microbial control. For example, some pathogens can be mass produced (as are chemical insecticides in chemical control), applied in a conventional manner at certain dosage levels to kill an existing infestation, and dissipated in the environment. In such cases, the microbial agent is essentially a "living insecticide," and no prolonged or residual effects of the application are expected.

Because microbial agents are living organisms, many of the principles that apply to other biological control agents, such as parasites and predators, apply equally as well to pathogens. For example, pathogens may be introduced into an environment to initiate a disease outbreak, but the main effects of the pathogens come from the reproduction and spread of the disease organisms in the pest population. Pathogens, like parasites and predators, are self-perpetuating and regulatory in nature. Some remain in the environment and become a permanent mortality factor in the pest population. Good examples of this are the milky (spore) disease bacteria *Bacillus popilliae* Dutky; *B. lentimorbus* Dutky, used to control the Japanese beetle; and *Nosema locustae*, a protozoan lethal to grasshoppers.

The bacterium *Bacillus thuringiensis* Berliner and the *Heliothis* polyhedrosis virus are pathogens used to control insects in a manner similar to that of conventional insecticides. They are the only microbial agents now registered for use on food crops in the United States. *B. thuringiensis* is available under several proprietary names for garden and ornamental use by homeowners and will control most species of caterpillars. The average user may not be particularly fond of these products because they are slower to act than the traditional insecticides, requiring several days to eradicate the pests. Because they are short-lived, repeated applications are necessary.

Disease-causing bacteria, viruses, fungi, nematodes, and protozoa affect a wide range of insects, beneficials and pests alike. In nature they play a large role in regulating insect pest numbers. For example, in most agroecosystems, cabbage loopers are annually decimated by a naturally occurring nuclear polyhedrosis virus. Similarly, a polyhedrosis virus of the alfalfa caterpillar is important in the natural control of this pest. Under favorable humidity and temperature conditions, pathogenic fungi play an important part in the natural control of a wide variety of insects. For example, several species of fungi are parasitic on the spotted alfalfa aphid.

Although many insect pests are subject to mortality from pathogenic agents occurring naturally in the environment, we cannot rely on them because of their unpredictable nature. Much research has been done in an effort to understand better the relationship of the three important components of a disease outbreak: the host insect, the pathogen, and the environment. Too, some pathogens exhibit high virulence against certain pests in the laboratory, but under field conditions relatively little effect is observed. The virus of the corn earworm is a case in point. The first attempts to utilize the nuclear polyhedrosis virus for field control of this pest were totally unsatisfactory, because ultraviolet radiation rendered the virus ineffective. Several formulations of this virus have been prepared, in an attempt to shield the virus particles from excessive radiation, and have increased its effectiveness in the field.

As scientists accumulate information on these organisms and better understand their ecological requirements, microbial control will surely become a major tool in the total management of insects. Pathogens, particularly the viruses, offer many potential new weapons for insect control, whether they are finally found to resemble the conventional chemical modes of action or to exert biological control similar to that of natural beneficials.

HERBICIDES

Herbicides, like all other pesticide groups, require new, more effective replacements for those that, for one reason or another, lose their utility. As new chemicals are introduced and groups of weeds are put under control, other weeds very soon, being freed of competition and being tolerant of the chemical, take over and become serious pests. With the advent of the chlorophenoxy compounds (2,4-D, etc.), which are highly selective for broad-leaf weeds, the grassy weeds became a more serious pest. Similar shifts have occurred when one chemical group of herbicides has been used continuously. In some instances, this problem has been solved by using a rotation of herbicide groups. In other situations, mixtures of herbicides are used to broaden the spectrum of weeds that can be controlled. (On weed tolerance and resistance to herbicides, see Chapter 11.)

The continuing increase in herbicide use and dollar investment indicates increased profits for agriculture and shows that chemical weed control, in addition to alleviating the tremendous burden of manual weeding, has increased the net income of growers around the world. Despite the advances of herbicide technology, improvements can be expected to continue for decades as herbicides become available and used in Third World countries.

Underlying this growing usage of herbicides in agriculture is a massive research effort involved in synthesis, testing, development, and production of new herbicides. Techniques from practically every aspect of biology are included in these activities. Biochemistry and plant physiology laboratories in universities and federal research facilities, as well as those of industry, carry out research on the absorption, translocation, and mode of action of herbicides; studies on the morphological effects of herbicides are conducted. Soil

TABLE 24-4
Herbicides, herbicide antidotes, defoliants, and plant growth regulators for the future.

Common name, trade name, and manufacturer	*Chemical structure, chemical name, and classification*	*Proposed uses*	*Oral LD_{50} to rats*	*Dermal LD_{50} to rabbits*
	HERBICIDES			
ACIFLUORFEN-SODIUM SALT Blazer® RH-6201 Rohm and Haas	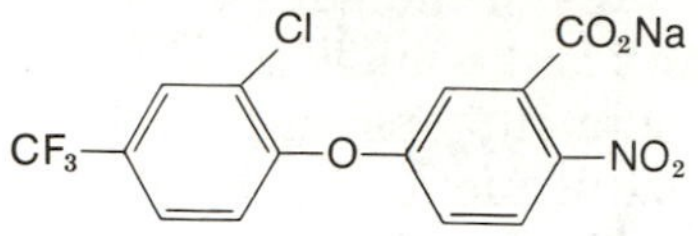 sodium 5-[2-chloro-4-(trifluoromethyl)-phenoxy]-2-nitrobenzoate Phenoxy herbicide	Selective soybean herbicide for pre- and post-emergence control of annual grasses and broad-leaf weeds	1540	>3680
BUTHIDAZOLE Ravage® VEL-5026 Velsicol	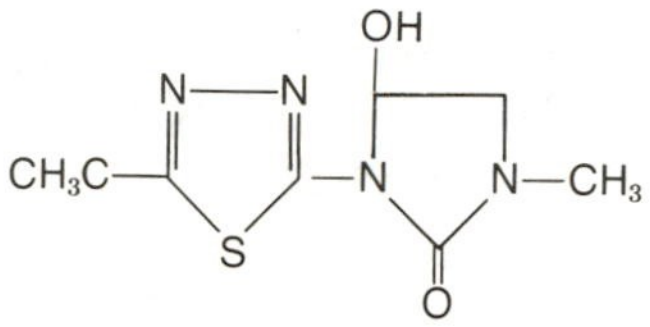 3[5-(1,1-dimethylethyl)-1,3,4-thiadiazol-2-yl]-4-hydroxy-1-methyl-2-imidazolidinone Thiazole herbicide	Broad-spectrum herbicide applied to soil or weed foliage to control most annual, biennial, and perennial weeds; uses for industrial and noncropland	1542	>2000
FLURIDONE Brake® Elanco	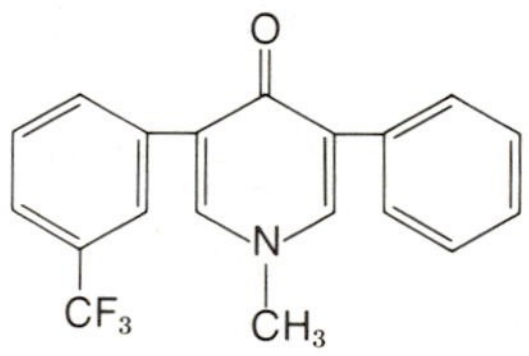 1-methyl-[3-phenyl-5-(3-trifluoromethyl) phenyl]-4(1H)-pyridinone	Pre-emergence selective herbicide for annual and perennial grass weeds in cotton, sugarcane, rice, trees, and vines	>10,000	—

science, microbiology, and pesticide residue laboratories are involved in studies of the fate of herbicides, including absorption, conjugation, chemical alteration, and photolytical and biological degradation. These studies represent the effort required to understand the functions of herbicides and secure their registrations from the EPA.

Herbicides that have produced residual problems, those that resist extensive degradation photolytically or biologically and remain in soils and plant products as intermediate breakdown products, are being phased out by the EPA or closely scrutinized. The history of the persistent insecticides provides many lessons for the future for herbicides.

The herbicides of the future will undoubtedly be developed rather heterogeneously from the many chemical groups already in existence, as have been those introduced within the last 3 years. Detailed information, including chemical structures where appropriate, is provided on 16 experimental, as yet unregistered, herbicides, defoliants, and plant growth regulators in Table 24-4.

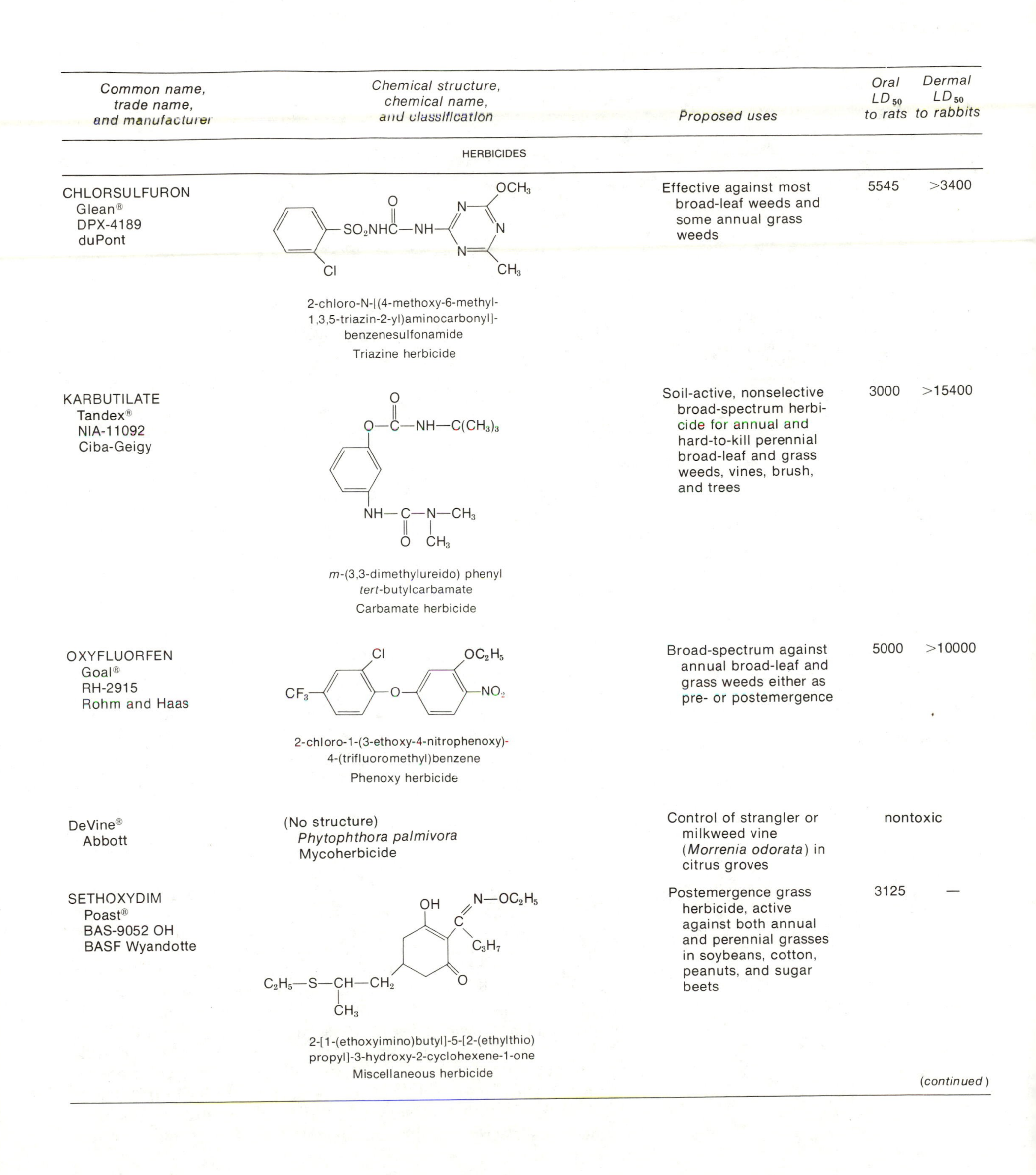

Common name, trade name, and manufacturer	Chemical structure, chemical name, and classification	Proposed uses	Oral LD_{50} to rats	Dermal LD_{50} to rabbits
HERBICIDES				
CHLORSULFURON Glean® DPX-4189 duPont	2-chloro-N-[(4-methoxy-6-methyl-1,3,5-triazin-2-yl)aminocarbonyl]-benzenesulfonamide Triazine herbicide	Effective against most broad-leaf weeds and some annual grass weeds	5545	>3400
KARBUTILATE Tandex® NIA-11092 Ciba-Geigy	*m*-(3,3-dimethylureido) phenyl *tert*-butylcarbamate Carbamate herbicide	Soil-active, nonselective broad-spectrum herbicide for annual and hard-to-kill perennial broad-leaf and grass weeds, vines, brush, and trees	3000	>15400
OXYFLUORFEN Goal® RH-2915 Rohm and Haas	2-chloro-1-(3-ethoxy-4-nitrophenoxy)-4-(trifluoromethyl)benzene Phenoxy herbicide	Broad-spectrum against annual broad-leaf and grass weeds either as pre- or postemergence	5000	>10000
DeVine® Abbott	(No structure) *Phytophthora palmivora* Mycoherbicide	Control of strangler or milkweed vine (*Morrenia odorata*) in citrus groves	nontoxic	
SETHOXYDIM Poast® BAS-9052 OH BASF Wyandotte	2-[1-(ethoxyimino)butyl]-5-[2-(ethylthio)propyl]-3-hydroxy-2-cyclohexene-1-one Miscellaneous herbicide	Postemergence grass herbicide, active against both annual and perennial grasses in soybeans, cotton, peanuts, and sugar beets	3125	—

(continued)

TABLE 24-4 *(continued)*

Common name, trade name, and manufacturer	Chemical structure, chemical name, and classification	Proposed uses	Oral LD_{50} to rats	Dermal LD_{50} to rabbits
	HERBICIDES			
THIOBENCARB (benthiocarb) Bolero® Saturn® Chevron	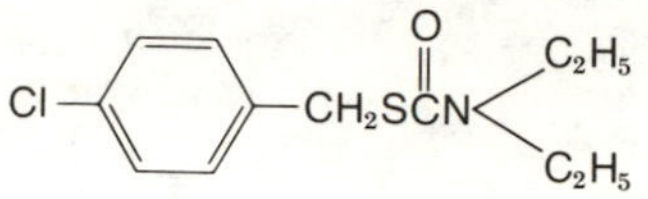 S-[(4-chorophenyl)] methyl diethylcarbamothioate Carbamate herbicide	Pre- and early post-emergence herbicide for grass and certain broad-leaf weeds in rice fields	1236	10000
UBI-S734 Uniroyal (Registration efforts were discontinued by the manufacturer in June 1982)	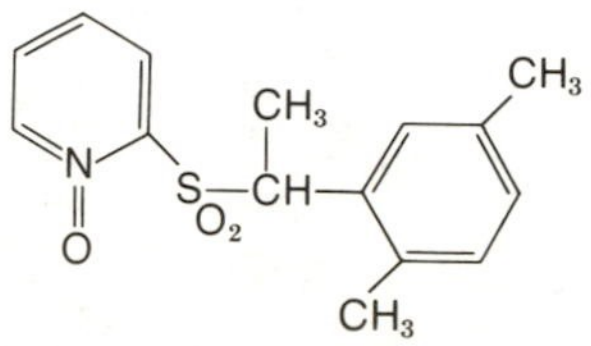 2-[[1-(2,5-dimethylphenyl)ethyl]sulfonyl]-pyridine 1-oxide Heterocyclic, pyridine herbicide	Preemergence herbicide primarily active against grasses and nutsedge	5200	>2000
FLUAZIFOP-BUTYL Fusilade® ICI Americas	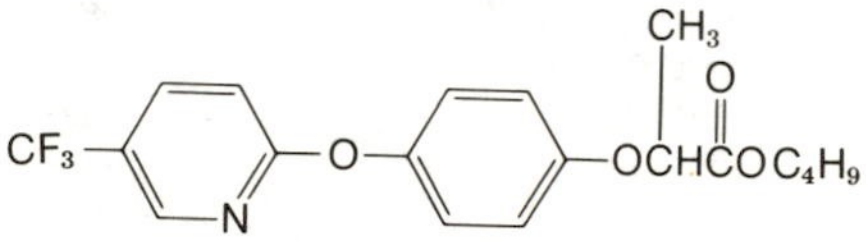 butyl 2-[4-(5-trifluoromethyl-2-pyridyloxy] phenoxy propionate Phenoxy herbicide	Highly selective for annual and perennial grasses. Systemic, translocating to roots, rhizomes, stolons, killing entire plant	3328	>2420
	HERBICIDE ANTIDOTE			
Concep® CGA-43089 Ciba-Geigy	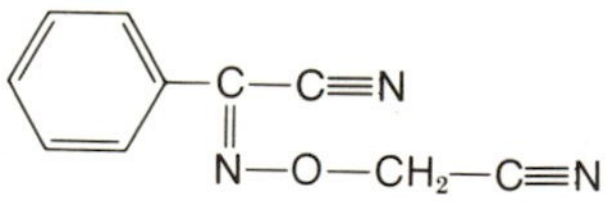 α-[(cyanomethoxy)imino] benzeneacetonitrile Substituted benzacetonitrile	Seed treatment to protect sorghum from phytotoxic effects of metolachlor (Dual®)	2227	>3100

FUNGICIDES

There are two approaches to chemical control of plant diseases: protecting plants from infection or curing plants after they become infected. Historically, prevention has been the only method of control until the recent appearances of the systemic fungicides, the oxathiins and benzimidazoles in 1966 and the pyrimidines in 1968. These are undoubtedly landmarks in the history of fungus control.

The systemics protect susceptible foliage and flowers more efficiently than the protectants because of their ability to translocate through the cuticle and across leaves. And, because of the great concern over agricultural chemical pollution, the systemics offer a

Common name, trade name, and manufacturer	Chemical structure, chemical name, and classification	Proposed uses	Oral LD_{50} to rats	Dermal LD_{50} to rabbits
	DEFOLIANTS			
Harvade® N252 Uniroyal	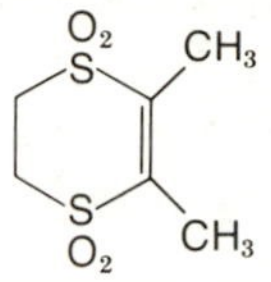 2,3-dihydro-5,6-dimethyl-1,4-dithiin 1,1,4,4-tetraoxide Dithiin defoliant	Cotton defoliant, also for nursery stock, grapes, and natural rubber	1175	>8000
THIDIAZURON Dropp® NOR-AM	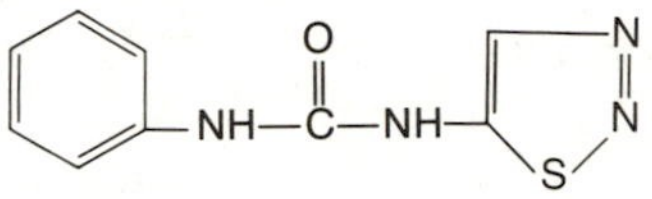 *N*-phenyl-*N*′-1,2,3-thiadiazol-5-ylurea Heterocyclic nitrogen, thiadiazole herbicide	Defoliant for cotton, inhibits regrowth of leaves on treated plants	>4000	>1000
	PLANT GROWTH REGULATORS			
GLYOXIME Pik-Off® CGA-22911 Ciba-Geigy	OH—N=CH—CH=N—OH glyoxal dioxime Ethanedial dioxime	Citrus abscission agent for removal of certain varieties of oranges destined for processing only	119	1580
MEPIQUAT CHLORIDE Pix® BAS 08300 BASF Wyandotte	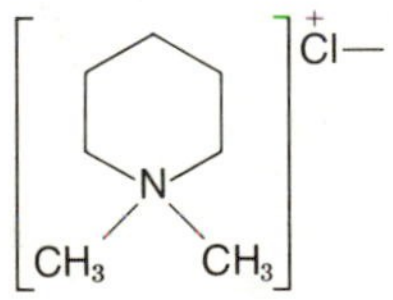 1,1-dimethyl-piperidiniumchloride Piperidinium growth regulator	Postemergence application to cotton to control vegetative growth	>6900	—

new path for specific placement of fungicides. If the total dosage and number of treatments needed for control can be reduced, excess chemical use can be avoided. Systemics may also replace certain more hazardous materials.

Unfortunately, the systemics developed to date are so specific that resistance is inevitable. This, of course, means that their overall life expectancies are much shorter than the traditional heavy-metal and broad-spectrum protectants. However, now that the principles of systemic fungicide structures and modes of action have been elucidated, it should be relatively simple to bring additional materials to commercial use.

In 1970 the EPA banned the use of alkyl mercurial fungicides for

TABLE 24-5
Fungicides for the future.

Common name, trade name, and manufacturer	*Chemical structure, chemical name, and fungicide classification*	*Proposed uses*	*Oral LD_{50} to rats*	*Dermal LD_{50} to rabbits*
FURMECYCLOX BAS 389F BASF Wyandotte	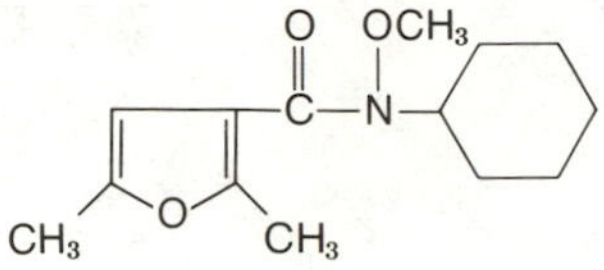 *N*-cyclohexyl-*N*-methoxy-2,5-dimethyl-furan-3-carboxylic acid-amid Furan fungicide	Seedling diseases, smuts	>3500	>5000
METALAXYL Ridomil® Subdue® Ciba-Geigy	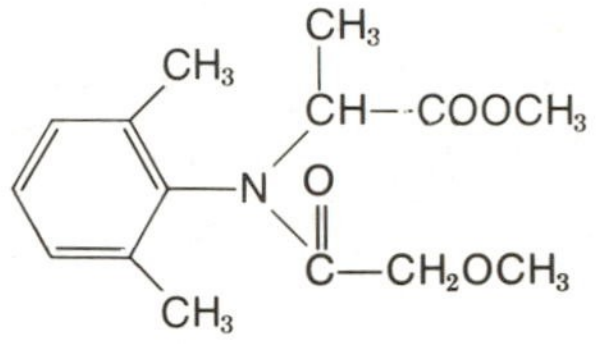 *N*-(2,6-dimethylphenyl)-*N*-(methoxyacetyl)-alanine methyl ester Systemic acylalanine fungicide	Ridomil®: soil-borne *Pythium* and *Phytophthora* and foliar diseases caused by Phycomycetes Subdue®: *Pythium* and *Phytophthora* on ornamentals and turf	669	>3100
PROPAMOCARB HYDROCHLORIDE Prevex® Schering AG and NOR-AM	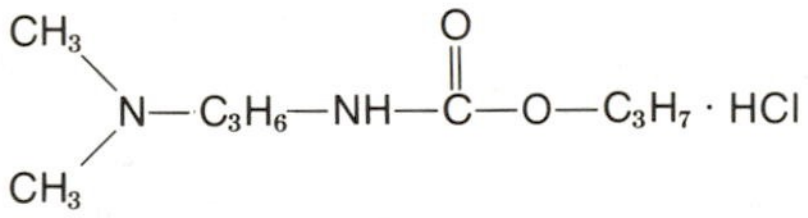 $(CH_3)_2N{-}C_3H_6{-}NH{-}C(=O){-}O{-}C_3H_7 \cdot HCl$ propyl [3-(dimethylamino)propyl] carbamate monohydrochloride Carbamate or alaphatic nitrogen fungicide	*Pythium* blight in turf	7860	>3600

seed treatment. Sweden had earlier banned the alkyl mercurials when it was established that the decline of seed-eating and predatory bird species in Sweden was caused by the use of alkyl mercurials for seed treatment. Immediately it became necessary to find substitutes for the banned chemical agents.

The needs in plant disease control are systemics and curatives for the downy mildew diseases, good systemic bactericides, and compounds that translocate to new foliage or roots following foliar treatment. As these are eventually provided, we should begin to lay aside those in current use that suffer from the usual inadequacies.

The future fungicides, then, will probably be based on the systemics currently in use, with similar basic chemical structures and their high order of specificity. This in no way will detract from use of protectants, since these are the "old reliables" and form the foundation for thorough and efficient disease control accepted as routine in today's cropping practices. Table 24-5 provides detailed information on six experimental fungicides, most of which are systemic and effective against the most difficult of plant diseases.

Common name, trade name, and manufacturer	*Chemical structure, chemical name, and fungicide classification*	*Proposed uses*	*Oral LD_{50} to rats*	*Dermal LD_{50} to rabbits*
Tilt® CGA-64250 Ciba-Geigy	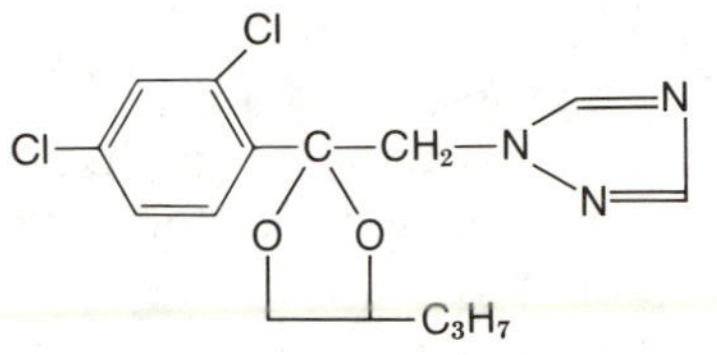 1-[[2(2,4-dichlorophenyl)-4-propyl-1,3-dioxolan-2-yl]methyl]1-*H*-1,2,4-triazole Triazole derivative fungicide	Systemic fungicide against Ascomycetes, Basidiomycetes, and Deuteromycetes	1517	>4000
Vangard® CGA-64251 Ciba-Geigy	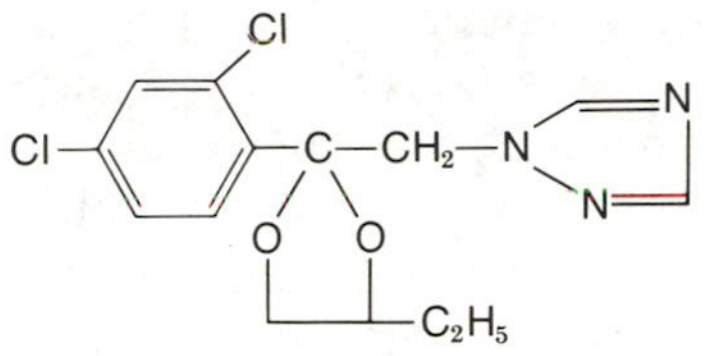 1-[[2-(2,4-dichlorophenyl)-4-ethyl-1,3-dioxolan-2-yl] methyl]-*1H*-1,2,4-triazole Triazole derivative fungicide	Systemic fungicide against Ascomycetes, Basidiomycetes, and Deuteromycetes	1343	>3100
VINCLOZOLIN Ronilan® BAS 352 04F BASF Wyandotte	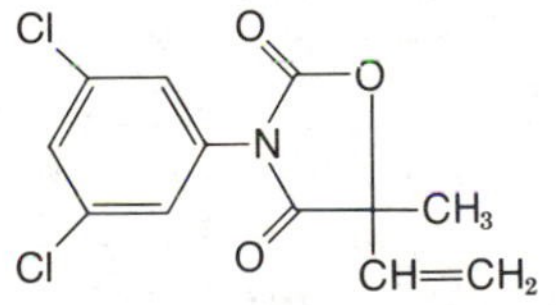 3-(3,5-dichlorophenyl)-5-ethenyl-5-methyl-2,4,oxazolidinedione	Contact fungicide against *Botrytis* spp., *Monilinia* spp., and *Sclerotinia* spp.	>10,000	2000

ALTERNATIVES TO PESTICIDES

It requires no wisdom to predict that we will continue to rely heavily on pesticides, at least throughout our generation. Pesticides are essential and will remain our first line of defense against pests when damage levels reach economic thresholds. However, to be completely dependent on chemical control and ignore all other methods of pest control has serious consequences, some of which we have already discussed.

First are the problems of persistence and biomagnification in relation to the chlorinated insecticides (see Chapter 4). Although persistence is a desirable characteristic of a pesticide in and of itself, assuring long-term effectiveness, the accumulation of fat-soluble compounds (such as DDT) in the food chain poses hazards to certain species of wildlife.

Second, overdependence on chemical control results in the pest populations' developing resistance to specific pesticides (see Chapter 20).

Third, we must consider the effects of chemical pesticides on nontarget organisms. For example, insecticides are normally applied to an agricultural crop for only a few pest insects. In most cases, these few key species require this artificial control measure to prevent economic losses to the crop. The history of insecticide usage, however, illustrates that additional problems may be created either by the rapid resurgence of the treated pest population or by the elevation of minor pests to the role of secondary or major pest status. Furthermore, insecticide applications reduce not only the pest population but also populations of natural enemies, with a resultant increase in pest populations.

There are many examples of unintentional damage to nontarget species: the drift of herbicides and insecticides onto sensitive crops or those intended for animal feeds; the reduction of beneficial soil microflora by application of fungicides or herbicides to the aboveground portions of the crop; the killing of pets and wildlife when baiting for rodent or predatory pests; the contamination of root crops from last year's insecticide or herbicide application; sickness and death of livestock feeding in pastures on which highly toxic materials drifted following application to adjacent crops; and, finally, the accidental poisoning of persons, including children, from improperly stored or secured pesticides.

Thus a total dependence on pesticides results in increased human-made pest problems, environmental damage, and potential hazard to humans themselves.

Long-range, intelligent pest control must address the management and manipulation of pests by using not just one but all available pest control methods. This combination of methods into one thoughtful program is referred to as *integrated pest management,* the practical manipulation of pests using any or all control methods in a sound ecological manner. This strategy brings together in a workable combination the best parts of all control methods that apply to a given problem created by the activities of pests.

Generally, control methods available today fall into six categories, though all may not apply to every pest form.

1. *Chemical control:* the control of pests by the use of pesticides, the method emphasized throughout this book.
2. *Biological control* (*biocontrol*): the reduction of pest numbers by predators, parasites, or pathogens.
3. *Cultural control:* the use of farming or cultural practices associated with crop production that make the environment less favorable for survival, growth, or reproduction of pest species.
4. *Host-plant resistance:* the use of and development of plants that are resistant to attack by insects, disease organisms, nematodes, or birds.
5. *Physical and mechanical control:* the application of direct or indirect measures that kill the pest, disrupt its physiology other than by chemical means, exclude it from an area, or adversely alter the pest's environment.
6. *Regulatory control:* the prevention of the entry and establishment of undesirable plant and animal pests in a country or area and the

eradication, containment, or suppression of pests already established in limited areas (quarantines).

Most methods of pest control fall within these classifications. Theoretically, all crops and their pests lend themselves readily to integrated pest management. In reality, this is not yet true. The greatest advances have been made in insect pest management, followed by the management of plant pathogens and weeds. We have a long way to progress in this area.

In summary, our pesticide arsenal is not keeping up with our pest problems, because of resistance, persistence, hazards, and environmental complications. Whatever the cause of the declining effectiveness or availability of these chemical tools, we need to refine our practice of integrated pest management to preserve their period of usefulness by using them when and only when they are needed. Our alternatives to this long-range plan are less than encouraging.

And, in closing, hear two more words of exhortation regarding chemicals and our environment: environmental respect. We have all seen or read about some of the undesirable effects that chemicals in excess—not just pesticides, any chemicals—can have on our precarious environment: the effects of food-chain-incorporated DDE, the metabolite of DDT most frequently found in the environment, and polychlorinated biphenyls on eggshell thickness in birds of prey; marine oil spills with resulting wildlife and beach contamination; the reduction of ozone in the upper atmosphere by fluorohydrocarbons; or contamination of rivers and streams by runoff of nitrogenous fertilizers and effluents from food-processing plants and the resulting reduction of aquatic life. These are but a few of the more bizarre, headline-rating episodes. We must, as a world culture, develop a greater respect for our delicate environment if the earth is to support the 6.4 billion residents anticipated by AD 2000.

APPENDIXES

APPENDIX A

Insecticides and Acaricides

Common, trade, and chemical names of insecticides and acaricides, their basic manufacturer(s), general use patterns, and oral and dermal LD_{50}s.

Common name, trade name, and basic manufacturer(s)	*Chemical name*	*General use pattern*	*Oral LD_{50} (rats)*	*Dermal LD_{50} (rabbits)*
Abate® (*see* temephos)				
Acaraben (*see* chlorobenzilate)				
acephate, Orthene® (Chevron)	*O,S*-dimethyl acetylphosphoramidothioate	Controls most insects on vegetables, cotton, soybeans. Residual and systemic activity.	866	2000
Actellic® (*see* pirimiphos-methyl)				
Advantage® (*see* carbosulfan)				
Agritox® (*see* trichloronat)				
aldicarb, Temik® (Union Carbide)	2-methyl-2-(methylthio)propionaldehyde *O*-(methylcarbamoyl)oxime	Systemic insecticide, acaricide and nematicide, soil-applied only. For cotton, sugar beets, potatoes, pecans, oranges, ornamentals, soybeans, peanuts.	0.9	>5
Aldrex® (*see* aldrin)				
aldrin, Aldrex® Aldrite® Octalene® (Shell International)	1,2,3,4,10,10-hexachloro-1,4,4a,5,8,8a-hexahydro-1,4-*endo-exo*-5,8-dimethanonapthalene	Used in U.S. exclusively as a termiticide. Gives long residual control of most soil insects.	39	65
Aldrite® (*see* aldrin)				
allethrin, Pynamin® (MGK; Sumitomo)	*cis, trans*-(±)-2,2-dimethyl-3-(2-methylpropenyl)-cyclopropanecarboxylic acid ester with (±)-2-allyl-4-hydroxy-3-methyl-2-cyclopenten-1-one	Aerosols, for controlling flying insects in homes.	680	11,200
d-*trans*-allethrin bioallethrin (MGK; Procida)	*trans*-(+)-2,2-dimethyl-3-(2-methylpropenyl)cyclopropanecarboxylic acid ester with (±)-2-allyl-4-hydroxy-3-methyl-2-cyclopenten-1-one	Sprays, aerosols against flying and crawling household insects.	425	4000
Altosid® (*see* methoprene)				
aluminum phosphide, Phostoxin® (Fumigators, Inc., Phostoxin Sales, Inc.)	AIP	Fumigant (hydrogen phosphide) kills insects in stored feed, grains, seeds, nuts.	2000 ppm gas in air	
Amaze® (*see* isofenphos)				
Ambush® (*see* permethrin)				

Common name, trade name, and basic manufacturer(s)	Chemical name	General use pattern	Oral LD_{50} (rats)	Dermal LD_{50} (rabbits)
Amdro® (American Cyanamid)	tetrahydro-5,5-dimethyl-2(1*H*)-pyrimidinone[3-[4-(trifluoromethyl)phenyl]-[1-[2-(4-trifluoro-methyl)phenyl]ethenyl]-2-propenylidene] hydrazone	Stomach insecticide used for fire ant control.	1131	>5000
Amethopterin® (*see* methotrexate)				
aminocarb, Matacil® (Mobay)	4-(dimethylamino)-3-methylphenol methylcarbamate	Controls forest insects (spruce budworm, jack pine budworm, etc.).	30	275
amitraz, Baam®, Mitac® (BFC Chemicals; TUCO)	*N'*-(2,4-dimethylphenyl)-*N*-[[(2,4-dimethyl-phenyl)imino]methyl]-*N*-methylmethan-imidamide	Controls pear psylla on pears; most mites on fruits, vegetables, ornamentals; lice, ticks on livestock.	600	1600
Ammo® (*see* cypermethrin)				
Animert V-101® (*see* tetrasul)				
Anthio® (*see* formothion)				
antiresistant DDT (discontinued)	*N,N*-di-*n*-butyl-*p*-chlorobenzene-sulfonamide added to DDT in 1:5 ratio.	For control of DDT-resistant flies.		
apholate	2,2,4,4,6,6-hexakis(1-aziridinyl)-2,2,4,4,6,6-hexa-hydro-1,3,5,2,4,6-triazatriphosphorine	Experimental chemosterilants for insects.	98	50
aphoxide (*see* tepa)				
Aramite® (discontinued)	2-(*p-tert*-butylphenoxy)-1-methylethyl-2-chloro-ethyl sulfite	General-use acaricide.	3900	—[a]
Aspon® (Stauffer)	*O,O,O,O*-tetra-*n*-propyl dithiopyrophosphate	For chinch bug control in lawns, turf.	891	—
azinphosethyl, Ethyl Guthion® (Bayer AG)	*O,O*-diethyl *S*-[[4-oxo-1,2,3-benzotriazin-3(4*H*)-yl]methyl] phosphorodithioate	Controls most foliage-feeding insects on several crops.	13	250
azinphosmethyl, Guthion® (Mobay)	*O,O*-dimethyl *S*-[[4-oxo-1,2,3-benzotriazin-3(4*H*)-yl]methyl]phosphorodithioate	Controls most foliage-feeding insects on many crops.	5	220
Azodrin® (*see* monocrotophos)				
Baam® (*see* amitraz)				
Bacillus thuringiensis Biotrol®K, Dipel®, Thuricide® (Abbott; Sandoz)	microbial insecticide for caterpillars	Controls a wide variety of caterpillar pests on most food crops.	nontoxic	
Bacillus thuringiensis, var. *israelensis* Sok-Bt® (TUCO) Teknar® (Sandoz) Bactimos® (Biochem Products)	microbial insecticide	Experimental aquatic larvicide for mosquitoes, black flies.	nontoxic	
Bacillus popilliae Doom®, Japidemic® (Fairfax)	microbial insecticide	Controls grub stage of Japanese beetle in soil.	nontoxic	
Barricade® (*see* cypermethrin)				
Baycarb® (*see* BPMC)				
Baygon® (*see* propoxur)				
Baygon MEB® (*see* plifenate)				

Common name, trade name, and basic manufacturer(s)	Chemical name	General use pattern	Oral LD_{50} (rats)	Dermal LD_{50} (rabbits)
Bayrusil® (*see* quinalphos)				
Baytex® (*see* fenthion)				
Baythion® (*see* phoxim)				
Baythion C® (*see* chlorphoxim)				
Belt® (*see* chlordane)				
bendiocarb, Ficam® Tattoo® (Fisons)	2,2-dimethyl-1,3-benzodioxol-4-yl-methylcarbamate	Controls the usual spectrum of household insects; turf, ornamentals; soil insects.	34	566
benzene hexachloride (several foreign) BHC, HCH	1,2,3,4,5,6-hexachlorocyclohexane (12 to 45% gamma isomer)	All uses canceled by EPA.	125	>4000
BHC (*see* benzene hexachloride)				
Bidrin® (*see* dicrotophos)				
binapacryl, Morocide® (Hoechst AG)	2-*sec*-butyl 4,6-dinitrophenyl 3-methyl-2-butenoate	Miticide with ovicidal action, fungicide for powdery mildews. For mites on tree fruits, nuts, cotton.	136	1010
bioallethrin (*see d-trans*-allethrin)				
bioresmethrin (*see d-trans*-resmethrin)				
Biotrol® K (*see Bacillus thuringiensis*)				
Biotrol VHZ® (*see Heliothis* virus)				
Bladafum® (*see* sulfotepp)				
Bo-Ana® (*see* famphur)				
Bollex® (U.S. Soil, Inc)	methyl α-eleostearate; methyl ester of (E,Z,E) 9,11,13-octadecatrienoic acid	Cotton boll weevil feeding deterrent; biorational control agent for integrated pest management.	5000	—
Bolstar® (*see* sulprofos)				
bomyl (Hopkins)	dimethyl 3-hydroxy glutaconate dimethyl phosphate	Fly baits.	31	20
BPMC, Baycarb® (Bayer AG)	2-*sec*-butylphenyl *N*-methylcarbamate	Effective against certain rice insects and bollworms, aphids on cotton.	340	4200
bromophos (Celamerck)	*O*-(4-bromo-2,5-dichlorophenyl) *O,O*-dimethyl phosphorothioate	Chinch bug control in lawns.	3750	2188
bromophos-ethyl (Celamerck)	*O*-(4-bromo-2,5-dichlorophenyl) *O,O*-diethyl phosphorothioate	Not registered in U.S.	52	1366
bufencarb, Bux® (Chevron)	3-(1-methylbutyl)phenyl methylcarbamate + 3-(1-ethylpropyl)phenyl methylcarbamate (3 : 1)	Larval control of corn rootworm, rice water weevil.	87	400
Butacide® (*see* piperonyl butoxide)				
Bux® (*see* bufencarb)				

Common name, trade name, and basic manufacturer(s)	Chemical name	General use pattern	Oral LD_{50} (rats)	Dermal LD_{50} (rabbits)
Carbamult® (*see* promecarb)				
carbaryl, Sevin® (Union Carbide)	1-naphthyl methylcarbamate	Has probably the greatest range of controlled pests of any insecticide; fruits, vegetables, field crops, ornamentals, pets.	307	2000
carbofuran, Furadan® (FMC)	2.3-dihydro-2,2-dimethyl-7-benzofuranyl methylcarbamate	Insecticide, miticide, nematicide. Wide range of soil and foliar pests on corn, alfalfa, tobacco, peanuts, rice, sugar cane, potatoes.	8	2550
carbophenothion, Trithion® (Stauffer)	*S*-[[(4-chlorophenyl)thio]methyl]-*O,O*-diethyl phosphorodithioate	Used on variety of fruit, nut, vegetable, fiber crops. Also acaricide with long residual.	6	22
carbosulfan, Advantage® (FMC)	2,3-dihydro-2,2-dimethyl-7-benzofuranyl-[(dibutylamino)thio]methyl carbamate	Experimental. Soil and foliar insects on alfalfa, citrus, corn, deciduous fruit, some nematodes.	209	>2000
Carzol® (*see* formetanate hydrochloride)				
chlordane, Belt® (Velsicol)	1,2,4,5,6,7,8,8-octachloro-3a,4,7,7a-tetrahydro-4,7-methanoindan	Used almost entirely for subterranean termite control; Belt® registered for cotton.	283	580
chlordecone (*see* Kepone®)				
chlordimeform, Fundal® Galecron® (Ciba-Geigy; NOR-AM)	*N*′-(4-chloro-o-tolyl)*N,N*-dimethylformamidine	Ovicide-insecticide for bollworm-budworm complex in cotton; ovicide-miticide for resistant mites.	170	225
chlorfenethol, Dimite® Qikron® (Nippon)	4,4′-dichloro-α-methylbenzhydrol	For spider mites on shrub trees, ornamentals.	926	—
chlorfenvinphos, several (several foreign)	2-chloro-1-(2,4-dichlorophenyl)vinyl diethyl phosphate	Soil insects, foliage feeders, livestock pests.	12	3200
chlorobenzilate, Acaraben® (Ciba-Geigy)	ethyl 4,4′-dichlorobenzilate	Mite control on citrus.	700	10,200
chloropicrin, several (several)	trichloronitromethane	Stored grain and soil fumigant. Usually combined with other fumigants as warning agent.		lachrymatory
chlorphoxim Baythion C® (Bayer AG)	(o-chlorophenyl)glyoxylonitrile oxime *O,O*-diethyl phosphorothioate	Controls mosquitoes, black flies.	>2500	>500
chlorpyrifos, Dursban® Lorsban® (Dow)	*O,O*-diethyl *O*-(3,5,6-trichloro-2-pyridyl) phosphorothioate	Household insects, ornamental pests, fire ants, turf insects; soil insects, fruit trees, cotton; termite control.	135	2000
chlorpyrifos-methyl Reidan® (Dow)	*O,O*-dimethyl *O*-(3,5,6-trichloro-2-pyridyl) phosphorothioate	Stored grain pests, flies, mosquitoes, aquatic larvae, several foliar pests.	941	2000
chlorthiophos (Celamerck)	*O*-[2,5-dichloro-4-(methylthio)phenyl]*O,O*-diethyl phosphorothioate and the 2,4,5 and 4,5,2 isomers	Controls a broad spectrum of biting, sucking and mining insects, spider mites, mosquito larvae, ticks.	7.8	50
Cidial® (*see* phenthoate)				
Ciodrin® (*see* crotoxyphos)				
Comite® (*see* propargite)				

Common name, trade name, and basic manufacturer(s)	Chemical name	General use pattern	Oral LD_{50} (rats)	Dermal LD_{50} (rabbits)
Co-ral® (*see* coumaphos)				
coumaphos, Co-Ral® (Bayer AG)	*O,O*-diethyl *O*-(3-chloro-4-methyl-2-oxo-2*H*-1-benzopyran-7-yl)phosphorothioate	Controls livestock pests; cattle grubs, lice, flies, ticks.	56	860
Counter® (*see* terbufos)				
Croneton® (*see* ethiofencarb)				
Cross Fire® (*see* resmethrin)				
crotoxyphos, Ciodrin® (Shell Intl.)	α-methylbenzyl(*E*)-3-hydroxycrotonate dimethyl phosphate	For flies, lice, ticks on lactating dairy and beef cattle, swine, goats, horses, sheep.	125	385
crufomate, Ruelene® (discontinued)	4-*tert*-butyl-2-chlorophenyl methyl methyl-phosphoramidate	Cattle grubs, lice, horn flies on cattle.	660	2000
cryolite, Kryocide® (Pennwalt)	sodium fluoaluminate (or sodium aluminofluoride)	Stomach insecticide usually for caterpillars on grapes, citrus, vegetables.	>10,000	—
cubé (*see* rotenone)				
Curacron® (*see* profenofos)				
cyanofenfos, Sureside® (Sumitomo)	*O*-(4-cyanophenyl)*O*-ethyl phenyl phosphonothioate	For rice borer, cotton bollworm, caterpillars on vegetables.	43	>2000
cyanophos (Sumitomo)	*O*-*p*-cyanophenyl *O,O*-dimethyl phosphorothioate	For caterpillar control on fruits, vegetables; residual household spray for cockroach control.	610	800
Cyflee® (*see* cythioate)				
Cygon® (*see* dimethoate)				
cyhexatin, Plictran® (Dow)	tricyclohexylhydroxystannane	Controls resistant and susceptible plant-feeding spider mites.	180	2000
Cymbush® (*see* cypermethrin)				
Cyolane® (*see* phosfolan)				
cypermethrin, Ammo® (FMC) Cymbush® (ICI) Ripcord®, Barricade® (Shell Intl.)	(±)-a-cyano (-3-phenoxyphenyl)methyl (±) *cis, trans* 3-(2,2-dichloroethenyl) -2,2dimethylcyclopropanecarboxylate.	Experimental. High-level activity against broad spectrum of insect pests.	247	>2000
cythioate, Cyflee® (American Cyanamid)	*O,O*-dimethyl *O*-*p*-sulfamoylphenyl phosphorothioate	Systemic for fleas, other ectoparasites of dogs and cats.	160	>2500
Cythion® (*see* malathion)				
Cytrolane® (*see* mephosfolan)				
Dacamox® (*see* thiofanox)				
Dasanit® (*see* fensulfothion)				
DBCP, dibromochloro-propane, Fumazone®, Nemafume®, Nemagon® (Amvac Chemical)	1,2-dibromo-3-chloropropane	Soil fumigant for nematode control. Now registered only for pineapple grown in Hawaii.	170	1420

Common name, trade name, and basic manufacturer(s)	Chemical name	General use pattern	Oral LD_{50} (rats)	Dermal LD_{50} (rabbits)
D-D® (*see* dichloropropane)				
DDD (*see* TDE)				
DDT (Montrose Chemical) (several foreign)	1,1,1-trichloro-2,2-bis(*p*-chlorophenyl)ethane	Not used in U.S. Some agricultural use in other countries, mostly in malaria eradication programs.	87	1931
DDVP (*see* dichlorvos)				
decamethrin, Decis® NRDC 161 (Roussel Uclaf)	(*S*)-α-cyano-*m*-phenoxybenzyl (1R,3R)-3-(2,2-dibromovinyl)-2,2-dimethylcyclo-propane-carboxylate	Experimental: wide range of effectiveness on field, fruit, vegetable crops.	128	>2000
Decis® (*see* decamethrin)				
deet, Delphene®, Off® (MGK)	*N,N*-diethyl-*m*-toluamide	Repellent for almost all biting arthropods.	2000	—
Delnav® (*see* dioxathion)				
Delphene® (*see* deet)				
Deltic® (*see* dioxathion)				
demeton, Systox® (Mobay)	*O,O*-diethyl-[2-(ethylthio)ethyl]phosphorothioate and *O,O*-diethyl S-[2-(ethylthio)-ethyl]phosphorothioate	Systemic insecticide applied to foliage or as soil drench for sucking insects, mites on wide variety of field, fruit, nut, vegetable crops.	2	8
demeton-methyl (*see* methyl demeton)				
derris (*see* rotenone)				
dialifor, Torak® (BFC Chemicals)	*O,O*-diethyl phosphorodithioate *S*-ester with *N*-(2-chloro-1-mercaptoethyl) phthalimide	Insecticide-acaricide used on grapes, apples, citrus, pecans.	5	145
diazinon, Knox-Out®, Spectracide® (Ciba-Geigy)	*O,O*-diethyl *O*-(2-isopropyl-6-methyl-4-pyrimidinyl) phosphorothioate	Broadly used insecticide against soil insects, pests of fruit, vegetables, field crops, ornamentals, household pets.	300	379
Dibrom® (*see* naled)				
dibromochloropropane (*see* DBCP)				
dibutyl phthalate (Rhone-Poulenc)	dibutyl phthalate	Clothing impregnate for chigger repellent.	8000	—
o-dichlorobenzene (discontinued)	*o*-dichlorobenzene	Insecticide, herbicide, soil fumigant; for bark beetles, grubs, borers.	500	—
p-dichlorobenzene, PDB (PPG)	*p*-dichlorobenzene	Used as moth crystals to protect fabrics; for peach tree borers.	500	2000
dichloropropane-dichloropropene, D-D® (Shell)	dichloropropane-dichloropropene mixture	Soil-injected fumigant to control nematodes, soil insects, certain diseases.	140	2100
dichloropropene, Telone II® (Dow)	1,3-dichloropropene	Preplant fumigant applications to soil for nematode, weed, disease control in many crops.	250	skin burns
dichlorvos, DDVP, Vapona® (Shell)	2,2-dichlorovinyl dimethyl phosphate	Contact, stomach, fumigant insecticide for household, public health, livestock, many crop insect and mite pests.	25	59

Common name, trade name, and basic manufacturer(s)	Chemical name	General use pattern	Oral LD_{50} (rats)	Dermal LD_{50} (rabbits)
dicofol, Kelthane® (Rohm & Haas)	4,4'dichloro-α-(trichloromethyl) benzhydrom	Wide use as acaricide on fruit, vegetable, field, ornamental crops.	575	4000
dicrotophos, Bidrin® (Shell)	dimethyl phosphate ester with (*E*)-3-hydroxy-*N,N*-dimethylcrotonamide	Systemic and contact insecticide, used on cotton, soybeans, ornamentals.	22	225
Dieldrex® (*see* dieldrin)				
dieldrin, Dieldrex® Dieldrite® (Shell Intl.)	1,2,3,4,10,10-hexachloro-6,7-epoxy 1,4,4a,5,6,7,8,8a-octahydro-1,4-endo-exo-5,8-dimethanophthalene	Used only for subterranean termite control in U.S.	40	65
Dieldrite® (*see* dieldrin)				
dienochlor, Pentac® (Hooker)	bis(pentachloro-2,4-cyclopentadien-1-yl)	Miticide effective against two-spotted spider mite on greenhouse ornamentals.	>3160	>3160
diflubenzuron, Dimilin® (Thompson-Hayward)	*N*-[[(4-chlorophenyl)amino]carbonyl]-2,6-difluorobenzamide	Insect growth regulator for cotton boll weevil, gypsy moth.	>4640	—
dimethoate, Cygon® (American Cyanamid)	*O,O*-dimethyl S-[2-(methylamino)-2-oxoethyl] phosphorodithioate	Residual fly spray for farm buildings; insect, mite control on wide variety of vegetable, field, fruit, ornamental crops.	250	150
Dimilin® (*see* diflubenzuron)				
Dimite® (*see* chlorfenethol)				
dinitrocresol, DNOC (Blue Spruce)	4,6-dinitro-*o*-cresol	Dormant spray for fruit trees to control mites, aphids, other pests; effective as ovicide.	20	200
dinitrophenol (several)	2,4-dinitrophenol	Insecticide, acaricide, fungicide as dormant fruit tree spray.	35	700
dinocap, Karathane® (Rohm & Haas)	2-(1-methylheptyl)-4,6-dinitrophenyl crotonate	For powdery mildew and mite control on fruits, vegetables, ornamentals.	980	4700
dinoseb, DNBP (several)	2-*sec*-butyl-4,6-dinitrophenol	Herbicide, desiccant, dormant fruit spray for controlling eggs, larvae, adults of various mites and insects.	37	80
dioxathion, Delnav® Deltic® (BFC Chemicals)	2,3-*p*-dioxanedithiol *S,S*-bis(*O,O*-diethyl phosphorodithioate)	Controls insects, mites on fruit and nut trees, ornamentals; also for many livestock pests.	19	53
Dipel® (*see Bacillus thuringiensis*)				
Dipterex® (*see* trichlorfon)				
disulfoton, Di-Syston® (Mobay)	*O,O*-diethyl *S*-[2-(ethylthio)ethyl] phosphorodithioate	Systemic insecticide, acaricide used on cotton, potatoes, other crops, usually applied as seed treatment or in furrow at planting.	2	6
Di-Syston® (*see* disulfoton)				
DNBP (*see* dinoseb)				
DNOC (*see* dinitrocresol)				
Doom® (*see Bacillus popilliae*)				

Common name, trade name, and basic manufacturer(s)	Chemical name	General use pattern	Oral LD_{50} (rats)	Dermal LD_{50} (rabbits)
Dri-Die® (Davison Chemical)	silica aerogel	Household insect control by desiccation.		nontoxic
Dursban® (*see* chlorpyrifos)				
Du-Ter® (*see* fentin hydroxide)				
Dyfonate® (*see* fonofos)				
Dylox® (*see* trichlorfon)				
Ectiban® (*see* permethrin)				
EDB (*see* ethylene dibromide)				
EDC (*see* ethylene dichloride)				
Elcar® (*see Heliothis* polyhedrosis virus)				
endosulfan, Thiodan® (several)	6,7,8,9,10,10-hexachloro-1,5,5a,6,9,9a-hexahydro-6,9-methano-2,4,3-benzodioxathiepin 3-oxide	Effective against many insect and mite pests on vegetable, fruit, forage, fiber crops, ornamentals.	18	74
endrin (Shell Intl.; Velsicol)	1,2,3,4,10,10-hexachloro-6,7-epoxy-1,4,4a,5,6,7,-8,8a-octahydro-1,4-*endo-endo*-5,8-dimethanonaphthalene	For cotton insect control and a few other crops; mouse control in orchards.	3	12
Entex® (*see* fenthion)				
EPN (duPont; Velsicol)	*O*-ethyl *O*-(4-nitrophenyl) phenylphosphonothioate	Used almost exclusively for cotton insect control.	14	110
epoxyethane (*see* ethylene oxide)				
ethiofencarb, Croneton® (Bayer AG)	2-ethyl-mercaptomethyl-phenyl-*N*-methylcarbamate	For aphid control on fruit, vegetables, ornamentals.	411	1000
ethion, Nialate® (FMC)	*O,O,O′,O′*-tetraethyl *S,S′*-methylene bis(phosphorodithioate)	Insecticide-acaricide for sucking and chewing pests on many crops.	27	915
ethoprop, Mocap® Moprop® (Mobil)	*O*-ethyl *S,S*-dipropyl phosphorodithioate	Soil insecticide-nematicide for root crops, corn, vegetables, commercial turf.	62	26
ethyl formate	ethyl formate	Fumigant for food products.	4000	—
Ethyl Guthion® (*see* azinphosethyl)				
ethyl parathion (*see* parathion)				
ethylan, Perthane® (discontinued)	1,1-dichloro-2,2-bis(4-ethylphenyl)ethane	Control of cabbage looper on vegetables, pear psylla; impregnate for moth and carpet beetle control in dry cleaning, textile industry.	6600	—
ethylene dibromide, EDB (Dow)	1,2-dibromoethane	Insecticide-nematicide fumigant for soil application; warehouse, mill, household fumigation.	146	200 ppm vapor
ethylene dichloride, EDC (Dow; PPG)	1,2-dichloroethane	Fumigant for stored products.	670	1000 ppm vapor

Common name, trade name, and basic manufacturer(s)	Chemical name	General use pattern	Oral LD_{50} (rats)	Dermal LD_{50} (rabbits)
ethylene oxide (several)	1,2-epoxyethane	Fumigant, sterilant for certain stored food products, soil.	irritant	50 ppm TLV
famphur, Bo-Ana®, Warbex® (American Cyanamid)	*O*-[*p*-(dimethylsulfamoyl)phenyl] *O,O*-dimethyl phosphorothioate	For louse and cattle grub control on cattle, reindeer.	35	1460
fenamiphos, Nemacure® (Mobay)	ethyl 3-methyl-4-(methylthio)phenyl (1-methylethyl)phosphoramidate	Systemic nematicide for pre- or postemergence application to soil.	8	72
fenbutatin-oxide, Vendex® (Shell)	hexakis (2-methyl-2-phenylpropyl)-distannoxane	Miticide for mites on deciduous fruits, citrus, greenhouse crops, ornamentals.	2631	>2000
fenchlorphos (*see* ronnel)				
fenitrothion, Sumithion® Folithion® (Bayer AG; Sumitomo)	*O,O*-dimethyl *O*-(4-nitro-*m*-tolyl) phosphorothioate	For chewing and sucking insects on wide variety of crops; public health insect control.	250	1300
fenson (Montedison)	*p*-chlorophenyl benzenesulfonate	Acaricide-ovicide for European red mite, other mites on tree fruits.	1560	2000
fensulfothion, Dasanit® (Mobay)	*O,O*-diethyl *O*-[*p*-methylsulfinyl)phenyl] phosphorothioate	Soil-applied nematicide-insecticide; limited systemic effects on foliage feeders.	2	3
fenthion, Baytex®, Entex®, Tiguvon® (Mobay)	*O,O*-dimethyl *O*-[4-(methylthio)-*m*-tolyl] phosphorothioate	Mosquitoes, flies, ornamentals; livestock insect pests.	255	330
fentin hydroxide, Du-Ter® (Thompson-Hayward)	triphenyltin hydroxide	A fungicide; has antifeeding effect on certain insects.	108	—
fenvalerate, Pydrin® (Shell)	cyano(3-phenoxyphenyl)methyl 4-chloro-α-(1-methylethyl)benzeneacetate	Broad spectrum of insecticidal activity on many crops. Currently registered only on cotton.	451	>5000
ferriamicide (Mississippi Agri. & Commerce Dept.)	mirex plus alkyl amines and ferrous chloride	Short-residual form of mirex for fire ant control.		
Ficam® (*see* bendiocarb)				
Flit MLO® (Exxon)	low odor, low viscosity petroleum oil	Mosquito larvicide.	nontoxic	
flucythrinate, Pay-Off® (American Cyanamid)	(*RS*)-*a*-cyano-3-phenoxybenzyl (*S*)-2-(4-difluromethoxyphenyl)-3-methylbutyrate	High level of activity against a broad spectrum of insect, mite, tick pests.	67	>1000
fluvalinate, Mavrik® (Zoecon)	*N*-[2-chloro-4-(trifluoromethyl)phenyl]-*DL*-valine (±) cyano (3-phenoxyphenyl) methyl ester	Broad-spectrum experimental pyrethroid insecticide for cotton, vegetables.	>6299	>20,000
Folimat® (*see* omethoate)				
Folithion® (*see* fenitrothion)				
fonofos, Dyfonate® (Stauffer)	*O*-ethyl *S*-phenyl ethyl phosphonodithioate	Soil insecticide for several species of soil pests.	8	25
formetanate hydrochloride, Carzol® (NOR-AM)	[3-dimethylamino-(methyleneiminophenyl)]-*N*-methylcarbamate hydrochloride	Insecticide-acaricide for tree fruits, citrus, alfalfa.	15	5600
formothion, Anthio® (Sandoz Ltd.)	*O,O*-dimethyl phosphorodithioate *S*-ester with *N*-formyl-2-mercapto-*N*-methylacetamide	Systemic and contact insecticide-acaricide for sucking insects, nonresistant mites.	365	>1000

Common name, trade name, and basic manufacturer(s)	Chemical name	General use pattern	Oral LD_{50} (rats)	Dermal LD_{50} (rabbits)
fosthietan, Nem-A-Tak® (American Cyanamid)	diethyl 1,3-dithietan-2-ylidenephosphoramidate	Experimental soil insecticide-nematicide for field crops.	4	27
Fumazone® (*see* DBCP)				
Fundal® (*see* chlordimeform)				
Furadan® (*see* carbofuran)				
Galecron® (*see* chlordimeform)				
gamma BHC (*see* lindane)				
Gardona® (*see* stirofos)				
Genite® (discontinued)	2,4-dichlorophenyl benzenesulfonate	Acaricide for deciduous fruit trees, some vegetables.	980	940
glyodin, Glyoxide® (Agway)	2-heptadecyl-2-imidazoline acetate	Acaricide-fungicide for deciduous fruit, ornamentals.	4600	—
Glyoxide® (*see* glyodin)				
Guthion® (*see* azinphosmethyl)				
HAG-107 (*see* tralomethrin)				
HCH (*see* benzene hexachloride)				
HCN (*see* hydrocyanic acid)				
Heliothis nuclear polyhedrosis virus Biotrol VHZ®, Elcar® (Sandoz)	viral insecticide specific for *Heliothis*	For control of bollworm, tobacco budworm on cotton.	nontoxic	
hellebore	poisonous alkaloids from white hellebore (*Veratrum album*)	No longer available; home garden and orchard insecticide, botanical in origin.		
hemel	hexamethylmelamine	Experimental chemosterilant for insects.	350	—
hempa	hexamethyl phosphoric triamide	Experimental chemosterilant for insects.	<2650	1500
heptachlor (Velsicol)	1,4,5,6,7,8,8-heptachloro-3a,4,7,7a-tetrahydro-4,7-methanoindene	For residual control of subterranean termites; a few agricultural uses.	40	119
Hirsutella thompsonii, Mycar® (Abbott)	naturally occurring disease fungus of mites	Controls certain mite species, especially citrus rust mite.	nontoxic	
Hostathion® (*see* triazophos)				
hydrocyanic acid, hydrogen cyanide, prussic acid, HCN	HCN	Highly toxic fumigant for buildings, premises, some stored products. (Extreme caution!)	<0.5	—
Imidan® (*see* phosmet)				
Indalone® (discontinued)	butyl 3,4-dihydro-2,2-dimethyl-4-oxo-2H-pyran-6-carboxylate	Insect repellent for skin, clothing.	7840	10,000
isobenzan, Telodrin® (discontinued)	1,3,4,5,6,7,8,8-octachloro-1,3,3a,4,7,7a-hexahydro-4,7-methanoisobenzofuran	Not available in U.S. Used on alfalfa, corn, tobacco.	8	5

Common name, trade name, and basic manufacturer(s)	Chemical name	General use pattern	Oral LD_{50} (rats)	Dermal LD_{50} (rabbits)
isodrin (discontinued)	1,2,3,4,10,10-hexachloro-1,4,4a,5,8,8a-hexahydro-1,4-*endo-endo*-5,8-dimethanonaphthalene	Isomer of aldrin; seldom used in the U.S.	24	85
isofenphos, Amaze® (Mobay)	1-methylethyl 2-[[ethoxy[(1-methylethyl)amino]-phosphinothioyl]oxy] benzoate	Soil insecticide for field crops, vegetables, turf.	32	162
isopropyl formate	isopropyl formate	Fumigant for packaged dry fruits, nuts.		
Japidemic® (*see Bacillus popilliae*)				
Karathane® (*see* dinocap)				
Kelthane® (*see* dicofol)				
Kepone®, chlordecone (discontinued)	decachlorooctahydro-1,3,4-methano-2*H*-cyclobuta(cd)pentalen-2-one	Originally used in area-wide control programs for fire ants.	95	345
Knox-Out® (*see* diazinon)				
Korlan® (*see* ronnel)				
Kryocide® (*see* cryolite)				
Lannate® (*see* methomyl)				
Larvin® (*see* thiodicarb)				
leptophos, Phosvel® (discontinued)	*O*-(4-bromo-2,5-dichlorophenyl)*O*-methyl phenylphosphonothioate	For export only; fruits, vegetables, rice, cotton.	43	10,000
lindane, gamma BHC (Hooker & several foreign)	1,2,3,4,5,6-hexachlorocyclohexane, gamma isomer of not less than 99%	Many uses including seed treatment; moderate fumigant action.	76	500
Lorsban® (*see* chlorpyrifos)				
malathion, Cythion® (American Cyanamid)	*O,O*-dimethyl phosphorodithioate of diethyl mercaptosuccinate	Controls very broad spectrum of insects in agriculture, public health, livestock, home, garden.	885	4,000
MAPS (*see* methiotepa)				
Marlate® (*see* methoxychlor)				
Matacil® (*see* aminocarb)				
Mavrik® (*see* fluvalinate)				
mecarbam (Crystal; Murphy)	*S*-[[(ethoxycarbonyl)methylcarbamoyl]-methyl] *O,O*-diethyl phosphorodithioate	Insecticide, acaricide-ovicide, for sucking insects, mites on fruit trees, citrus, certain vegetables.	36	>1220
mephosfolan, Cytrolane® (American Cyanamid)	diethyl (4-methyl-1,3-dithiolan-2-ylidene)-phosphoramidate	Systemic insecticide for leaf-eating larvae, stem borers.	9	100
mercaptodimethur (*see* methiocarb)				
Mesurol® (*see* methiocarb)				

Common name, trade name, and basic manufacturer(s)	Chemical name	General use pattern	Oral LD_{50} (rats)	Dermal LD_{50} (rabbits)
metam-sodium, Vapam® (Stauffer)	sodium *N*-methyldithiocarbamate	General-purpose soil fumigant; nematicide, fungicide, herbicide.	820	800
Metasystox® (*see* methyl demeton)				
Metasystox-R® (*see* oxydemetonmethyl)				
metepa, methaphoxide	tris(2-methyl-1-aziridinyl)phosphine oxide	Experimental insect chemosterilant.	93	156
methamidophos, Monitor® (Chevron; Mobay)	*O,S*-dimethyl phosphoramidothioate	Controls most common insect pests on vegetables, cotton, potatoes.	13	110
metaphoxide (*see* metepa)				
methidathion, Supracide® (Ciba-Geigy)	*O,O*-dimethyl phosphorodithioate *S*-ester with 4-(mercaptomethyl)-2-methoxy-Δ^2-1,3,4-thiadiazolin-5-one	Controls wide spectrum of insects in alfalfa, cotton, tree fruits, nuts.	25	375
methiocarb, mercapto-dimethur, Mesurol®, (Mobay)	4-(methylthio)-3,5-xylyl methylcarbamate	Contact insecticide with bird-repellent qualities; used on cherries, blueberries, other bird-problem crops.	15	2000
methiotepa, MAPS	tris(2-methyl-1-aziridinyl)phosphine sulfide	Experimental insect chemosterilant.	—	—
methomyl, Lannate® Nudrin® (duPont; Shell)	*S*-methyl *N*-[(methylcarbamoyl)oxy]-thioacetimidate	Controls wide range of insects in fruit crops, vegetables, field crops, ornamentals.	17	1000
methoprene, Altosid® (Zoecon)	isopropyl (*E,E*)-11-methoxy-3,7,11-trimethyl-2,4-dodecadienoate	Insect growth regulator, used as mosquito larvicide.	>34,600	>3000
methoxychlor, Marlate® (several)	1,1,1-trichloro-2,2-bis(*p*-methoxyphenyl)ethane	Many uses on fruit and shade trees, vegetables, home gardens, livestock.	5000	2820
methotrexate Amethopterin®	*N*-[*p*-[[(2,4-diamino-6-pteridinyl)methyl]methyl-amino]benzoyl]glutamic acid	Experimental insect chemosterilant.	—	—
methyl apholate	2,2,4,4,6,6-hexahydro-2,2,4,4,6,6-hexakis (2-methyl-1-aziridinyl)-1,3,5,2,4,6-triazatriphosphorine	Experimental insect chemosterilant.	—	—
methyl bromide (several)	bromomethane	Fumigant action controls all living matter; used as space, soil, stored products, drywood termite fumigant. (Extreme caution!)	200 ppm vapor	15 ppm TLV
methyl demeton, Metasystox® (Bayer AG; Shell Intl.)	mixture of *O,O*-dimethyl *S*(and *O*)-(2-(ethylthio)-ethyl)phosphorothioates	Controls sucking insects, mites on vegetables, fruit, hops.	64	302
methyl formate	methyl formate	Fumigant for food products.	4000	—
methyl parathion, Penncap-M® (several)	*O,O*-dimethyl *O*-(*p*-nitrophenyl) phosphorothioate	Controls most cotton pests including nonresistant bollworms, tobacco budworms.	9	63
methyl trithion (discontinued)	*S*-[[(*p*-chlorophenyl)thio]methyl] *O,O*-dimethyl phosphorodithioate	Nonsystemic acaricide, also effective against cotton boll weevil.	157	2420
mevinphos, Phosdrin® (Amvac; Shell)	methyl (*E*)-3-hydroxycrotonate dimethyl phosphate	Contact and systemic insecticide-acaricide with broad range of use on vegetable, fruit, field, forage crops.	3	16
mexacarbate, Zectran® (discontinued)	4-(dimethylamino)-3,5-xylyl methylcarbamate	Broad range of control for nonfood crops; ornamentals, turf, shrubs; some forest use.	15	500

Common name, trade name, and basic manufacturer(s)	Chemical name	General use pattern	Oral LD_{50} (rats)	Dermal LD_{50} (rabbits)
Mirex® (discontinued)	dodecachlorooctahydro-1,3,4-metheno-1*H*-cyclobuta [*cd*]pentalene	Was used to control all ant species.	235	800
Mitac® (*see* amitraz)				
Mocap® (*see* ethoprop)				
Monitor® (*see* methamidophos)				
monocrotophos, Azodrin® (Shell)	dimethyl phosphate ester of (*E*)-3-hydroxy-*N*-methylcrotonamide	Controls most insect pests of cotton, tobacco, sugarcane, potatoes, peanuts, ornamentals.	8	354
Moprop® (*see* ethoprop)				
Morestan® (*see* oxythioquinox)				
Morocide® (*see* binapacryl)				
Multicide® (*see* *d*-phenothrin)				
Mycar® (*see Hirsutella thompsonii*)				
naled, Dibrom® (Chevron)	1,2-dibromo-2,2-dichloroethyl dimethyl phosphate	Effective against chewing and sucking pests on many crops; fly control in barns, stables, poultry houses, kennels; mosquito control projects.	430	1100
naphthalene	naphthalene	Fumigant for nonfood uses.		
Nemacure® (*see* fenamiphos)				
Nemafume® (*see* DBCP)				
Nemagon® (*see* DBCP)				
Nem-A-Tak® (*see* fosthietan)				
Neo-Pynamin® (*see* tetramethrin)				
Nialate® (*see* ethion)				
nicotine	3-(1-methyl-2-pyrrolidyl)pyridine	Contact, fumigant; used as greenhouse fumigant.	50	50
nicotine sulfate	3-(1-methyl-2-pyrrolidyl)pyridine sulfate	Marketed as aqueous solution with 40% nicotine equivalent; used mostly against aphids, scales on ornamentals.	60	140
Niran® (*see* parathion)				
NOLOC® (*see Nosema locustae*)				
Nosema locustae, NOLOC® (Sandoz)	naturally occurring protozoan spores	Grasshopper control on rangeland.	nontoxic	
NRDC 161 (*see* decamethrin)				
Nudrin® (*see* methomyl)				
Octalene® (*see* aldrin)				

Common name, trade name, and basic manufacturer(s)	Chemical name	General use pattern	Oral LD_{50} (rats)	Dermal LD_{50} (rabbits)
Off® (*see* deet)				
omethoate, Folimat® (Bayer AG)	*O,O*-dimethyl *S*-[2-(methylamino)-2-oxoethyl] phosphorothioate	For sucking insects, mites on fruits, vegetables, hops, ornamentals.	50	1400
Omite® (*see* propargite)				
OMPA (*see* schradan)				
Orthene® (*see* acephate)				
ovex, Ovotran® (Nippon)	*p*-chlorophenyl *p*-chlorobenzene sulfonate	Acaricide-ovicide for cotton, fruits, nuts, ornamentals.	2000	—
Ovotran® (*see* ovex)				
oxamyl, Vydate® (du Pont)	methyl *N′,N′*-dimethyl-*N*-[(methylcarbamoyl)-oxy]-1-thiooxamimidate	Insecticide, nematicide, acaricide; controls some of each on field crops, vegetables, fruits, ornamentals.	5	710
oxydemetonmethyl, Metasystox-R® (Mobay)	*S*-[2-(ethylsulfinyl)ethyl]*O,O*-dimethyl phosphorothioate	Systemic insecticide-acaricide for vegetables, fruits, field crops, flowers, shrubs, trees.	65	100
oxythioquinox Morestan® (Mobay)	6-methyl-1,3-dithiolo(4,5-b)quinoxalin-2-one	Insecticide, acaricide, fungicide; controls mites, mite eggs, powdery mildew, pear psylla; used on fruits, vegetables, citrus, ornamentals.	2500	>2000
paradichlorobenzene (*see p*-dichloro-benzene)				
parathion, ethyl parathion, Niran®, Penncap-E® (Bayer AG; Monsanto)	*O,O-diethyl O*-(4-nitrophenyl) phosphorothioate	Broad-spectrum insecticide used on wide variety of crops.	3	6.8
Pay-Off® (*see* flucythrinate)				
PDB (*see p*-dichlorobenzene)				
Penncap-E® (*see* parathion)				
Penncap-M® (*see* methyl parathion)		Microencapsulated, slow-release formulation	>270	>5400
Pentac® (*see* dienochlor)				
permethrin, Ambush®, Ectiban®, Pounce®, Pramex® (FMC; ICI Americas; Penick)	3-(phenoxyphenyl)methyl (±)-cis,trans-3-(2,2-dichloroethenyl)-2,2-dimethyl cyclopropane-carboxylate (60% trans, 40% cis isomers)	Ambush® and Pounce® for cotton insects; Ectiban® for animal health; Pramex® for greenhouse ornamentals.	>4000	>4000
Perthane® (*see* ethylan)				
d-phenothrin, Multicide®, Sumithrin® (MGK; Sumitomo)	(3-phenoxyphenyl)methyl 2,2-dimethyl-3-(2-methyl-1-propenyl)cyclopropanecarboxylate	Space sprays for homes, industry, aircraft.	>10,000	>5000
penthoate, Cidial® (Montedison; Nissan)	ethyl α-[(dimethoxyphosphinothioyl)thio] benzeneacetate	Insecticide-acaricide; used as mosquito adulticide; controls citrus scales.	200	4000
phorate, Thimet® (American Cyanamid)	*O,O*-diethyl *S*-[(ethylthio)methyl] phosphorodithioate	Systemic insecticide used on cotton, corn, some small grains, vegetables.	1.6	2.5

Common name, trade name, and basic manufacturer(s)	Chemical name	General use pattern	Oral LD_{50} (rats)	Dermal LD_{50} (rabbits)
phosalone, Zolone® (Rhone-Poulenc)	*O,O*-diethyl *S*-[(6-chloro-2-oxobenzoxazolin-3-yl)methyl]phosphorodithioate	Insecticide-acaricide; used on tree fruit, nuts, citrus, potatoes, ornamentals.	125	1500
phosfolan, Cyolane® (American Cyanamid)	diethyl 1,3-dithiolan-2-ylidenephosphoramidate	Systemic control of leaf-feeding caterpillars, whiteflies, aphids, thrips, mites on cotton.	8.9	23
Phosdrin® (*see* mevinphos)				
phosmet, Imidan® Prolate® (Stauffer)	*N*-(mercaptomethyl)phthalimide *S*-(*O,O*-dimethylphosphorodithioate)	Used for a wide variety of pests on alfalfa, tree fruits, nuts, citrus, corn, cotton.	147	3160
phosphamidon (Chevron)	2-chloro-3-(diethylamino)-1-methyl-3-oxo-1-propenyl dimethyl phosphate	Systemic insecticide used for sucking insects in small grains, cotton, other field crops.	15	125
Phostoxin® (*see* aluminum phosphide)				
Phosvel® (*see* leptophos)				
phoxim, Baythion® (Bayer AG)	phenylglyoxylonitrile oxime *O,O*-diethyl phosphorothioate	For stored products, insects in granaries, mills, ships; public health insect control.	1845	1126
phthalthrin (*see* tetramethrin)				
piperonyl butoxide, Butacide® (several)	α-[2-(2-butoxy)ethoxy]-4,5-(methylene-dioxy)-2-propyltoluene	Synergist for pyrethrins, some of the synthetic pyrethroids.	>7500	>7500
piprotal	piperonal, bis[2-(2-butoxyethoxy)ethyl]acetal	Synergist for pyrethrins, some carbamate insecticides.	4400	—
pirimicarb, Pirimor® (discontinued)	2-(dimethylamino)-5,6-dimethyl-4-pyrimidinyl dimethylcarbamate	Useful against OP-resistant aphids on potatoes, ornamentals.	147	>500
pirimiphos-ethyl, Primicid® (ICI Americas)	*O*,[2-(diethylamino)-6-methyl-4-pyrimidinyl] *O,O*-diethyl phosphorothioate	Registered in Florida for sod webworms, chinch bugs.	170	1000
pirimiphos-methyl, Actellic® (ICI Americas)	*O*-[2-(diethylamino)-6-methyl-4-pyrimidinyl] *O,O*-dimethyl phosphorothioate	Wide range of pests of stored products; fruit, vegetable, field crops.	2050	>2000
Pirimor® (*see* pirimicarb)				
Plictran® (*see* cyhexatin)				
plifenate, Baygon MEB® (Bayer AG)	2,2,2-trichloro-1-(3,4-dichlorophenyl)-ethyl acetate	Household insecticide for flies, mosquitoes, clothes moths, carpet beetles.	>10,000	>1000
Pounce® (*see* permethrin)				
Pramex® (*see* permethrin)				
Primicid® (*see* pirimiphos-ethyl)				
profenofos, Curacron® (Ciba-Geigy)	*O*-(4-bromo-2-chlorophenyl)-*O*-ethyl-S-propyl phosphorothioate	Controls most caterpillar pests on cotton, soybeans.	400	472
Prolate® (*see* phosmet)				

Common name, trade name, and basic manufacturer(s)	Chemical name	General use pattern	Oral LD_{50} (rats)	Dermal LD_{50} (rabbits)
promecarb, Carbamult® (Schering AG)	3-methyl-5-(1-methylethyl)phenyl methylcarbamate	For Colorado potato beetle, caterpillars, leaf miners of fruit; corn rootworm.	61	>1000
propargite, Comite® Omite® (Uniroyal)	2-[4-(1,1-dimethylethyl)phenoxy]cyclohexyl 2-propynyl sulfite	Acaricide used on fruit, nuts, corn, cotton, citrus, some vegetables, ornamentals.	2200	5000
propetamphos, Safrotin® (Sandoz Ltd.)	(*E*)-1-methylethyl 3-[[(ethylamino)methoxy-phosphinothioyl]oxy]-2-butenoate	Effective against cockroaches, mosquitoes, fleas.	82	2300
propoxur, Baygon® (Mobay)	*o*-isopropoxyphenyl methylcarbamate	Used against household crawling insects, flies, mosquitoes, lawn insects.	95	>1000
propylene dichloride (discontinued)	1,2-dichloropropane	Fumigant mixed with other materials for stored grain; controls some soil insects.	2000	—
propylene oxide (Jefferson Chemical)	1,2-epoxypropane	Fumigant used to sterilize packaged products.	irritant	100 ppm TLV
propyl isome (discontinued)	dipropyl 5,6,7,8-tetrahydro-7-methylnaptho-(2,3-*d*)-1,3-dioxole-5,6-dicarboxylate	Synergist for pyrethrins.	1500	—
Proxol® (*see* trichlorfon)				
Pydrin® (*see* fenvalerate)				
Pynamin® (*see* allethrin)				
pyrethrins, pyrethrum (several)	mixture of pyrethrins and cinerins	Fast knockdown natural insecticide; safe for household aerosols, pets, livestock, stored products.	1500	1800
pyrethrum (*see* pyrethrins)				
Qikron® (*see* chlorfenethol)				
quinalphos, Bayrusil® (Bayer AG; Sandoz Ltd.)	*O,O*-diethyl *O*-2-quinoxalinyl phosphorothioate	Broad range of insect pests on cotton, peanuts, vegetables, fruit trees.	65	340
Rabon® (*see* stirofos)				
Reldan® (*see* chlorpyrifos-methyl)				
resmethrin, Cross Fire® Synthrin®, SBP-1382® (Fairfield American; Penick)	(5-phenylmethyl-3-furanyl)methyl 2,2-dimethyl-3-(2-methyl-1-propenyl)-cyclopropanecarboxylate	Uses similar to pyrethrins but longer lasting, more effective.	2000	2500
d-trans-resmethrin, bioresmethrin	(5-phenylmethyl-3-furanyl)methyl (1*R-trans*)-2,2-dimethyl-3-(2-methyl-1-propenyl) cyclopropanecarboxylate	Uses similar to pyrethrins (see above) but longer lasting, more effective.	7070	>10,000
Ripcord® (*see* cypermethrin)				
ronnel, fenchlorphos, Korlan®, Trolene® (discontinued)	*O,O*-dimethyl *O*-(2,4,5-trichlorophenyl) phosphorothioate	Systemic animal and residual premise insecticide; Trolene® administered orally for variety of internal, external parasites.	906	1000
rotenone, cubé, derris (several)	1,2,12,12a-tetrahydro-2-isopropenyl-8,9-dimethoxy[1]benzopyrano[2,4-*b*] furo[2,3-*b*][1]benzopyran-6(6a*H*)-one	Used in home gardens, on pets; fish control in ponds, lakes.	60	>1000

Common name, trade name, and basic manufacturer(s)	Chemical name	General use pattern	Oral LD_{50} (rats)	Dermal LD_{50} (rabbits)
Ruelene® (*see* crufomate)				
Rutgers 612 (discontinued)	2-ethyl-1,3-hexanediol	Early insect repellent for skin, clothing.	6500	—
ryania (discontinued)	ground stemwood of *Ryania speciosa*	Stomach and contact insecticide used in apple production; selective for worm pests, leaving beneficials.	750	4000
sabadilla (Prentiss Drug)	ground seeds of *Schoenocaulon officinale*	Historic botanical insecticide useful in home gardens.	4000	—
Safrotin® (*see* propetamphos)				
SBP-1382 (*see* resmethrin)				
schradan, OMPA (Wacker-Chemie GmbH)	octamethylpyrophosphoramide	Systemic insecticide-acaricide; used as soil drench for transplanted shrubs, trees.	9	15
sesamex, Sesoxane® (Shulton Inc.)	2-(2-ethoxyethoxy)ethyl 3,4-(methylenedioxy) phenylacetal of acetaldehyde	Synergist for pyrethrins, allethrin.	2000	11,000
sesamin	2,6-bis[3,4-(methylenedioxy)phenyl]-3,7-dioxabicyclo(3.3.0)octane	Synergist for pyrethrins.	2000	11,000
sesamolin	6-[3,4-(methylenedioxy)phenoxy]-2-[3,4-methylenedioxy)-phenyl]-3,7-dioxabicyclo-[3.3.0]octane	Synergist for pyrethrins.	2000	10,000
Sesoxane® (*see* sesamex)				
Sevin® (*see* carbaryl)				
Sok-Bt® (see *Bacillus thuringiensis* var. *israelensis*)				
Spectracide® (*see* diazinon)				
stirofos, tetrachlorvinphos, Gardona®, Rabon® (Shell)	2-chloro-1-(2,4,5-trichlorophenyl)-vinyl dimethylphosphate	Wide range of insects on many crops, stored products, livestock, residual premise, sprays.	4000	>2500
Strobane® (discontinued)	polychlorinates of camphene, pinene, and related terpenes	Used almost exclusively on cotton.	220	
Strobane-T® (*see* toxaphene)				
sulfotepp, Bladafum® (Bayer AG)	*O,O,O,O*-tetraethyl dithiopyrophosphate	Greenhouse fumigant for aphids, mites, thrips, whiteflies.	5	—
sulfoxide (discontinued)	1,2-(methylenedioxy)-4-[2-(octylsulfinyl) propyl]benzene	Synergist for pyrethrins, allethrin.	2000	9000
sulfuryl fluoride, Vikane® (Dow)	sulfuryl fluoride	Fumigant used to control structural and household pests, especially drywood termites.		5 ppm TLV
sulprofos, Bolstar® (Mobay)	*O*-ethyl *O*-[4-(methylthio)phenyl] *S*-propyl phosphorodithioate	Controls several pests on cotton, corn, tobacco, tomatoes, peanuts, alfalfa.	107	820
Sumithion® (*see* fenitrothion)				

Common name, trade name, and manufacturer	Chemical name	General use pattern	Oral LD_{50} (rats)	Dermal LD_{50} (rabbits)
Sumithrin® (*see* *d*-phenothrin)				
Supracide® (*see* methidathion)				
Sureside® (*see* cyanofenfos)				
Synthrin® (*see* resmethrin)				
Systox® (*see* demeton)				
Tabatrex® (Glen Chemical)	di-*n*-butylsuccinate	Repels biting flies on cattle, cockroaches, ants in dwellings.	8000	—
Tattoo® (*see* bendiocarb)				
TDE, DDD (discontinued)	1,1-dichloro-2,2-bis(*p*-chlorophenyl)ethane	Used in the past on many fruits, vegetables.	3400	4000
Tedion® (*see* tetradifon)				
Teknar® (*see Bacillus thuringiensis,* var. israelensis)				
Telodrin® (*see* isobenzan)				
Telone II® (*see* dichloropropene)				
temephos, Abate® (American Cyanamid)	*O,O*′-(thiodi-4,1-phenylene)bis(*O,O*-dimethyl phosphorothioate)	Aquatic larvicide for mosquitoes, black flies, midges.	2030	970
Temik® (*see* aldicarb)				
tepa, aphoxide	tris(1-aziridinyl)phosphine oxide	Experimental insect chemosterilant.	37	—
tepp (discontinued)	tetraethyl pyrophosphate	Formerly used for control of non-caterpillar pests on vegetables, cotton.	0.2	2
terbufos, Counter® (American Cyanamid)	*S*-[[(1,1-dimethylethyl)thio]methyl] *O,O*-diethyl phosphorodithioate	Soil insecticide, nematicide for corn rootworm, nematodes in corn, sugar beet maggots.	1.6	1.0
tetrachlorvinphos (*see* stirofos)				
tetradifon, Tedion® (Philips-Duphar)	4-chlorophenyl 2,4,5-trichlorophenyl sulfone	Acaricide used on many fruits, nuts, citrus, vegetables, cotton, ornamentals.	>14,700	10,000
tetramethrin, phthalthrin Neo-Pynamin® (Fairfield American; Sumitoma)	1-cyclohexane-1,2-dicarboximidomethyl 2,2-dimethyl-3-(2-methylpropenyl) cyclopropanecarboxylate	Fast knockdown insecticide for aerosols, sprays; controls flying insects, garden pests, livestock insects, stored products pests.	>4640	>15,000
tetrasul, Animert V-101® (Philips-Duphar)	4-chlorophenyl 2,4,5-trichlorophenyl sulfide	Controls spider mites, which hibernate in the egg stage on plants.	>10,800	2000
Thanite® (discontinued)	isobornyl thiocyanoacetate	Older fast-knockdown insecticide for household, pet sprays, aerosols; human louse control.	1000	6000
Thimet® (*see* phorate)				

Common name, trade name, and manufacturer	Chemical name	General use pattern	Oral LD_{50} (rats)	Dermal LD_{50} (rabbits)
Thiodan® (*see* endosulfan)				
thiodicarb, Larvin® (Union Carbide)	dimethyl-*N,N*-[thiobis[(methylimino)-carbonyloxy]]bis(ethanimidothioate)	Experimental insecticide for control of caterpillars of various crops.		
thiofanox, Dacamox® (Diamond Shamrock)	3,3-dimethyl-1-(methylthio)-2-butanone *O*-[(methylamino)carbonyl]oxime	Systemic soil insecticide for sucking insects on cotton, soybeans, sugar beets, peanuts, potatoes, some cereals.	9	39
Thuricide® (*see Bacillus thuringiensis*)				
Tiguvon® (*see* fenthion)				
Torak® (*see* dialifor)				
toxaphene, Strobane-T® (BFC Chemicals; Idacon, Inc.; Vertac Chemical)	chlorinated camphene containing 67–69% chlorine	Once used heavily on cotton; effective for grasshoppers, range caterpillars; cattle dips for mange control.	40	600
tralomethrin (*see* HAG-107)				
tretamine	2,4,6-tris(1-aziridinyl)-1,3,5-triazine	Experimental insect chemosterilant.	1	—
triazophos, Hostathion® (Hoechst AG)	*O,O*-diethyl *O*-(1-phenyl-H-1,2,4-triazol-3-yl phosphorothioate	Insecticide, miticide, nematicide; very broad spectrum of activity.	64	1100
trichlorfon, Dipterex® Dylox®, Neguvon®, Proxol® (Mobay; TUCO)	dimethyl (2,2,2-trichloro-1-hydroxyethyl) phosphonate	Controls many insects on field crops, vegetables, alfalfa, cotton, seed crops, ornamentals; cattle grubs, lice on beef; several fruits, forestry, sugarcane.	144	>2000
trichloronat, Agritox® (Bayer AG)	*O*-ethyl *O*-(2,4,5-trichlorophenyl) ethyl phosphonothioate	For vegetable fly larvae in field, garden crops; soil pests in meadows.	16	135
Trithion® (*see* carbophenothion)				
Trolene® (*see* ronnel)				
Vapam® (*see* metam-sodium)				
Vapona® (*see* dichlorvos)				
Vendex® (*see* fenbutatin-oxide)				
Vidden D® (*see* dichloropropane)				
Vikane® (*see* sulfuryl fluoride)				
Vydate® (*see* oxamyl)				
Warbex® (*see* famphur)				
Zectran® (*see* mexacarbate)				
Zolone® (*see* phosalone)				

Herbicides and Plant Growth Regulators

Common, trade, and chemical names of herbicides and plant growth regulators, their basic manufacturer(s), general use patterns, and oral and dermal LD_{50}s.

Common name, trade name, and basic manufacturer(s)	*Chemical name*	*General use pattern*	*Oral LD_{50} (rats)*	*Dermal LD_{50} (rabbits)*
acifluorfen, Blazer® (Rohm and Haas)	sodium 5-[2-chloro-4-(trifluoromethyl)-phenoxy]-2-nitrobenzoate	Selective pre-, post emergence control of broad-leaf, grass weeds in soybeans, peanuts.	1300	450
acrolein, Aqualin® (discontinued)	2-propenal	Aquatic weed and slime control.	46	skin burns
alachlor, Lasso® (Monsanto)	2-chloro-2′,6′-diethyl-*N*-(methoxymethyl) acetanilide	Preemergence control of annual grass, some broadleaf weeds in soybeans, corn, peanuts, potatoes.	1200	>1200
alloxydim-sodium, Clout® (Schering)	sodium salt of 2-(1-allyl-oxyaminobutylidene)-5,5-dimethyl-4-methoxycarbonylcyclohexane-1,3-dione	Postemergence control of grass weeds in broad-leaf crops.	2322	—
Alanap® (*see* naptalam)				
ametryn, Evik® (Ciba-Geigy)	2-(ethylamino)-4-(isopropylamino)-6-(methylthio)-*s*-triazine	Broad-leaf, grass weeds in pineapple, sugarcane, banana; potato vine desiccant.	1100	>10,200
Amex® (*see* butralin)				
Amiben® (*see* chloramben)				
amitrole (several) (American Cyanamid; Union Carbide)	3-amino-1,2,4-triazole	Noncropland use only; for annual, perennial grass; broad-leaf weeds.	1100	>10,000
Ammate® (*see* AMS)				
ammonium thiocyanate, Trans-Aid® (Union Carbide)	ammonium thiocyanate	Contact herbicide and soil sterilant.	repellent	
AMS, Ammate® (duPont)	ammonium sulfamate	Controls woody plants; also short-term weed control.	3900	—
Antor® (*see* diethatyl ethyl)				
Aqualin® (*see* acrolein)				
Aquazine® (*see* simazine)				
asulam, Asulox® (May & Baker) (Rhone-Poulenc)	methyl sulfanilylcarbamate	Postemergent grass weed control in sugarcane; bracken control in reforestation areas; crabgrass control in turf, certain ornamentals.	>5000	—

Common name, trade name, and basic manufacturer(s)	*Chemical name*	*General use pattern*	*Oral* LD_{50} *(rats)*	*Dermal* LD_{50} *(rabbits)*
Asulox® (*see* asulam)				
atrazine several (several)	2-chloro-4-(ethylamino)-6-(isopropylamino)-*s*-triazine	Season-long weed control mainly in corn, sorghum.	1869	>3100
Avadex® (*see* diallate)				
Avadex BW® (*see* triallate)				
Avenge® (*see* difenzoquat)				
Azak® (*see* terbucarb)				
Balan® (*see* benefin)				
Banvel® (*see* dicamba)				
Banex®, (*see* dicamba)				
barban, Carbyne® (Velsicol)	4-chloro-2-butynyl-*m*-chlorocarbanilate	Postemergence control of wild oats and canarygrass in many crops.	600	23,000
Basagran® (*see* bentazon)				
Basalin® (*see* fluchloralin)				
benazolin (many) (BFC Chemicals)	4-chloro-2-oxobenzothiazolin-3yl-acetic acid	Systemic, synergistic with dicamba; selective for certain broad-leaf weeds in rape, cereals, grassland.	4800	3200
benefin, Balan® (Elanco)	*N*-butyl-*N*-ethyl-a-a-a-trifluoro-2,6-dinitro-*p*-toluidine	Preemergence control of annual grass, broad-leaf weeds in several crops.	>10,000	—
bensulide, Betasan®, Prefar® (Stauffer)	*S*-(*O,O*-diisopropyl phosphorodithioate) ester of *N*-(2-mercaptoethyl)benzenesulfonamide	Preemergence control of crabgrass, annual bluegrass, broad-leaf weeds in dichondra, grass lawns.	770	3950
bentazon, Basagran® (BASF Wyandotte)	3-isopropyl-1*H*-2,1,3-benzothiadiazin-(4) 3*H*-one-2,2-dioxide	Postemergence control of broadleaf weeds in soybeans, rice, corn, peanuts.	1100	>2500
benthiocarb (*see* thiobencarb)				
Betanal® (*see* phenmedipham)				
Betanex® (*see* desmidipham)				
Betasan® (*see* bensulide)				
Bexton® (*see* propachlor)				
bifenox, Modown® (Mobil)	methyl 5-(2,4-dichlorophenoxy)-2-nitrobenzoate	Preemergence control of weeds in soybeans, sorghum, rice.	>6400	>20,000
Bladex® (*see* cyanazine)				
Blazer® (*see* acifluorfen)				
Bolero® (*see* thiobencarb)				
bromacil, Hyvar® (duPont)	5-bromo-3-*sec*-butyl-6-methyluracil	Weed and brush control in noncrop area; selective weed control in citrus, pineapple.	5200	>5000
Brominal® (*see* bromoxynil)				

Common name, trade name, and basic manufacturer(s)	Chemical name	General use pattern	Oral LD_{50} (rats)	Dermal LD_{50} (rabbits)
bromoxynil, Brominal® Buctril® (several)	3,5-dibromo-4-hydroxybenzonitrile	Postemergence broad-leaf weed control in small grains, noncrop areas.	190	>3660
Buctril® (*see* bromoxynil)				
Bush Buster® (*see* dicamba)				
butam (Gulf)	2,2-dimethyl-*N*-(1-methylethyl)-*N*-(phenyl-methyl)propanamide	Grass and broad-leaf weeds in peanuts, soybeans, tomatoes, sugar beets, alfalfa.	6000	>2000
butralin, Amex® (Union Carbide)	4-(1,1-dimethylethyl)-*N*-(1-methylpropyl)-2,6-dinitrobenzenamine	Annual grass, broad-leaf weeds in cotton, soybeans.	1000	10,200
butylate, Sutan® (Stauffer)	*S*-ethyl diisobutylthiocarbamate	Incorporated preplant for controlling grass weeds especially nutgrass in corn.	4000	>2000
cacodylic acid, several (Crystal; Vineland)	hydroxydimethylarsine oxide	For killing trees in forestry, as a nonselective herbicide; cotton defoliant.	700	—
Caparol® (*see* prometryn)				
Carbyne® (*see* barban)				
Casoron® (*see* dichlobenil)				
CDAA, Randox® (Monsanto)	*N-N*-diallyl-2-chloroacetamide	Preemergence grass, broad-leaf weed control in corn, sorghum, soybeans, some vegetable crops.	700	—
CDEC, Vegadex® (Monsanto)	2-chloroallyl diethyldithiocarbamate	Preemergence grass, broad-leaf weeds in vegetable crops, ornamentals, shrubbery.	850	—
CGA-43089 (*see* Concep®)				
Chem-Hoe® (*see* propham)				
chloramben, Amiben® Vegiben® (Union Carbide)	3-amino-2,5-dichlorobenzoic acid	Preemergence weed control in many vegetable, field crops.	3500	>3160
chloridazon, pyrazon, Pyramin® (BASF Wyandotte)	5-amino-4-chloro-2-phenyl-3-(2*H*)-pyridazinone	Pre and postemergence weed control in sugar, red, and fodder beets.	3030	—
Chloro IPC® (*see* chlorpropham)				
chlorpropham, Chloro-IPC®, Furloe® (PPG)	isopropyl *m*-chlorocarbanilate	Preemergence herbicide for weeds in several crops; potato sprout inhibitor.	3800	10,300
Clout® (*see* alloxydim-sodium)				
Concep®, CGA-43089 (Ciba-Geigy)	α-(cyanomethoximino)benzacetonitrile	Herbicide safener seed treatment for sorghum and certain herbicides.	2277	>3100
Cotoran® (*see* fluometuron)				
cyanazine, Bladex® (Shell)	2-[(4-chloro-6-(ethylamino)-*s*-triazin-2-yl) amino]-2-methylpropionitrile	Pre-, postemergence herbicide for cotton, corn.	182	>1200

Common name, trade name, and basic manufacturer(s)	Chemical name	General use pattern	Oral LD_{50} (rats)	Dermal LD_{50} (rabbits)
cycloate, Ro-Neet® (Stauffer)	*S*-ethyl *N*-ethylthiocyclohexanecarbamate	Preplant herbicide for grass, broad-leaf weeds in sugar beets, table beets, spinach.	3160	>4640
2,4-D, several (several)	(2,4-dichlorophenoxy)acetic acid	Selective broad-leaf weed control in monocots (small grains, corn, sorghum, sugarcane), noncrop areas.	375	—
Dacthal® (*see* DCPA)				
dalapon, Dowpon® (Dow; Diamond Shamrock)	2,2-dichloropropionic acid (sodium salt)	Systemic herbicide for various grasses and rushes in crop, noncrop areas.	6500	10,000
dazomet, Mylone® (Hopkins; Stauffer)	tetrahydro-3,5-dimethyl-2*H*-1,3,5-thiadiazine-2-thione	Preplanting seedbed treatment for tobacco, turf, ornamentals; anti-microbial in glues for paper products; nematicide, slimicide.	500	—
2,4-DB, several (several)	4-(2,4-dichlorophenoxy)butyric acid	Selective control of certain broad-leaf weeds in alfalfa, clover, soybeans, peanuts.	500	>10,000
DCPA, Dacthal® (Diamond Shamrock)	dimethyl tetrachloroterephthalate	Preemergence herbicide for grass, broad-leaf weeds in soybeans, cotton, seeded vegetables.	>3000	>10,000
DEF® (Mobay)	*S,S,S*-tributylphosphorotrithioate	Defoliant for cotton.	200	>1000
desmidipham, Betanex® (NOR-AM)	ethyl *m*-hydroxycarbanilate carbanilate (ester)	Broad-leaf weed control in sugar beets.	>9600	>2000
Destun® (*see* perfluidone)				
Devine® (*see* *Phythophora palmivora*)				
Devrinol® (*see* napropamide)				
diallate, Avadex® (Monsanto)	*S*-(2,3-dichloroallyl)diisopropylthiocarbamate	Soil-incorporated preemergence control of wild oats in several crops.	395	2000
dicamba, Banvel®, Banex®, Bush Buster® (Velsicol)	3,6-dichloro-*o*-anisic acid	Controls annual, perennial weeds in corn, sorghum, small grains; pasture, rangeland, noncropland.	1040	>2000
dichlobenil, Casoron® (Thompson-Hayward)	2,6-dichlorobenzonitrile	Selective weed control in ornamentals, orchards, vineyards; total weed control for industrial sites.	3160	1350
dichlorprop, several (several)	2-(2,4-dichlorophenoxy)propionic acid	Systemic brush control on rights-of-way, rangeland; broad-leaf weeds in cereals.	400	1400
diclofop methyl, Hoelon® (American Hoechst)	methyl 2-[4-(2′,4′-dichlorophenoxy)-phenoxy] propanoate	For annual weed grasses (wild oats) in wheat, barley, soybeans.	679	>2000
diethatyl ethyl, Antor® (BFC Chemicals)	*N*-chloroacetyl-*N* (2,6-diethylphenyl)glycine ethyl ester	Preplant preemergence for control of annual grass, broad-leaf weeds in sugar beets, soybeans.	2300	4000
difenzoquat, Avenge® (American Cyanamid)	1,2-dimethyl-3,5-diphenyl-1*H*-pyrazolium methyl sulfate	Postemergence for controlling wild oats in wheat, barley.	270	3540
Dinitro® (*see* dinoseb)				
dinoseb, DNBP, Dinitro® (several)	2-*sec*-butyl-4,6-dinitrophenol	General contact herbicide, desiccant.	40	75

Common name, trade name, and basic manufacturer(s)	*Chemical name*	*General use pattern*	*Oral LD_{50} (rats)*	*Dermal LD_{50} (rabbits)*
diphenamid, Dymid® Enide® (TUCO)	*N,N*-dimethyl-2,2-diphenylacetamide	Preemergence control of annual, broad-leaf weeds in many crops.	1000	>6320
dipropetryn, Sancap® (Ciba-Geigy)	2-ethylthio-4-6-bis-isopropylamino-*s*-triazine	Preemergence control of pigweed, Russian thistle in cotton.	4050	>10,000
diquat (Chevron)	6,7-dihydrodipyridol(1,2-α:2′,1′-*c*)-pyrazinediium ion	Industrial and aquatic weed control; seed crop desiccant.	400	irritant
diuron, Karmex® Krovar® (duPont)	3-(3,4-dichlorophenyl)-1,1-dimethylurea	Controls most germinating weeds in many crops; soil sterilant, general weed killer at high rates.	3400	—
DNBP (*see* dinoseb)				
DNOC (Blue Spruce)	4,6-dinitro-*o*-cresol	General weed killer; also insecticide, fungicide, defoliant.	20	200
Dowco® 290 (*see* Lontrel®)				
Dowpon® (*see* dalapon)				
Dozer® (*see* fenuron TCA)				
Dropp® (*see* thidiazuron)				
DSMA, several (several)	disodium methanearsonate	Postemergence grass weed control in cotton, noncrop areas.	600	—
Dual® (*see* metolachlor)				
Dymid® (*see* diphenamid)				
Embark® (*see* mefluidide)				
endothall, several (Pennwalt)	7-oxabicyclo(2.2.1)heptane-2,3-dicarboxylic acid	Weed control in sugar beets, turf; aquatic herbicide; cotton defoliant; potato, seedcrop desiccant.	38	irritant
Enide® (*see* diphenamid)				
Eptam® (*see* EPTC)				
EPTC, Eptam® (Stauffer)	*S*-ethyldipropylthiocarbamate	Controls most grass weeds in several crops.	1367	10,000
ethalfluralin, Sonalan® (Elanco)	*N*-ethyl-*N*-(2-methyl-2-propenyl)-2,6-dinitro-4-(trifluoromethyl) benzenamine	For most annual grasses and many broad-leaf weeds; preemergence, soil incorporated.	>10,000	—
ethephon, Ethrel® (Union Carbide)	(2-chloroethyl) phosphonic acid	Plant growth regulator, produces ethylene; registered for many crops.	4229	5730
ethidimuron, Ustilan® (Mobay)	3-(5-ethyl-sulphonyl-1,3,4-thiadiazol-2-yl)-1,3-dimethyl urea	Total weed control on noncropland and sugarcane.	>5000	>1000
ethofumesate, Nortron® (Fisons)	2-ethoxy-2,3-dihydro-3,3-dimethyl-5-benzofuranyl methanesulphonate	Preplant, pre- or postemergence on sugar beets for annual broad-leaf, grass weeds.	5650	>1440
Ethrel® (*see* ethephon)				
Evik® (*see* ametryn)				
Far-Go® (*see* triallate)				
fenac (Union Carbide)	(2,3,6-trichlorophenyl)acetic acid	Preemergence control of weeds in sugar cane, most weeds in noncrop areas.	1760	>3160

Common name, trade name, and basic manufacturer(s)	Chemical name	General use pattern	Oral LD_{50} (rats)	Dermal LD_{50} (rabbits)
fenuron (Crewe; Murphy)	1,1-dimethyl-3-phenylurea	Weed brush killer.	6400	—
fenuron TCA, Dozer® (Hopkins)	1,1-dimethyl-3-phenylurea mono(trichloroacetate)	Nonselective weed, brush control in noncrop areas.	4000	—
fluchloralin, Basalin® (BASF Wyandotte)	n-(2-chloroethyl)-2,6-dinitro-*N*-propyl-4-(trifluoromethyl)aniline	Preplant preemergence control of weeds in cotton, soybeans.	1550	—
fluometuron, Cotoran® (Ciba-Geigy)	1,1-dimethyl-3-(α,α,α-trifluoro-*m*-tolyl)urea	Pre-, postemergence control of weeds in cotton, sugarcane.	7880	>10,000
fluridone, Sonar® (Elanco)	1-methyl-3-phenyl-5-[3-(trifluoromethyl)phenyl]-4(1*H*)-pyridinone	Experimental soil-incorporated preemergence herbicide for most grass and broad-leaf weeds in cotton.	>10,000	—
Folex® (*see* merphos)				
fosamine ammonium Krenite® (duPont)	ammonium ethyl carbamoylphosphonate	Fall-applied brush control agent on noncropland areas.	24,000	>4000
Furloe® (*see* chlorpropham)				
Garlon® (*see* triclopyr)				
glyoxime, Pik-Off® (Ciba-Geigy)	ethanedial dioxime	Growth regulator, enhances abscission of fruit (oranges and pineapple).	185	1580
glyphosate, Roundup® (Monsanto)	*N*-(phosphonomethyl)glycine, isopropylamine salt	Highly versatile translocated, non-selective herbicide—many uses on many crops.	4320	>7940
glyphosine, Polaris® (Monsanto)	*N,N*-bis(phosphonomethyl)glycine	Chemical ripener for sugarcane.	3925	—
Goal® (*see* oxyfluorfen)				
hexazinone, Velpar® (duPont)	3-cyclohexyl-6-(dimethylamino)-1-methyl-1,3,5-triazine-2,4-(1*H*, 3*H*)-dione	Contact and residual control of every kind of weed on noncropland areas.	1690	5278
Hoelon® (*see* diclofop methyl)				
Hyvar® (*see* bromacil)				
Igran® (*see* terbutryn)				
IPC (*see* propham)				
isocil (DuPont)	5-bromo-3-isopropyl-6-methyluracil	Nonselective herbicide.	3400	—
isopropalin, Paarlan® (Elanco)	2,6-dinitro-*N,N*-dipropylcumidine	Preemergence control of weeds in transplant tobacco.	5000	>2000
karbutilate, Tandex® (discontinued)	*tert*-butylcarbamic acid ester with 3(*m*-hydroxy-phenyl)-1,1-dimethylurea	Weed, brush, vine control in industrial, noncrop areas.	3000	>15,400
Karmex® (*see* diuron)				
Kerb® (*see* pronamide)				
Kloben® (*see* neburon)				
Krenite® (*see* fosamine ammonium)				
Krovar® (*see* diuron)				
Lasso® (*see* alachlor)				

Common name, trade name, and basic manufacturer(s)	Chemical name	General use pattern	Oral LD_{50} (rats)	Dermal LD_{50} (rabbits)
lenacil, Venzar® (duPont)	3-cyclohexyl-6,7-dihydro-1*H*-cyclopenta-pyrimidine-2,4(3*H*,5*H*)-dione	Weed control in sugar beets, cereal grains.	11,000	—
Lexone® (*see* metribuzin)				
linuron, Lorox® (duPont)	3-(3,4-dichlorophenyl)-1-methoxy-1-methylurea	Selective weed control in corn, sorghum, cotton, soybeans, wheat; annual weeds in noncrop area.	1500	—
Lontrel®, Dowco® 290 (Dow)	3,6-dichloro-2-pyridinecarboxylic acid	Broad-leaf weeds in wheat and barley.	>5000	>2000
Lorox® (*see* linuron)				
MAA, several (several)	methanearsonic acid	Grass weeds in cotton, noncrop areas.	1300	—
MAMA, several (several)	monoammonium methanearsonate	Dallis grass, nutgrass in turf.	750	—
MCPA, several (several)	[(4-chloro-*o*-tolyl)oxy]acetic acid	Translocated herbicide used in small grains, rice for postemergence control of wide weed spectrum.	700	—
MCPB, several (Rhone-Poulenc)	4-(4-chloro-2-methylphenoxy)butanoic acid	Weed control in peas.	680	—
MCPP (*see* mecoprop)				
mecoprop, MCPP (several)	2-(4-chloro-*o*-tolyl)oxypropionic acid	Turf, cereal herbicide for broad-leaf weeds.	930	—
mefluidide, Embark® (3M)	*N*-[2,4-dimethyl-5-[[(trifluoromethyl)-sulfonyl]amino]phenyl]acetamide	Plant growth regulator to prevent seedhead formation in turf grasses.	1920	>4000
mepiquat chloride, Pix® (BASF Wyandotte)	1,1-dimethyl-piperidiniumchloride	Cotton growth regulator to reduce vegetative growth.	>6900	—
merphos, Folex® (Mobil)	tributyl phosphorotrithioite	Defoliant for cotton.	1272	>4600
metam-sodium, Vapam®, SMDC (Stauffer)	sodium methyldithiocarbamate	Soil fumigant for weeds, weed seeds, nematodes, fungi.	820	2000
methazole, Probe® (Velsicol)	2-(3,4-dichlorophenyl)-4-methyl-1,2,4-oxadiazolidine-3,5-dione	Pre-, postemergence herbicide for cotton.	1350	>12,500
metolachlor, Dual® (Ciba-Geigy)	2-chloro-*N*-(2-ethyl-6-methylphenyl)-*N*-(2-methoxy-1-methylethyl) acetamide	Preemergence, preplant in corn, peanuts, soybeans, others.	2780	>10,000
metribuzin, Lexone® Sencor® (duPont; Mobay)	4-amino-6-*tert*-butyl-3-(methylthio)-1,2,4-triazin-5(4*H*)one	Controls wide spectrum of weeds in several crops.	1937	>20,000
MH (maleic hydrazide) (several)	1,2-dihydro-3,6-pyridazinedione	Growth retardant in trees, shrubs, grasses; sprout inhibitor for stored onions, potatoes.	2200	>8000
Milogard® (*see* propazine)				
Modown® (*see* bifenox)				
molinate, Ordram® (Stauffer)	*S*-ethyl hexahydro-1*H*-azepine-1-carbothioate	Watergrass control in rice.	584	>2000
monuron (Hopkins)	3-(*p*-chlorophenyl)-1,1-dimethylurea	Complete weed control in noncrop-land areas.	3600	—
monuron TCA, Urox® (Hopkins)	3-(*P*-chlorophenyl)1,1-dimethylurea mono(trichloroacetate)	Complete weed control in noncrop-land areas.	2300	irritant
MSMA, several (several)	monosodium methanearsonate	Preplant for cotton, grass weeds in turf, noncropland.	700	—

Common name, trade name, and basic manufacturer(s)	Chemical name	General use pattern	Oral LD_{50} (rats)	Dermal LD_{50} (rabbits)
Mylone® (*see* dazomet)				
napropamide, Devrinol® (Stauffer)	2-(α-naphthoxy)-*N,N*-diethylpropionamide	Selective weed control in orchards, vineyards, several vegetables, ornamentals.	>5000	>4640
naptalam, Alanap® (Crystal; Uniroyal)	*N*-1-naphthylphthalamic acid	Broad-leaf weed control in soybeans, peanuts, vine crops.	1770	—
neburon, Kloben® (duPont)	1-butyl-3-(3,4-dichlorophenyl)-1-methylurea	Selective weed control in woody ornamental nurseries.	>11,000	
nitralin, Planavan® (Shell Intl.)	4-(methylsulfonyl)-2,6-dinitro-*N,N*-dipropylaniline	Weed control in cotton, soybeans, peanuts, alfalfa, vegetables.	>2000	>2000
nitrapyrin, N-Serve® (Dow)	2-chloro-6-(trichloromethyl)pyridine	Inhibits nitrification in soil by *Nitrosomonas* bacteria.	1230	—
nitrofen, Tok® (Rohm & Haas)	2,4-dichlorophenyl-*p*-nitrophenyl ether	Pre-, postemergence control of broad weed spectrum in vegetables, sugar beets, rice, ornamentals.	2630	—
norflurazon, Solicam® Zorial® (Sandoz)	4-chloro-5-(methylamino)-2-(α,α,α-trifluro-*m*-tolyl)-3(2*H*)-pyridazinone	Preemergence control of weeds in cotton, cranberries, fruit, nut trees.	>8000	>20,000
Nortron® (*see* ethofumesate)				
N-Serve® (*see* nitrapyrin)				
Ordram® (*see* molinate)				
oryzalin, Surflan® (Elanco)	3,5-dinitro-N^4,N^4-dipropylsulfanilamide	Preemergence control of many weeds in soybeans, cotton, nonbearing fruit, nut trees, ornamentals.	>10,000	>2000
oxadiazon, Ronstar® (Rhone-Poulenc)	2-*tert*-butyl-4-(2,4-dichloro-5-isopropoxy-phenyl)-Δ2-1,3,4-oxadiazolin-5-one	Pre- postemergence weed control in rice; for weeds in turf, ornamentals.	>8000	>8000
oxyfluorfen, Goal® (Rohm & Haas)	2-chloro-1-(3-ethoxy-4-nitrophenoxy)-4-(trifluoromethyl)benzene	Pre-, postemergence control of broad weed spectrum in corn, cotton, soybeans, fruit, nut trees, ornamentals.	>5000	>10,000
Paarlan® (*see* isopropalin)				
paraquat, several (Chevron)	1,1′-dimethyl-4,4′-bipyridinium ion	Contact herbicide and desiccant with broad scope of uses.	150	236
PCP, penta (several)	pentachlorophenol	Preharvest defoliant, wood preservative, molluscicide.	50	irritant
PEBC (*see* pebulate)				
pebulate, PEBC, Tillam® (Stauffer)	*S*-propyl butylethylthiocarbamate	Preplant control of grass, weeds in sugar beets, tobacco, tomatoes.	1120	>2936
pendimethalin, Prowl® (American Cyanamid)	*N*-(1-ethylpropyl)-3,4-dimethyl-2,6-dinitrobenzenamine	Pre- or postemergence use in corn; preplant for cotton, soybeans, tobacco.	1250	>5000
penta (*see* PCP)				
perfluidone, Destun® (3M)	1,1,1-trifluoro-*N*-[2-methyl-4-(phenylsulfonyl)-phenyl]methanesulfonamide	Preemergence weed control in cotton.	633	>4000
phenmedipham, Betanal® (NOR-AM)	methyl-*m*-hydroxycarbanilate-methylcarbanilate	Broad-leaf weed control in sugar beets, table beets.	>8000	>4000
Phytophthora palmivora Devine® (Abbott)	spores of a naturally occurring fungus (mycoherbicide)	Milkweed vine control in citrus groves.	nontoxic	

Common name, trade name, and basic manufacturer(s)	Chemical name	General use pattern	Oral LD_{50} (rats)	Dermal LD_{50} (rabbits)
picloram, Tordon® (dow)	4-amino-3,5,6-trichloropicolinic acid	Brush control on industrial sites, pastures, rangeland; broad-leaf weed control in small grains.	8200	>4000
Pik-Off® (*see* glyoxime)				
Pix® (*see* mepiquat chloride)				
Planavan® (*see* nitralin)				
Polaris® (*see* glyphosine)				
Pramitol® (*see* prometon)				
Prefar® (*see* bensulide)				
Prep® (Union Carbide)	sodium *cis*-3-chloroacrylate	Defoliant for cotton and potatoes.	320	—
Princep® (*see* simazine)				
Probe® (*see* methazole)				
profluralin, Tolban® (Ciba-Geigy)	*N*-(cyclopropylmethyl)-α,α,α-trifluoro-2,6-dinitro-*N*-propyl-*p*-toluidine	Controls most grass, broad-leaf weeds in cotton, soybeans, sunflower, certain vegetables.	2200	>10,000
prometon, Pramitol® (Ciba-Geigy)	2,4-bis(isopropylamino)-6-methoxy-*s*-triazine	Nonselective pre-, postemegence control for most weeds in noncropland.	2980	>2000
prometryn, Caparol® (Ciba-Geigy)	2,4-bis(isopropylamino)-6-(methylthio)-*s*-triazine	Versatile pre-, postemergence cotton herbicide.	3150	>3100
pronamide, Kerb® (Rohm & Haas)	3,5-dichloro(*N*-1,1-dimethyl-2-propynyl)benzamide	Pre- or postemergence weed control in legumes; also used on turf, woody ornamentals.	5620	>3160
propachlor, Bexton® Ramrod® (Dow; Monsanto)	2-chloro-*N*-isopropylacetanilide	Controls most annual grass, broad-leaf weeds in corn, sorghum, soybeans (grown for seed).	710	irritant
propanil, several (several)	3′-4′-dichloropropionalide	Postemergence control of grasses, certain other weeds in rice, wheat.	1384	4830
propazine, Milogard® (Ciba-Geigy)	2-chloro-4,6-bis(isopropylamino)-*s*-triazine	Preemergence control of most annual weeds in milo, sorghum.	5000	>10,200
propham, Chem-Hoe®, IPC (PPG)	isopropyl carbanilate	Pre- and postemergence weed control in forage legumes, sugar beets, vegetables.	9000	6800
prosulfalin, Sward® (Elanco)	*N*-[(4-dipropylamino)-3,5-dinitrophenyl]-sulfonyl)-*S,S*-dimethylsulfilimine	Selective preemergence herbicide for use against crabgrass, bluegrass in turf.	2000	>500
Prowl® (*see* pendimethalin)				
Pyramin® (*see* chloridazon)				
pyrazon (*see* chloridazon)				
Ramrod® (*see* propachlor)				
Randox® (*see* CDAA)				
Ro-Neet® (*see* cycloate)				
Ronstar® (*see* oxadiazon)				

Common name, trade name, and basic manufacturer(s)	Chemical name	General use pattern	Oral LD_{50} (rats)	Dermal LD_{50} (rabbits)
Roundup® (*see* glyphosate)				
Rowmate® (Union Carbide)	3,4- and 2,3-dichlorobenzyl *N*-methylcarbamate	Controls grass and broad-leaf weeds in soybeans, peanuts, potatoes, beans, peas, ornamentals.	1879	—
Sancap® (*see* dipropetryn)				
Sencor® (*see* metribuzin)				
siduron, Tupersan® (duPont)	1-(2-methylcyclohexyl)-3-phenylurea	Weed grass control in bluegrass, ryegrass, other lawn grasses.	>7500	—
silvex, several (Dow; Thompson-Hayward; Vertac)	2-(2,4,5-trichlorophenoxy)propionic acid	For control of woody plants, broad-leaf herbaceous weeds, aquatic weeds.	375	>3940
simazine, Princep® Aquazine® (Ciba-Geigy)	2-chloro-4,6-bis(ethylamino)-*s*-triazine	Most annual grasses, broad-leaf weeds in corn, tree fruits, nuts, ornamentals, turf; aquatic weed, algae control.	5000	>3100
Sinbar® (*see* terbacil)				
sodium arsenite, several (several)	sodium arsenite	Weed control in industrial areas; tree, stump destruction.	10–50	—
sodium chlorate, several (Kerr-McGee; Occidental)	sodium chlorate	Soil sterilant herbicide, defoliant, desiccant, harvest aid.	1200	—
Solicam® (*see* norflurazon)				
Sonalan® (*see* ethalfluralin)				
Sonar® (*see* fluridone)				
Spike® (*see* tebuthiuron)				
Surflan® (*see* oryzalin)				
Sutan® (*see* butylate)				
Sward® (*see* prosulfalin)				
swep (Nissan)	methyl (3,4-dichlorophenyl)carbamate	Soil treatment and foliage contact weed control in rice, corn, peanuts.	522	—
2,4,5-T, many (Dow; Thompson-Hayward; Vertac)	(2,4,5-trichlorophenoxy)acetic acid	Woody plant control on industrial sites, rangeland. Amine form used on rice.	300	—
Tandex® (*see* karbutilate)				
2,3,6-TBA (Fisons Ltd.)	2,3,6-trichlorobenzoic acid	Control of deep-rooted perennials.	750	>1000
TCA (Akzo Zout)	trichloroacetic acid	Soil sterilant for perennial weed grass control in noncropland.	3200	—
tebuthiuron, Spike® (Elanco)	*N*-[5-(1,1-dimethylethyl)-1,3,4-thiadiazol-2-yl]-*N,N'*-dimethylurea	Total vegetation control in noncropland areas.	644	>200
terbacil, Sinbar® (duPont)	3-*tert*-butyl-5-chloro-6-methyluracil	Selective control of annual, some perennial weeds in a wide range of crops.	>5000	>5000
terbucarb, terbutol, Azak® (BFC chemicals)	2,6-di-*tert*-butyl-*p*-tolyl methyl carbamate	Selective preemergence herbicide.	>34,000	—

Common name, trade name, and basic manufacturer(s)	Chemical name	General use pattern	Oral LD_{50} (rats)	Dermal LD_{50} (rabbits)
terbutol (*see* terbucarb)				
terbutryn, Igran® (Ciba-Geigy)	2-(*tert*-butylamino)-4-(ethylamino)-6-(methylthio)-*s*-triazine	Preemergence weed control in grain sorghum; postemergence control in winter wheat, barley.	2100	>2000
thidiazuron, Dropp® (NOR-AM)	*N*-phenyl-*N'*-(1,2,3-thiadiazol-5yl) urea	Defoliant for cotton.	>4000	>1000
thiobencarb, benthiocarb Bolero® (Chevron)	*S*-(4-chlorophenyl)methyl diethylcarbamothioate	Pre-, postemergent herbicide for grass; broad-leaf weeds in rice fields.	1903	2900
Tillam® (*see* pebulate)				
Tok® (*see* nitrofen)				
Tolban® (*see* profluralin)				
Tordon® (*see* picloram)				
Trans-Aid® (*see* ammonium thiocyanate)				
Treflan® (*see* trifluralin)				
triallate, Avadex® BW Far-Go® (Monsanto)	*S*-(2,3,3-trichloroallyl)diisopropylthiocarbamate	Wild oat control in barley, wheat, peas, lentils.	1675	2225
triclopyr, Garlon® (Dow)	[(3,5,6-trichloro-2-pyridinyl)oxy]acetic acid	Systemic control of woody plants, broad-leaf weeds in rights-of-way, industrial sites, forests.	713	—
trifluralin, Treflan® (Elanco)	α,α,α-trifluoro-2,6-dinitro-*N,N*-dipropyl-*p*-toluidine	Selective preemergence incorporated herbicide registered for many crops.	>10,000	—
Tupersan® (*see* siduron)				
Urox® (*see* monuron TCA)				
Ustilan® (*see* ethidimuron)				
Vapam® (*see* metam-sodium)				
Vegadex® (*see* CDEC)				
Vegiben® (*see* chloramben)				
Velpar® (*see* hexazinone)				
Venzar® (*see* lenacil)				
Vernam® (*see* vernolate)				
vernolate, Vernam® (Stauffer)	*S*-propyl dipropylthiocarbamate	For most common weed grasses in soybeans, peanuts.	1780	4640
Zorial® (*see* norflurazon)				

APPENDIX C

Fungicides and Bactericides

Common, trade, and chemical names of fungicides, their basic manufacturer(s), general use patterns, and oral and dermal LD_{50}s.

Common name, trade name, and basic manufacturer(s)	*Chemical name*	*General use pattern*	*Oral LD_{50} (rats)*	*Dermal LD_{50} (rabbits)*
Acti-Dione® (*see* cycloheximide)				
Afugan® (*see* pyrazophos)				
Algae-Rhap® (*see* copper-triethanolamine)				
Algimycin PLLC® (*see* copper chelates)				
anilazine, Dyrene® (Mobay)	2,4-dichloro-6-(*o*-chloroanilino)-*s*-triazine	Fungus diseases of lawns, turf, vegetables; potato, tomato leafspots.	2710	>9400
Antracol® (*see* propineb)				
Arbotect® (*see* thiabendazole)				
basic copper sulfate (several)	basic copper sulfate	Bacterial and fungal diseases on fruit, vegetable, nut, field crops.	low	
Baycor® (*see* bitertanol)				
Bayleton® (*see* triadimefon)				
Baytan® (*see* triadimenol)				
Beam® (*see* tricyclazole)				
Benlate® (*see* benomyl)				
benomyl, Benlate® (du Pont)	methyl-1-(butylcarbamoyl)-2-benzamidazole carbamate	Systemic control of fruit, vegetable, nut, field crop, turf, ornamental diseases.	9590	—
binapacryl, Morocide® (Hoechst)	2-*sec*-butyl-4,6-dinitrophenyl 3-methyl-2-butenoate	Mostly powdery mildews; also a miticide.	136	750
biphenyl, diphenyl (several)	diphenyl	Citrus wrap impregnate for rot fungi.	3280	—
bitertanol, Baycor® (Mobay)	*β*[(1,1′-biphenyl)-4-yloxy]-*a*-(1,1 dimethylethyl)-1*H*-1,2,4 triazole-1-ethanol	Diseases of fruits, field crops, vegetables, ornamentals.	>5000	>5000
blasticidin-S (*Streptomyces griseochromogenes*) (Kaken Kagaku Co., Japan)	(*S*)-4[[3-amino-5[(aminoiminomethyl)methyl-amino]-1-oxopentyl]amino]-1-[4- amino-2-oxo-1(2*H*)-pyrimidinyl]-1,2,3,4-tetradeoxy-*β*-D-erythrohex-2-enopyranuronic acid	Systemic antibiotic effective against rice blast.	16.3	75.5

Common name, trade name, and basic manufacturer(s)	Chemical name	General use pattern	Oral LD_{50} (rats)	Dermal LD_{50} (rabbits)
bluestone (*see* copper sulfate)				
Bordeaux mixture (Chemical Formulators)	mixture of copper sulfate and calcium hydroxide forming basic copper sulfates	For many diseases of vegetables, fruits, nuts; also acts as insecticide, repellent to some species.	nontoxic	
Botran® (*see* dicloran)				
Bravo® (*see* chlorothalonil)				
Brestan® (*see* fentin acetate)				
bronopol (BFC Chemicals)	2-bromo-2-nitro-1,3-propanediol	Bacterial diseases of cotton.	400	>1600
bupirimate, Nimrod® (ICI)	5-butyl-2-ethylamino-6-methyl-pyrimidin-4-yl-dimethylsulfamate	Systemic control of powdery mildew of fruit.	>4000	>500
Busan-72®, TCMTB (Buckman)	2-(thiocyanomethylthio)benzothiazole	Soak treatment for bulbs and seeds.	1590	—
butrizol (*see* Indar®)				
Cadminate® (Mallinckrodt)	cadmium succinate	Fungal diseases of turf.	660	>200
cadmium chloride (Cleary)	cadmium chloride	Fungal diseases of turf.	88	—
Calixin® (*see* tridemorph)				
captafol, Difolatan® (Chevron)	*cis-N*(1,1,2,2-tetrachloroethyl)thio-4-cyclohexene-1,2-dicarboximide	Many diseases of fruits, vegetables, nuts, seeds.	6200	>15,400
captan (many names) (Chevron, Hopkins, & Stauffer)	*N*-[(trichloromethyl)thio]-4-cyclohexene-1,2-dicarboximide	Many diseases of fruits, vegetables, nuts, seeds.	9000	—
Carbamate® (*see* ferbam)				
carboxin, DCMO, Vitavax® (Uniroyal)	5,6-dihydro-2-methyl-*N*-phenyl-1,4-oxathiin-3-carboxamide	Systemic seed treatment for grains, peanuts, other field crops.	3820	>8000
Ceresan®, MEMC (discontinued)	2-methoxyethylmercuric chloride	Seed treatment and bulb dip.	22	—
Chipco 26019® (*see* iprodione)				
chloranil, Spergon® (uniroyal)	2,3,5,6-tetrachloro-1,4-benzoquinone	Foliar diseases of fruit, vegetables, ornamentals.	4000	—
chloroneb, Demosan® Tersan® (du Pont)	1,4-dichloro-2,5-dimethoxybenzene	Systemic seed treatment; in-furrow soil treatment; turf diseases.	>11,000	>5000
chlorothalonil, Daconil® 2787, Bravo® (Diamond Shamrock)	tetrachloroisophthalonitrile	Variety of vegetable, fruit, turf, ornamental diseases.	10,000	>10,000
copper ammonium carbonate (Mineral Research)	copper ammonium carbonate	Protectant for fruits, vegetables tolerant to copper.	low	—
copper chelates Algimycin PLLC® (Great Lakes)	copper chelates of citrate and gluconate	Most common algae in various water sources.	low	—
copper hydroxide, Kocide® (Kocide)	copper hydroxide	Protectant for fruit, vegetable, tree, field crops.	1000	—

Common name, trade name, and basic manufacturer(s)	Chemical name	General use pattern	Oral LD_{50} (rats)	Dermal LD_{50} (rabbits)
copper naphthenates (several)	cupric cyclopentanecarboxylate	Rot, mildew in fabrics and wood.	low	—
copper oxides (several)	cuprous oxide and cupric oxide	Seed treatment; various fruit, vegetable diseases.	low	—
copper oxychloride (several)	basic copper chloride	Protectant for fruit, vegetable crops.	700	—
copper oxychloride sulfate (CP Chemicals)	mixture of basic copper chloride and basic copper sulfate	Downy mildew, leafspots of row, vine, field, and tree crops.	low	—
copper quinolinolate (several)	copper-8-quinolinolate	Treatment of fruit-handling equipment.	10,000	—
copper sulfate, bluestone (several)	copper sulfate pentahydrate	Mixed with lime for Bordeaux mixture; almost never used alone due to phytotoxicity.	300	—
copper-triethanolamine Algae-Rhap® (CP Chemicals)	copper-triethanolamine complex in liquid form	Filamentous and planktonic algae in various water sources.	low	—
cycloheximide, Acti-Dione® (TUCO)	3-[2-(3,5-dimethyl-2-oxocyclohexyl)-2-hydroxyethyl]-glutarimide	Mostly diseases of ornamentals, turf.	2.5	—
Cyprex® (*see* dodine)				
Daconil® 2787 (*see* chlorothalonil)				
dazomet, Mylone® (Hopkins, Stauffer)	tetrahydro-3,5-dimethyl-2*H*-1,3,5-thiadiazine-2-thione	Seed-bed soil fungi, nematodes, soil insects, weeds; slimicide; industrial fungicide.	500	—
DCMO (*see* carboxin)				
DCNA (*see* dicloran)				
Demosan® (*see* chloroneb)				
dichlone, Quintar® (Hopkins)	2,3-dichloro-1,4-naphthoquinone	Protectant for fruit, vegetables, field crops, ornamentals.	1300	5000
dichlozoline, Sclex® (discontinued)	3-(3,5-dichlorophenyl)-5,5-dimethyl-2,4-oxazolidinedione	General fungicide.	3000	—
dicloran, Botran®, DCNA (TUCO)	2,6-dichloro-4-nitroaniline	Many fruit, vegetable, field crop, greenhouse, ornamental diseases.	5000	—
Difolatan® (*see* captafol)				
dimethirimol, Milcurb® (ICI)	5-*n*-butyl-2-dimethylamino-4-hydroxy-6-methylpyrimidine	Systemic control of powdery mildew on cucurbits, ornamentals.	2350	7500
dinocap, Karathane® (Rohm and Haas)	mixture of 2,4-dinitro-6-octylphenyl crotonate and 2,6-dinitro-4-octylphenyl crotonate	Powdery mildew on fruits, vegetables, ornamentals: also a good miticide.	980	—
diphenyl (*see* biphenyl)				
dodemorph, Milban® (BASF, Mallinckrodt)	4-cyclododecyl-2,6-dimethylmorpholimium acetate	Powdery mildew of ornamentals and greenhouse roses.	4180	—
dodine, Cyprex®, Melprex® (American Cyanamid)	dodecylquanidine monoacetate	Many diseases of tree fruits.	660	>1500
Dowicide A®, sodium phenyl phenate (Dow)	sodium-*o*-phenyl phenate	Fruit wrapper impregnate and seed box disinfectant.	1160	—
Dowicide 1® (*see* phenyl phenol)				

Common name, trade name, and basic manufacturer(s)	Chemical name	General use pattern	Oral LD_{50} (rats)	Dermal LD_{50} (rabbits)
Du-Ter® (*see* fentin hydroxide)				
Dyrene® (*see* anilazine)				
ethazol, etridiazole, Terrazole®, Koban®; Truban® (Olin)	5-ethoxy-3-trichloromethyl-1,2,4-thiadiazole	Seedling diseases of field and vegetable crops.	1077	—
etridiazole (*see* ethazol)				
ethirimol, Milcurb® Super (ICI)	5-butyl-2-(ethylamino)-6-methyl-4(1*H*)-pyrimidinone	Systemic control of powdery mildew of cereals.	6340	—
fenaminosulf, Lesan® (Mobay)	Sodium [4-(dimethylamino)phenyl] diazenesulfonate	Seed treatment for field crops, some vegetables, ornamentals, pineapple, sugar cane.	75	>100
fentin acetate, Brestan® (Hoechst)	triphenyltin acetate	Vegetable diseases, especially potato late blight.	125	500
fentin chloride, Tinmate® (Nitto Kasei)	triphenyltin chloride	*Cercospora* leafspot on sugar beets; potato late blight.	18	—
fentin hydroxide, Du-Ter® (Thompson-Hayward)	triphenyltin hydroxide	Vegetable diseases, especially potato late blight.	108	—
ferbam, Carbamate® (FMC)	tris(dimethylcarbamodithioato-*S,S'*)iron	Apple and tobacco diseases; protectant for other crops.	>4000	—
folpet, Phaltan® (Chevron, Stauffer)	2-[(trichloromethyl)thio]-1*H*-isoindole-1,3(2*H*)-dione	Broad uses on fruits, vegetables, ornamentals; seed-, plant-bed treatments.	10,000	>22,600
Frucote® (discontinued)	2-aminobutane	Citrus diseases.	380	—
Funginex® (*see* triforine)				
Glyodin® (Agway)	2-heptadecyl-2-imidazoline acetate	Wide range of tree fruit, ornamental diseases.	4600	—
hexachlorobenzene (Compania Quimica)	1,2,3,4,5,6-hexachlorobenzene	Seed treatment for small grains.	10,000	—
hexachlorophene (Kalo)	2,2'-methylene bis (3,4,6-trichlorophenol)	Fungal and bacterial diseases of tomatoes, peppers and cucumbers; cotton seedling diseases.	2700	>10,000
Indar®, butrizol (Rohm & Haas)	4-N-butyl-4*H*-1,2,4-triazole	Wheat leaf rust.	50	315
iprodione, Chipco 26019® Rovral® (Rhone-Poulenc)	3-(3,5-dichlorophenyl)-N-(1-methyl-ethyl)-2,4-dioxo-1-imidazolidinecarboxamide	Spring and summer turf diseases. Experimental for fruits, vegetables, field crops.	4000	—
Karathane® (*see* dinocap)				
kasugamycin (*Streptomyces kasugaenis*) (Hokko Chemical Industry Co. (Japan))	D-3-*O*-[2-amino-4-[(1-carboxyiminomethyl)-amino]-2,3,4,6-tetradeoxy-α-D-arabino-hexopyranosyl] D-chiro-inositol	Systemic antibiotic effective against rice blast.	22,000	—
Koban® (*see* ethazol)				
Kocide® (*see* copper hydroxide)				
Lesan® (*see* fenaminosulf)				
lime sulfur (several)	calcium polysulfide	Apple scab, powdery skin irritation mildews.		
mancozeb several (several)	coordination product of zinc ion and manganese ethylene bisdithiocarbamate	Wide spectrum of fruit, vegetable, field crop, nut diseases; also seed treatment for field crops.	>8000	

Common name, trade name, and basic manufacturer(s)	Chemical name	General use pattern	Oral LD_{50} (rats)	Dermal LD_{50} (rabbits)
maneb (many) (several)	manganese ethylenebisdithiocarbamate	Many diseases of fruits, vegetables, turf.	6750	—
Melprex® (*see* dodine)				
MEMC (*see* Ceresan®)				
metalaxyl, Ridomil® Subdue® (Ciba-Geigy)	*N*-(2,6-dimethylphenyl)-*N*-(methoxyacetyl)-alanine methyl ester	Systemic control of *Pythium* and *Phytophthora* soil-borne diseases, downy mildews.	669	>3100
metiram, Polyram-Combi® (BASF)	mixture of ammoniates of ethylenebis (dithiocarbamate)-zinc and ethyelnebisdithiocarbamic acid cyclic anhydrosulfides	Many diseases of fruits, vegetables ornamentals, turf.	6200	—
Milban® (*see* dodemorph)				
Milcurb® (*see* dimethirimol)				
Milcurb® Super (*see* ethirimol)				
Morestan® (*see* oxythioquinox)				
Morocide® (*see* binapacryl)				
Mylone® (*see* dazomet)				
nabam (several) (Chemical Formulators, Rohm & Haas)	disodium ethylenebisdithiocarbamate	Industrial applications only; not for food crops.	395	—
Nimrod® (*see* bupirimate)				
oxycarboxin, Plantvax® (Uniroyal)	5,6-dihydro-2-methyl-*N*-phenyl-1,4-oxathiin-3-carboxamide 4,4-dioxide	Systemic control of rusts on greenhouse flowers.	2000	>16,000
oxythioquinox, Morestan® (Mobay)	6-methyl-1-3-dithiolo(4,5-*b*)quinoxalin-2-one	Powdery mildews on fruits, vegetables, ornamentals.	2500	—
parinol, Parnon® (discontinued)	α,α-bis(*p*-chlorophenyl)-3-pyridine-methanol	Powdery mildew on ornamentals.	5000	—
PCNB, Terraclor® (Olin)	pentachloronitrobenzene	Many seedlings, vegetable, ornamental, turf diseases.	12,000	—
PCP (several)	pentachlorophenol	Wood preservative.	50	—
Phaltan® (*see* folpet)				
phenyl phenol, Dowicide 1® (Dow)	*o*-phenyl phenol	Wax treatment or impregnated wraps for harvested fruit.	2700	—
piperalin, Pipron® (Elanco)	3-(2-methylpiperidinyl)propyl 3,4-dichlorobenzoate	Powdery mildew on flowers, ornamentals.	2500	—
Plantvax® (*see* oxycarboxin)				
PMA (many) (discontinued)	phenylmercury acetate	Turf; seed-, soil-borne cereal diseases.	25	—
polyoxin B (*Streptomyces cacaoi*) (Hokko Chemical Industry Co., Kaken Kagaku Co., Kumiai Chemical Industry Co., (Japan))	5-[[2-amino-5-*O*-(aminocarbonyl)-2-deoxy-L-xylonoyl]amino]-pyrimidinyl]-β-D-allofura-1,5-dideoxy-1-[3,4-dihydro-5-(hydroxymethyl)-2,4-dioxo-1(2*H*)-nuronic acid	Systemic antibiotic effective against *Alternaria* on apples & pears, *Botrytis* on tomatoes & rice sheath blight.	21,000	—
Polyram-Combi® (*see* metiram)				

Common name, trade name, and basic manufacturer(s)	Chemical name	General use pattern	Oral LD_{50} (rats)	Dermal LD_{50} (rabbits)
Previcure® (*see* prothiocarb)				
Previcur-N® (*see* propamocarb hydrochloride)				
propamocarb hydro-chloride, Previcur-N® (Nor-Am)	propyl[3-(dimethylamino)propyl]carbamate-monohydrochloride	Soil or foliar application to control downy mildews.	2000	>5000
propineb, Antracol® (Bayer)	zinc-propylenebis(dithiocarbamate)	Grape, tobacco diseases; early/late blight of potatoes, tomatoes.	8500	>1000
prothiocarb, Previcure® (discontinued)	*S*-ethyl-*N*-(3-dimethylaminopropyl)-thiocarbamate hydrochloride	Seedling diseases, soil treatment.	1300	>980
pyrazophos, Afugan® (Hoechst AG)	ethyl 2-[(diethoxyphosphinothioyl)oxy]-5-methylpyrazolo(1,5-*a*)pyrimidine-6-carboxylate	Powdery mildew on cusurbits, fruits, cereals.	415	>2000
Quintar® (*see* dichlone)				
Ridomil® (*see* metalaxyl)				
Ronilan® (*see* vinclozolin)				
Rovral® (*see* iprodione)				
Sclex® (*see* dichlozoline)				
sodium phenyl phenate (*see* Dowicide A®)				
Spergon® (*see* chloranil)				
streptomycin (Merck, Pfizer)	2,4-diguanidino-3,5,6-trihydroxycyclohexyl-5-deoxy-2-*O*-2-deoxy-2-methylamino-α-glucopyranosyl)-3-formyl pentofuranoside	Bacterial blights; cankers of pome, stone fruits, ornamentals.	9000	—
Subdue® (*see* metalaxyl)				
sulfur (several)	elemental sulfur in many formulations	Fruits rots, powdery mildews, rusts on vegetables.	nontoxic irritant	
TBZ (*see* thiabendazole)				
TCMTB (*see* Busan-72®)				
Terraclor® (*see* PCNB)				
Terrazole® (*see* ethazol)				
Tersan® (*see* chloroneb)				
thiabendazole, TBZ Arbotec® (Merck)	2-(4′-thiazoyl)benzimidazole	Systemic for blue and green molds; soybean pod, stem blight; others.	3100	—
thiophanate, Topsin E® (Pennwalt, Cleary)	1,2-bis(3-ethoxycarbonyl-2-thioureido) benzene	Systemic, controls many diseases of fruits, vegetables, turf, cereals.	15,000	—
thiophanate-methyl, Topsin-M® (Pennwalt)	dimethyl [(1,2-phenylene)bis-(imino-carbonothioyl)]bis-(carbamate)	Systemic, controls wide spectrum of diseases of fruits, vegetables, field crops, turf.	7500	—
thiram, several (several)	tetramethylthiuramdisulfide	Seed treatment; diseases of fruits, vegetables, turf.	780	—
Tinmate® (*see* fentin chloride)				
Topsin E® (*see* thiophanate)				

Common name, trade name, and basic manufacturer(s)	*Chemical name*	*General use pattern*	*Oral* LD_{50} *(rats)*	*Dermal* LD_{50} *(rabbits)*
Topsin M® (*see* thiophanate-methyl)				
triadimefon, Bayleton® (Mobay)	1-(4-chlorophenoxy)-3,3-dimethyl-1-(1*H*-1,2,4-triazol-1-yl)-2-butanone	Systemic control of powdery mildews of fruits, vegetables, cereals; rusts of cereals, coffee.	363	>1000
triadimenol, Baytan® (Mobay)	β-(4-chlorophenoxy)-α-(1,1-dimethylethyl)-1*H*-1,2,4-triazole-1-ethanol	Systemic control of cereal diseases; smuts, powdery mildew, rust.	1105	>5000
tricyclazole, Beam® (Elanco)	5-methyl-1,2,4-triazole(3,4-b) benzothiazole	Systemic control of rice blast disease.	250	—
tridemorph, Calixin® (BASF)	2,6-dimethyl-4-tridecyl-morpholine	Variety of field crop, vegetable, fruit, ornamental diseases.	1112	1350
triforine, Funginex® (Celamerck, Shell Ltd.)	*N,N'*-[piperazinediyl-bis(2,2,2-trichloro-ethylidene)]-bis-(formamide)	Systemic control of several fungal diseases of fruits, vegetables, cereals, ornamentals.	6000	>10,000
Truban® (*see* ethazol)				
Urbacide® (discontinued)	methylarsenic-dimethyl-dithiocarbamate	Apple scab, coffee diseases.	175	—
vinclozolin, Ronilan® (BASF Wyandotte)	3-(3,5-dichlorophenyl)-5-ethenyl-5-methyl-2,4-oxazolidinedione	*Botrytis* and *Monilia* spp. in fruits, vegetables, ornamentals.	10,000	—
Vitavax® (*see* carboxin)				
zineb, several (several)	zinc ethylenebis(dithiocarbamate)	Many diseases of fruits, vegetables.	5200	—
ziram, several (several)	zinc dimethyldithiocarbamate	Many diseases of vegetables, some fruits.	1400	—

APPENDIX D

Rodenticides

Common, trade, and chemical names of rodenticides, their basic manufacturer(s), and their oral LD_{50}s to rats.

Common name, trade name, and basic manufacturer(s)	*Chemical name*	*Oral LD_{50} (rats)*
antu	α-naphthylthiourea	6
Baran® (*see* Compound 1081)		
barium carbonate	barium carbonate	630
brodifacoum, Talon® (ICI Americas)	3-[3-[4′-bromo(1-1′-biphenyl)-4-yl]-1,2,3,4-tetrahydro-1-naphthalenyl]-4-hydroxy-*2H*-1-benzopyran-2-one	0.27
bromadiolone, Maki® (Chempar)	3-[3-[4′-Bromo(1,1′-biphenyl)-4-yl]-3-hydroxy-1-phenylpropyl]-4-hydroxy-*2H*-1-benzopyran-2-one	1.13
Castrix® (*see* crimidine)		
chlorophacinone, Rozol® (Chempar)	2-[(p-chlorophenyl)phenylacetyl]-1*H*-indene-1,3(2*H*)-dione	20.5
Compound 1080 (discontinued)	sodium fluoroacetate or sodium monofluoroacetate	0.22
Compound 1081, Rodex® (Jewnin-Joffe) Baran® (Tamogen)	fluoroacetamide	5.75
coumachlor, Tomorin® Ratilan® (Ciba-Geigy Ltd.)	3-(α-acetonyl-4-chlorobenzyl)-4-hydroxycoumarin	900
coumafuryl, Fumarin® (Union Carbide)	3-(α-acetonylfurfuryl)-4-hydroxycoumarin	25
coumatetralyl, Racumin® (Bayer AG)	3-(α-tetralyl)-4-hydroxycoumarin	16.5
crimidine, Castrix® (Bayer AG)	2-chloro-4-dimethylamino-6-methylpyrimidine	1.25
Cymag® (*see* sodium cyanide)		
Dethdiet® (*see* red squill)		
difenacoum, Ratak® (ICI)	3-{3-1,1′-biphenyl-4yl-1,2,3,4-tetrahydro-1-naphthylenyl}-4-hydroxy-2*H*-1-benzopyran-2-one	1.8
Diphacin® (*see* diphacinone)		
diphacinone, Diphacin® Promar®, Ramik® (Velsicol) (Hopkins)	2-diphenylacetyl-1,3-indandione	
endrin (Velsicol)	1,2,3,4,10,10-hexachloro-7-epoxy-1,4,4a,5,6,7,8,8a-octahydro exo-1,4-exo-5,8-dimethanonaphthalene	10
Fumarin® (*see* coumafuryl)		
Gophacide® (*see* phorazetim)		
Maki® (*see* bromadiolone)		
norbormide, Raticate® (discontinued)	5-(α-hydroxy-α-2-pyridyl-benzyl)-7-(α-2-pyridyl-benzylidene)-5-norbornene-2,3-dicarboximide	5.3

Common name, trade name, and basic manufacturer(s)	Chemical name	Oral LD_{50} (rats)
nux vomica (*see* strychnine)		
phorazetim, Gophacide® (discontinued)	*O,O*-bis(4-chlorophenyl)acetimidoylphosphoramidothioate	3.7
phosphorus (*see* yellow phosphorus)		
pindone, Pival® (Motomco)	2-pivaloylindane-1,3-dione	280
Pival® (*see* pindone)		
Prolin® (Prentiss) (Hopkins)	warfarin plus sulfaquinoxaline	1000
Promar® (*see* diphacinone)		
pyriminil, Vacor® (discontinued)	1-(3-pyridylmethyl)-3-(4-nitrophenyl)-urea	4.75
Racumin® (*see* coumatetralyl)		
Ratak® (*see* difenacoum)		
Ratilan® (*see* coumachlor)		
Ratox® (*see* thallium sulfate)		
Raticate® (*see* norbormide)		
Ramik® (*see* diphacinone)		
red squill, Dethdiet® Rodine®	scilliroside = (3β, 6β-6-acetyloxy-3-(β-D-glucopyranosyloxy)-8,14-dihydroxybufa-1,20,22-trienolide)	0.7
Rodex® (*see* Compound 1081)		
Rodine® (*see* red squill)		
Rozol® (*see* chlorophacinone)		
sodium cyanide, Cymag® (ICI)	sodium cyanide	0.5
strychnine, nux vomica	alkaloid from tree, *Strychnos nux-vomica*	1 to 30
Talon® (*see* brodifacoum)		
Tomorin® (*see* coumachlor)		
thallium sulfate, Ratox® (Bayer AG)	thallium sulfate	16
Vacor® (*see* pyriminil)		
Valone® (Motomco)	2-isovaleryl-1,3-indandione	50
warfarin (many) (several)	3-(α-acetonylbenzyl)-4-hydroxycoumarin	Varies greatly from 1 to 186 (in daily doses)
warfarin sodium salt	water soluble sodium salt of warfarin	Same as warfarin
yellow phosphorus	elemental phosphorus	<6
zinc phosphide (Hooker)	zinc phosphide	45

Glossary

AAPCO. Association of American Pesticide Control Officials, Inc.

Abscission. Process by which a leaf or other part is separated from the plant.

Absorption. Process by which pesticides are taken into plant tissues by roots or foliage (stomata, cuticle, etc.).

Acaricide (miticide). An agent that destroys mites and ticks.

Acceptable daily intake (ADI). The daily exposure level of a pesticide residue (expressed as mg/kg body weight per day) that, over the entire lifetime of a human being, appears to be without appreciable risk on the basis of all facts known at a given time.

Acetylcholine (ACh). Chemical transmitter of nerve and nerve-muscle impulses in animals.

Activator. Material added to a fungicide to increase toxicity.

Active ingredient (AI). Chemicals in a product that are responsible for the pesticidal effect.

Acute toxicity. The toxicity of a material determined at the end of 24 hours; toxicity that causes injury or death from a single dose or exposure.

Additive, pesticide. *See* Adjuvant.

ADI. *See* Acceptable daily intake.

Adjuvant. An ingredient that improves the properties of a pesticide formulation. Includes wetting agents, spreaders, emulsifiers, dispersing agents, foam suppressants, penetrants, and correctives.

Adsorption. Chemical and/or physical attraction of a substance to a surface. Refers to gases, dissolved substances, or liquids on the surface of solids or liquids.

Aerosol. Colloidal suspension of solids or liquids in air.

Adulterated pesticide. A pesticide that does not conform to the professed standard or quality as documented on its label or labeling.

Agroecosystem. An agricultural area sufficiently large to permit long-term interactions of all the living organisms and their nonliving environment.

Algicide. Chemical used to control algae and aquatic weeds.

Alkylating agent. Highly active compound (chemosterilant) that replaces hydrogen atoms with alkyl groups, usually in cells undergoing division.

Annual. Plant that completes its life cycle in one year, that is, germinates from seed, produces seed, and dies in the same season.

Antagonism. Decreased activity arising from the effect of one chemical on another (opposite of synergism).

Antibiotic. Chemical substance produced by a microorganism that is toxic to other microorganisms.

Anticoagulant. A chemical that prevents normal blood-clotting. The active ingredient in some rodenticides.

Antidote. A practical treatment, including first aid, used in the treatment of pesticide poisoning or some other poison in the body.

Antidrift agent. A compound used to reduce the number of fine droplets produced at the spray nozzle.

Antimetabolite. Chemical that is structurally similar to biologically active metabolites, and that may take their place detrimentally in a biological reaction.

Antitranspirant. A chemical applied directly to a plant that reduces the rate of transpiration or water loss by the plant.

Apiculture. Care and culture of bees.

Aromatics. Solvents containing benzene or compounds derived from benzene.

Atropine (atropine sulfate). An antidote used to treat organophosphate and carbamate poisoning.

Attractant, insect. A substance that lures insects to trap or poison-bait stations. Usually classed as food, oviposition, and sex attractants.

Attracticide. Combination of an insect attractant (usually a sex pheromone) with an insecticide designed to kill attracted insects.

Auxin. Substance found in plants that stimulates cell growth in plant tissues.

Avicide. Lethal agent used to destroy birds; also refers to materials used for repelling birds.

Aziridine. Chemical classification of chemosterilants containing three-membered rings composed of one nitrogen and two carbon atoms.

Bactericide. Any bacteria-killing chemical.

Bacteriostat. Material used to prevent growth or multiplication of bacteria.

Band application. Application to a continuous restricted band such as in or along a crop row, rather than over the entire field area.

Bentonite. A colloidal native clay (hydrated aluminum silicate) that has the property of forming viscous suspensions (gels) with water; used as a carrier in dusts to increase a pesticide's adhesion.

Biennial. Plant that completes its growth in two years. The first year it produces leaves and stores food; the second year it produces fruit and seeds.

Biocide. Usually used as a synonym for pesticide, suggesting a chemical substance that kills living things or tissue; *not in current use among scientists.*

Biological control agent. Any biological agent that adversely affects pest species.

Biomagnification. The increase in concentration of a pollutant in animals as related to their position in a food chain, usually referring to the persistent, organochlorine insecticides and their metabolites.

Biorational pesticides. Biological pesticides such as bacteria, viruses, fungi, and protozoa; includes pest control agents and chemical analogues of naturally occurring biochemicals (pheromones, insect growth regulators, etc.).

Biota. Animals and plants of a given habitat.

Biotic insecticide. Microorganism, known as an *insect pathogen*, applied in the same manner as conventional insecticides to control pest species.

Biotype. Subgroup within a species differing in some respect from the species, such as a subgroup that is capable of reproducing on a resistant variety.

Blast. Plant disease similar to blight.

Blight. Common name for a number of different diseases on plants, especially those causing sudden collapse—for example, leaf blight, blossom blight, shoot blight.

Botanical pesticide. A pesticide produced from naturally occurring chemicals found in some plants. Examples are nicotine, pyrethrum, strychnine, and rotenone.

Brand. The name, number, or designation of a pesticide.

Broadcast application. Application over an entire area rather than only on rows, beds, or middles.

Broad-spectrum insecticide. Nonselective insecticide having about the same toxicity to most insects.

Calibrate. To determine the amount of pesticide that will be applied to the target area.

Cancelation, pesticide registration. Statement by the EPA of reasons for judging that the risks of a pesticide outweigh its benefits and its uses should be canceled.

Canker. A lesion on a stem.

Carbamate insecticide. One of a class of insecticides derived from carbamic acid.

Carcinogen. A substance that causes cancer in animal tissue.

Carrier. An inert material that serves as a diluent or vehicle for the active ingredient or toxicant.

Cataractogen. A substance that causes cataracts in the eyes of animals.

Causal organism. The organism (pathogen) that produces a given disease.

Certified applicator. Commercial or private person qualified to apply restricted-use pesticides as defined by the EPA.

Chelating agent. Organic chemical (e.g., ethylenediaminetetraacetic acid) that combines with metal to form soluble chelates and prevent conversion to insoluble compounds.

Chemical name. Scientific name of the active ingredient(s) found in the formulated product. The name is derived from the chemical structure of the active ingredient.

Chemosterilant. A chemical compound that causes sterilization or prevents effective reproduction.

Chemotherapy. Treatment of a diseased organism with chemicals to destroy or inactivate a pathogen without seriously affecting the host.

CHEMTREC. A toll-free, long-distance, telephone service that provides 24-hour emergency pesticide information (800-424-9300).

Chlorosis. Loss of green color in foliage.

Cholinesterase (ChE). An enzyme of the body necessary for proper nerve function; it is inhibited or damaged by organophosphate or carbamate insecticides taken into the body by any route.

Chronic toxicity. The toxicity of a material determined beyond 24 hours and usually after several weeks of exposure.

Common pesticide name. A common chemical name given to a pesticide by a recognized committee on pesticide nomenclature. Many pesticides are known by a number of trade or brand names but have only one recognized common name. For example, the common name for Sevin® insecticide is *carbaryl*.

Compatible materials. Two materials that can be mixed together with neither affecting the action of the other.

Concentration. Content of a pesticide in a liquid or dust; for example, pounds/gallon or percentage by weight.

Contact herbicide. Herbicide that causes localized injury to plant tissue upon contact.

Contamination. The presence of an unwanted pesticide or other material in or on a plant, animal, or their by-products. *See also* Residue.

Criteria met or exceeded. Reason(s) why the EPA is investigating a pesticide or placing it under RPAR.

Cultivar. An accepted term for a variety of a man-made selection of a particular plant.

Cumulative pesticides. Those chemicals that tend to accumulate or build up in the tissues of animals or in the environment (soil, water).

Curative pesticide. A pesticide that can inhibit or eradicate a disease-causing organism after it has become established in the plant or animal.

Cutaneous toxicity. *See* Dermal toxicity.

Cuticle. Outer covering of insects.

Days-to-harvest. The least number of days between the last pesticide application and the harvest date, as set by law. Also called *harvest intervals.*

Deciduous plant. Plant that loses its leaves during the winter.

Decontamination. Removal or breakdown of any pesticide chemical from any surface or piece of equipment.

Deflocculating agent. Material added to a spray preparation to prevent aggregation or sedimentation of the solid particles.

Defoliant. A chemical that initiates abscission in leaves.

Deposit. Quantity of a pesticide deposited on a unit area.

Dermal toxicity. Toxicity of a material as tested on the skin, usually on the shaved belly of a rabbit; the property of a pesticide to poison an animal or human when absorbed through the skin. Also called *cutaneous toxicity.*

Desiccant. A chemical that induces rapid desiccation of a leaf or plant part.

Desiccation. Accelerated drying of a plant or plant parts.

Detoxify. To make an active ingredient in a pesticide or other poisonous chemical harmless and incapable of being toxic to plants and animals.

Diapause. A period of arrested development or suspended animation.

Diatomaceous earth. A very abrasive, white powder prepared from naturally occurring deposits formed by the silicified skeletons of diatoms; used as a diluent in dust formulations.

Diluent. Component of a dust or spray that dilutes the active ingredient.

Directed application. Precise application to a specific area or plant organ, such as to a row or bed or to the leaves or stems of plants.

Disinfectant. A chemical or other agent that kills or inactivates disease-producing microorganisms in animals, seeds, or other plant parts. Also commonly refers to chemicals used to clean or surface-sterilize inanimate objects.

DNA. Deoxyribonucleic acid.

Dormant spray. Chemical applied in winter or very early spring before treated plants have started active growth.

Dose, dosage. The amount of toxicant given or applied per unit of plant, animal, or surface. Same as *rate.*

Drift, spray. Movement of air-borne spray droplets from the spray nozzle beyond the intended contact area.

EC_{50}. The median effective concentration (ppm or ppb) of the toxicant in the environment (usually water) that produces a designated effect in 50 percent of the test organisms exposed.

Ecdysone. Hormone secreted by insects essential to the process of molting from one stage to the next.

Ecology. Derived from the Greek *oikos,* "house or place to live." A branch of biology concerned with organisms and their relation to the environment.

Economic injury level. The lowest population density of a pest that will cause economic damage (loss).

Economic level. The insect pest level at which additional management practices must be employed to prevent economic losses.

Economic poison. *See* Pesticide.

Economic threshold. The density of a pest at which control measures should be initiated to prevent an increasing pest population from reaching the economic injury level.

Ecosystem. The interacting system of all the living organisms of an area and their nonliving environment.

Ectoparasite. A parasite that feeds on a host from the host's exterior or outside.

ED_{50}. The median effective dose, expressed as mg/kg of body weight, which produces a designated effect in 50 percent of the test organisms exposed.

Emergence. The visible emerging phase of a specified crop or weed.

Emulsifiable concentrate. Concentrated pesticide formulation containing organic solvent and emulsifier to facilitate emulsification with water.

Emulsifier. Surface active substance used to stabilize suspensions of one liquid in another; for example, oil in water.

Emulsion. Suspension of minuscule droplets of one liquid in another.

Encapsulated formulation. Pesticide enclosed in capsules (or beads) of thin polyvinyl or other material, to control the rate of release of the chemical and extend the period of diffusion.

Endangered species. A species of animal or plant threatened with extinction.

Endoparasite. A parasite that enters the host's tissues and feeds from within.

Environment. All the organic and inorganic features that surround and affect a particular organism or group of organisms.

Environmental Protection Agency (EPA). The federal agency responsible for pesticide rules and regulations, and all pesticide registrations.

EPA. See Environmental Protection Agency.

EPA Establishment Number. A number assigned to each pesticide production plant by the EPA. The number indicates the plant at which the pesticide product was produced and must appear on all labels of that product.

EPA Registration Number. A number assigned to a pesticide product by the EPA when the product is registered by the manufacturer or his designated agent. The number must appear on all labels for a particular product.

Eradicant. Fungicide in which a chemical is used to eliminate a pathogen from its host or environment.

Extermination. Complete extinction of a species over a large continuous area such as an island or a continent.

FAO. Food and Agricultural Organization of the United Nations.

FDA. *See* Food and Drug Administration.

FEPCA. The Federal Environmental Pesticide Control Act of 1972.

Fetotoxin. A substance that can poison an unborn fetus.

Field scout. A person who samples fields for insect infestations.

FIFRA. The Federal Insecticide, Fungicide, and Rodenticide Act of 1947.

Filler. Diluent in powder form.

Fixed coppers. Insoluble copper fungicides whose copper is in a combined form. Usually finely divided, relatively insoluble powders.

Flowable. A type of pesticide formulation in which a very finely divided pesticide is mixed in a liquid carrier.

Foaming agent. A chemical that causes a pesticide preparation to produce a thick foam, which aids in reducing drift.

Fog treatment. The application of a pesticide as a fine mist for the control of pests.

Foliar application. Application of a pesticide to the leaves or foliage of plants.

Food and Drug Administration (FDA). Federal agency responsible for purity and wholesomeness of food; safety of drugs, cosmetics, and food additives; and enforcement of pesticide tolerances for food, as set by the EPA.

Food chain. Sequence of species within a community, each member of which serves as food for the species next higher in the chain.

Formamidine insecticide. New group of insecticides with a mode of action highly effective against insect eggs and mites.

Formulation. Way in which basic pesticide is prepared for practical use. Includes preparation as wettable powder, granular, emulsifiable concentrate, and the like.

Full-coverage spray. Spray applied thoroughly over a crop to the point of runoff or drip.

Fumigant. A volatile material that forms vapors that destroy insects, pathogens, and other pests.

Fungicide. A chemical that kills fungi.

Fungistatic. Action of a chemical that inhibits the germination of fungus spores while in contact.

Gallonage. Number of gallons of finished spray mix applied per acre, tree, hectare, square mile, or other unit.

General-use pesticide. A pesticide that can be purchased and used by the general public without undue hazard to the applicator and environment as long as the instructions on the label are followed carefully. *For comparison, see* Restricted-use pesticide.

Genestatic. Pertaining to the action of a chemical that prevents sporulation.

Germicide. A substance that kills "germs" (microorganisms); *not in current use among scientists.*

Granular. A dry formulation of pesticide and other components in discrete particles generally less than 10 mm^3 in size.

Growth regulator. Organic substance effective in minute amounts for controlling or modifying (plant or insect) growth processes.

Harvest-aid chemical. Material applied to a plant before harvest to facilitate harvesting by reducing plant foliage.

Harvest intervals. Period between last application of a pesticide to a crop and the harvest as permitted by law.

Hepatoxin. A substance that causes damage to the liver.

HHS. U.S. Department of Health and Human Services (formerly Department of Health, Education and Welfare).

Hormone. A product of living cells that circulates in the animal or plant fluids and that produces a specific effect on cell activity remote from its point of origin.

Host. Any plant or animal attacked by a parasite.

Hydrolysis. Splitting of a molecule with the addition of a water molecule.

Hyperplasia. Abnormal increase in the number of cells of a tissue.

Hypertrophy. Abnormal increase in the size of cells of a tissue.

Imminent hazard. A situation in which the continued use of a pesticide during the time required for cancelation by the EPA would likely result in unreasonable adverse effects on the environment or to an endangered species.

Incompatible materials. Two or more materials that cannot be mixed or used together.

Inert ingredient. Any substance in a pesticide product having no pesticidal action; it may have nonpesticidal actions.

Ingest. To eat or swallow.

Ingredient statement. That portion of the label on a pesticide container that gives the name and amount of each active ingredient and the total amount of inert ingredients in the formulation.

Inhalation. Exposure of test animals either to vapor or dust for a predetermined time.

Inhalation toxicity. Toxicity of a material to man or animals when breathed into the lungs.

Insect-growth regulator (IGR). Chemical substance that disrupts the action of insect hormones controlling molting, maturity from pupal stage to adult, and other growth functions.

Insect pest management. The practical manipulation of insect (or mite) pest populations, using any or all control methods in a sound ecological manner.

Integrated control. The integration of the chemical and biological control methods.

Integrated pest management. A management system that uses all suitable techniques and methods in as compatible a manner as possible to maintain pest populations at levels below those causing economic injury.

Intramuscular. Injected into the muscle.

Intraperitoneal. Injected into the viscera but not into the organs.

Intravenous. Injected into the vein.

Invert emulsion. An emulsion in which the water is dispersed in oil rather than oil in water; usually a thick mixture, like salad dressing.

IR-4. Interregional Research Project No. 4. A national program devoted to the registration of pesticides for minor crops, headquartered at Rutgers University, New Jersey.

Juvenoid. Insect growth regulator that affects embryonic, larval, and nymphal development.

Label. All printed material attached to or part of the pesticide container.

Labeling. Supplemental pesticide information that complements the information on the label but is not necessarily attached to or part of the container.

Larva (plural, **larvae**). Immature stage of insects that undergo complete metamorphosis.

Larvicide. More commonly refers to chemicals used for controlling mosquito larvae (wiggle tails), but also to chemicals for controlling caterpillars on crops.

Layby pesticide. Pesticide applied with or after the last cultivation of a crop, usually a herbicide.

LC_{50}. The median lethal concentration, the concentration that kills 50 percent of the test organisms, expressed as milligrams (mg) or cubic centimeters (cc, if liquid) per animal. It is also the concentration expressed as parts per million (ppm) or parts per billion (ppb) in the environment (usually water) that kills 50 percent of the test organisms exposed.

LD_{50}. A lethal dose for 50 percent of the test organisms. The dose of toxicant producing 50 percent mortality in a population. A value used in presenting mammalian toxicity, usually oral toxicity, expressed as milligrams of toxicant per kilogram of body weight (mg/kg).

Leaching. The movement of a pesticide chemical or other substance downward through soil as a result of water movement.

Low-volume spray. Concentrate spray, applied to uniformly cover the crop, but not as a full coverage to the point of runoff.

Mesh size. Number of grids per inch through which described particles will pass, for example, 60-mesh granules.

Metabolite, pesticide. A breakdown product of a pesticide resulting from biological, chemical, or physical action on the pesticide within a living organism.

mg/kg (milligrams per kilogram). Measurement used to designate the amount of toxicant per kilogram of body weight of a test organism required to produce a designated effect, usually the amount necessary to kill 50 percent of the test animals.

Microbial insecticide. A microorganism applied in the same way as conventional insecticides to control an existing pest population.

Mildew. Fungus growth on a surface.

Minimum tillage. Practices that utilize minimum cultivation for seedbed preparation and may reduce labor and fuel costs; may also reduce damage to soil structure.

Miscible liquids. Two or more liquids capable of being mixed in any proportions and of remaining mixed under normal conditions.

Mist spraying. Method in which concentrated spray is atomized into an air stream.

MLD. Median lethal dose; same as LD_{50}.

Moiety. Part or component of a molecule.

Molluscicide. A chemical used to kill or control snails and slugs.

Mosaic. Leaf pattern of yellow and green or light-green and dark-green produced by certain virus infections.

Mutagen. Substance causing genes in an organism to mutate or change.

Mutagenicity. The quality of causing birth defects in future generations.

Mycoplasma. A microorganism intermediate in size between viruses and bacteria possessing many virus-like properties and not visible with a light microscope.

NCI. National Cancer Institute.

Necrosis. Death of tissue, plant or animal.

Negligible residue. A tolerance set on a food or feed crop permitting a minuscule amount of pesticide at harvest as a result of indirect contact with the chemical.

Nematicide. Chemical used to kill nematodes.

Nephrotoxin. A substance that causes damage to the kidney.

Neurotoxin. A substance that causes defects in nerve tissue.

No observable effect level (NOEL). The dosage of a pesticide that results in no discernible harm to experimental animals in chronic toxicity studies that include the close examination of all body organs for abnormalities.

Nonselective herbicide. A chemical that is generally toxic to plants without regard to species; toxicity may be a function of dosage, method of application, and the like.

Nuclear polyhedrosis virus (NPV). A disease virus of insects, cultured commercially and sold as a biological insecticide.

Oncogenic. Tending to cause tumors (not necessarily cancerous) in tissue. *See also* Carcinogen.

OPP. Office of Pesticide Programs, EPA.

Oral toxicity. Toxicity of a compound when given by mouth. Usually expressed as number of milligrams of chemical per kilogram of body weight of animal (white rat) when given orally in a single dose that kills 50 percent of the animals. The smaller the number, the greater the toxicity.

Organochlorine insecticide. Insecticide that contains carbon, chlorine, and hydrogen, for example, DDT, chlordane, lindane.

Organophosphate pesticides. Insecticides (also one or two herbicides and fungicides) derived from phosphoric acid esters.

Organotins. A classification of miticides and fungicides containing tin as the nucleus of the molecule.

OTS. Office of Toxic Substances, EPA.

Overtop application. Application over the top of transplanted or growing plants, such as by airplane or raised spray boom of ground rigs; a broadcast or banded application above the plant canopy.

Ovicide. A chemical that destroys an organism's eggs.

Pathogen. Any disease-producing microorganism or virus.

Penetrant. An additive or adjuvant that aids the pesticide in moving through the outer surface of plant tissues.

Perennial. Plant that continues to live from year to year. Plants may be herbaceous or woody.

Persistence. The quality of an insecticide to persist as an effective residue due to its low volatility and chemical stability, for example, certain organochlorine insecticides.

Persistent herbicide. Herbicide that, when applied at the recommended rate, will harm susceptible crops planted in normal rotation after harvesting the treated crop, or that interferes with regrowth of native vegetation in noncrop sites for an extended period of time. *See also* Residual herbicide.

Pest Control Advisor (PCA). Individual who is qualified to make recommendations for use of agricultural chemicals, including pesticides.

Pesticide. An "economic poison" defined in most state and federal laws as any substance used for controlling, preventing, destroying, repelling, or mitigating any pest. Includes fungicides, herbicides, insecticides, nematicides, rodenticides, desiccants, defoliants, plant growth regulators, and the like.

Pheromone. Highly potent chemical substance produced by insects for communication. For some species, laboratory-synthesized pheromones have been developed for trapping purposes.

Phosphorylation. (1) The attachment of the phosphate moiety of an organophosphate insecticide to one of the OP-sensitive enzymes (for example, cholinesterase, acetylcholinesterase, aliesterase, carboxyesterase), rendering it inactive. (2) The coupling of inorganic phosphate to adenosine diphosphate (ADP) to form the high-energy molecule adenosine triphosphate (ATP).

Photolysis. The splitting or degradation of a molecule by light, usually ultraviolet; thus, the chemical decomposition of molecules by sunlight.

Photosensitizer. A chemical that causes increased sensitivity to light.

Physical selectivity. Selective action of broad-spectrum insecticides through timing, dosage, formulation, and the like.

Physiological selectivity. Selective action of insecticides that are inherently more toxic to some insects than to others.

Phytotoxic. Injurious to plants.

Piscicide. Chemical used to kill fish.

Plant regulator (growth regulator). A chemical that increases, decreases, or changes the normal growth or reproduction of a plant.

Plasmolysis. Shrinkage of cell protoplasm away from its wall due to removal of water from its large central vacuole.

Poison. Any chemical or agent that can cause illness or death when eaten, absorbed through the skin, or inhaled by humans or animals.

Poison control center. Information sources for human poisoning cases, including pesticides, usually located at major hospitals.

Position Document 1 (PD 1). A statement of the EPA's position on risk criteria that have been met or exceeded. If an RPAR is issued, the notice is published in the *Federal Register* along with the PD 1 and announces the rebuttal period during which the public may comment.

Position Document 2/3 (PD 2/3). The proposed regulatory action by the EPA to conclude the RPAR statement of the rebuttal analysis and risk-benefit analysis. The proposed Notice of Determination is published in the *Federal Register* with a notice of the availability of the document upon request.

Position Document 4 (PD 4). A statement of the EPA's final decision on the resolution of the RPAR along with an evaluation of the comments from interested parties.

Postemergence. After emergence of the specified weed or crop.

Potentiation. The greatly increased toxicity of two toxicants when administered together, usually organophosphates.

ppb. Parts per billion (parts in 10^9 parts) is the number of parts of toxicant per billion parts of the substance in question.

ppm. Parts per million (parts in 10^6 parts) is the number of parts of toxicant per million parts of the substance in question. They may include residues in soil, water, or whole animals.

Predacide. Chemical used to poison predators.

Preemergence. Prior to emergence of the specified weed or planted crop.

Preplant application. Treatment applied on the soil surface before seeding or transplanting.

Preplant soil incorporated. Herbicide applied and tilled into the soil before seeding or transplanting.

Pre-RPAR. Single study indicating that a compound meets risk criteria; may place pesticide on candidate RPAR list.

Propellant. An inert ingredient in self-pressurized products that produces the force necessary to dispense the active ingredient from the container. *See also* Aerosol.

Protectant. Fungicide applied to plant surface before pathogen attack to prevent penetration and subsequent infection.

Protective clothing. Clothing to be worn in pesticide-treated fields under certain conditions as required by federal law, e.g., reentry intervals.

Protopam chloride (2-pam). An antidote for certain organophosphate pesticide poisoning, but not for carbamate poisoning.

Rate. Amount of active ingredient applied to a unit area regardless of percentage of chemical in the carrier (dilution).

Raw agricultural commodity. Any food in its raw and natural state, including fruits, vegetables, nuts, eggs, raw milk, and meats.

RCRA. *See* Resource Conservation and Recovery Act.

Rebuttable Presumption Against Registration (RPAR). A regulatory investigation process used by the EPA when a pesticide shows potentially dangerous characteristics. At the conclusion of the investigation, the chemical is (1) returned to full registration, or (2) some or all uses become restricted, or (3) the intent to cancel or suspend some or all uses is announced, or (4) a combination of these.

Recirculating sprayer. A sprayer system with the nozzle aimed at a catchment device, which recovers and recirculates herbicide that does not hit plants or weeds.

Reentry interval. Waiting interval required by federal law between application of certain hazardous pesticides to crops and the entrance of workers into those crop fields without protective clothing.

Registered pesticides. Pesticide products that have been approved by the Environmental Protection Agency for the uses listed on the label.

Repellent (insects). Substance used to repel ticks, chiggers, gnats, flies, mosquitoes, and fleas.

Residual. Having a continued lethal effect over a period of time.

Residual herbicide. Herbicide that persists in the soil and injures or kills germinating weed seedlings over a relatively short period of time. *See also* Persistent herbicide.

Residue. Trace of a pesticide and its metabolites remaining on and in a crop, soil, or water.

Resistance (insecticide). Natural or genetic ability of an organism to tolerate the poisonous effects of a toxicant.

Resource Conservation and Recovery Act (RCRA). This act (PL 94-580) became effective January 1, 1977, and is administered by the Environmental Protection Agency (EPA). It deals primarily with solid and hazardous waste disposal through appropriate state and local agencies, and recommends model codes, ordinances, and statutes providing for sound solid-waste management. Overall, it provides technical and financial assistance for development of management plans and facilities for the recovery of energy and other resources from discarded materials and for the safe disposal of discarded materials, and regulates the management of hazardous waste.

Restricted-use pesticide. One of several pesticides, designated by the EPA, that can be applied only by certified applicators, because of their inherent toxicity or potential hazard to the environment.

RNA. Ribonucleic acid.

Rodenticide. Pesticide applied as a bait, dust, or fumigant to destroy or repel rodents and other animals, such as moles and rabbits.

Ropewick applicator. A rope saturated with a foliage-applied translocated herbicide solution that is wiped across the surface of weed foliage. The rope utilizes forces of capillary attraction and conducts the herbicide from a reservoir.

RPAR. *See* Rebuttable Presumption Against Registration.

Rust. A disease with symptoms that usually include reddish-brown or black pustules; a group of fungi in the Basidiomycetes.

Safener. Chemical that reduces the phytotoxicity of another chemical.

Scientific name. Name (genus and species) of a plant or animal used throughout the international scientific community.

Secondary pest. A pest that usually does little if any damage but can become a serious pest under certain conditions; for example, when insecticide applications destroy a given insect's predators and parasites.

Selective pesticide. Pesticide that, while killing the pest individuals, spares most of the other fauna or flora, including beneficial species, either through differential toxic action or through the manner in which the pesticide is used (formulation, dosage, timing, placement, etc.).

Senescence. Process or state of growing old.

Sensitivity, herbicide. Susceptibility of plants to effects of toxicant at low dosage; for example, many broadleaf plants are sensitive to 2,4-D.

Sex lure. Synthetic chemical that acts as the natural lure (pheromone) for one sex of an insect species.

Signal word. A required word that appears on every pesticide label to denote the relative toxicity of the product. The signal words are *Danger–Poison* for highly toxic compounds, *Warning* for moderately toxic, or *Caution* for slightly toxic.

Silvicide. Herbicide used to control undesirable brush and trees, as in wooded areas.

Slimicide. Chemical used to prevent slimy growth, as in wood-pulping processes for manufacture of paper and paperboard.

Slurry. Thin, watery mixture, such as liquid mud or cement. Fungicides and some insecticides are applied to seeds as slurries to produce thick coating and reduce dustiness.

Smut. A fungus with sooty spore masses; a group of fungi in the Basidiomycetes.

Soil application. Application of pesticide made primarily to soil surface rather than to vegetation.

Soil incorporation. Mechanical mixing of herbicide with the soil.

Soil persistence. Length of time that a pesticide application on or in soil remains effective.

Soluble powder. A finely ground, solid material that will dissolve in water or some other liquid carrier.

Spermatogenicity. The quality of causing a reduction in sperm.

Spore. A single- to many-celled reproductive body in fungi that can develop a new fungus colony.

Spot treatment. Application to localized or restricted areas, as differentiated from overall, broadcast, or complete coverage.

Spreader. Ingredient added to spray mixture to improve contact between pesticide and plant surface.

Sterilize. To treat with a chemical or other agent to kill every living thing in a certain area.

Sticker. Ingredient added to spray or dust to improve its adherence to plants.

Stomach poison. A pesticide that must be eaten in order to kill or control the insect or other animal.

Structural pests. Pests that attack and destroy buildings and other structures, clothing, stored food, and manufactured and processed goods; for example, termites, cockroaches, clothes moths, rats, and dry-rot fungi.

Stupefacient or soporific. Drug used as a pesticide to cause birds to enter a state of stupor so they can be captured and removed, or to frighten other birds away from the area.

Subcutaneous toxicity. The toxicity determined following its injection just below the skin.

Sulfhydryl radical. The —SH group found in many plant and animal enzymes.

Surfactant. Ingredient that aids or enhances the surface-modifying properties of a pesticide formulation (wetting agent, emulsifier, or spreader).

Suspension. Finely divided particles dispersed in a liquid, including emulsions.

Suspension, pesticide registration. EPA finds that an "imminent hazard" exists when the risks from the use of a pesticide during the time required to complete cancelation proceedings outweigh the benefits that may be derived from such use. Suspended products cannot continue to be sold.

Swath. The width of the area covered by an airplane, sprayer, or duster making one sweep.

Synergism. Increased activity resulting from the effect of one chemical on another.

Synthesize. To produce a compound by joining various elements or simpler compounds.

Systemic. Compound that is absorbed and translocated throughout the plant or animal.

Tank mix. Mixture of two or more pesticides in the spray tank at time of application. Such mixture must be cleared by the EPA.

Target. The plants, animals, structures, areas, or pests to be treated with a pesticide application.

Temporary tolerance. A tolerance established on an agricultural commodity by the EPA to permit a pesticide manufacturer or his agent time, usually one year, to collect additional residue data to support a petition for a permanent tolerance; in essence, an experimental tolerance. *See also* Tolerance.

Teratogen. Substance that causes physical birth defects in the offspring following exposure of the pregnant female.

Threshold limit value (TLV). The maximum air concentration of a chemical, expressed as milligrams per cubic meter (mg/m^3), in which workers may perform their duties 8 hours a day, 40 hours per week, with no adverse health effects, as established by the American Conference of Governmental Hygienists.

Tolerance. Amount of pesticide residue permitted by federal regulation to remain on or in a crop. Expressed as parts per million (ppm).

Tolerant. Capable of withstanding effects.

Topical application. Treatment of a localized surface site, such as a single leaf blade, as opposed to oral application.

Toxic. Poisonous to living organisms.

Toxicant. A poisonous substance, such as the active ingredient in pesticide formulations, that can injure or kill plants, animals, or microorganisms.

Toxic Substances Control Act (TSCA). This act (PL 94-469) became effective January 1, 1977, and gives broad authority to the EPA to obtain information from industry on the production, use, health effects, and other matters concerning chemical substances and mixtures. Chemicals already regulated by other laws are exempt from this act (pesticides, tobacco, ammunition, food, food additives, drugs, cosmetics, and nuclear material).

Toxin. A naturally occurring poison produced by plants, animals, or microorganisms; for example, the poison produced by the black widow spider, the venom produced by snakes, and the botulism toxin.

Toxogonin®. An antidote used in the treatment of organophosphate pesticide poisoning. Its action is identical to that of 2-PAM, and is not to be used in the treatment of carbamate poisoning. Toxogonin® is manufactured by Merck (Germany) and is neither sold nor used in the United States. Chemical name: bis-(4-hydroxy-iminomethyl-pyridinum-(1)-methyl) ether dichloride.

Toxophore. The toxic component of a toxic molecule, or that portion of a molecule responsible for its toxic action.

Trade name (trademark name, proprietary name, brand name). Name given a product by its manufacturer or formulator, distinguishing it as being produced or sold exclusively by that company.

Translocation. Transfer of food or other materials such as herbicides from one plant part to another.

Trigger, RPAR. The major factor that alerts federal authorities to the potential environmental threat presented by a compound.

Trivial name. Name in general or commonplace usage; for example, nicotine.

TSCA. *See* Toxic Substances Control Act.

Ultralow volume (ULV). Sprays that are applied at 0.5 gallon or less per acre or sprays applied as the undiluted formulation.

Vector. An organism, as an insect, that transmits pathogens to plants or animals.

Viricide. A substance that completely and permanently inactivates a virus.

Virustatic. Pertaining to the action of a chemical that inhibits the multiplication of a virus.

Volatilize. To vaporize.

Weed. Plant growing where it is not desired.

Wettable powder. Pesticide formulation of toxicant mixed with inert dust and a wetting agent that mixes readily with water and forms a short-term suspension (requires tank agitation).

Wetting agent. Compound that causes spray solutions to contact plant surfaces more thoroughly.

Winter annual. Plant that starts germination in the fall, lives over winter, and completes its growth, including seed production, the following season.

Bibliography

Agriculture Council of America. 1982. *Food Globe*. Agriculture Council of America, Washington, D.C.

Ahrens, W. H., C. J. Arntzen, and E. W. Stoller. 1981. Chlorophyll fluorescence assay for the determination of triazine resistance. *Weed Science* 29(3): 316–322.

Ashton, F. M., and A. S. Crafts. 1973. *Mode of Action of Herbicides*. Wiley, New York. 504 pp.

Aspelin, A. L., and G. L. Ballard. 1980. *Pesticide Industry Sales and Usage: 1980 Market Estimates*. Economic Analysis Bureau, Environmental Protection Agency, Washington, D.C. 12 pp.

Baily, J. B., and J. E. Swift. 1968. *Pesticide Information and Safety Manual*, University of California Agricultural Extension Service, Berkeley. 147 pp.

Bergquist, L. M. 1981. *Microbiology for the Hospital Environment*. Harper & Row, New York. 719 pp.

Beroza, M. (Ed.). 1970. *Chemicals Controlling Insect Behavior*. Academic Press, New York. 170 pp.

Borkovec, A. B. 1972. *Safe Handling of Insect Chemosterilants in Research and Field Use*. ARS-NE-2. Agricultural Research Service, U.S. Department of Agriculture, Washington, D.C.

Brazelton, R. W., N. B. Akesson, and W. E. Yates (Eds.). 1972. *The Safe Application of Agricultural Chemicals–Equipment and Calibration. Study Guide for Agricultural Pest Control Advisers*. Agricultural Publications, University of California, Berkeley. 58 pp.

Brown, A. W. A. 1963. "Chemical Injuries." In *Insect Pathology: An Advanced Treatise*, E. A. Steinhaus (Ed.). Vol. 1, pp. 65–131. Academic Press, New York.

Burges, H. D., and N. W. Hussey (Eds.). 1971. *Microbial Control of Insects and Mites*. Academic Press, New York. 861 pp.

Chemical and Engineering News. 1982. U.S. output of aerosols rose 1.5% in 1981 over 1980 production. May 24, p. 8.

Chemical Week. 1980. Pesticides: $6 billion by 1990. May 7, p. 45.

Corbett, J. R. 1974. *The Biochemical Mode of Action of Pesticides*. Academic Press, New York. 330 pp.

Council for Agricultural Science and Technology. 1980. *Organic and Conventional Farming Compared*. CAST Report No. 84. Iowa State University, Ames.

Cummings, N. W. (Ed.). 1973. *Vertebrate Pests. Study Guide for Agricultural Pest Control Advisers*. Agricultural Publications, University of California, Berkeley. 125 pp.

Davies, J. E. 1976. *Pesticide Protection: A Training Manual for Health Personnel*. University of Miami School of Medicine, Miami, Florida. 71 pp.

Deal, A. S. (Ed.). 1972. *Insects, Mites, and Other Invertebrates and Their Control in California. Study Guide for Agricultural Pest Control Advisers*. Agricultural Publications, University of California, Berkeley. 138 pp.

Delp, C. J. 1980. Coping with resistance to plant disease control agents. *Plant Disease* 64(7): 652–657.

Eichers, T. R. 1981. *Farm Pesticide Economic Evaluation, 1981*. Agr. Econ. Report No. 464. U.S. Drug Administration, Washington, D.C. 21 p.

Erwin, D. C. 1973. Systemic fungicides: Disease control, translocation, and mode of action. *Annual Review of Phytopathology* 11: 389–422.

Falcon, L. A. 1971. "Microbial Control as a Tool in Integrated Control Programs." In *Biological Control*, C. B. Huffaker (Ed.). Pp. 346–364. Plenum Press, New York.

Finch, W. (Ed.). 1980. *Arizona Study Guide for Agricultural Pest Control Advisors*. Arizona Agricultural Chemicals Association, Glendale.

Fowler, D. L., 1982. Private communication. Pesticide specialist, U.S. Department of Agriculture. Washington, D.C.

Fowler, D. L., and J. N. Mahan. 1980. *The Pesticide Review 1978*. U.S. Department of Agriculture, Washington, D.C. 42 pp.

Gleason, M. N., R. E. Gosselin, and H. C. Hodge. 1976. *Clinical Toxicology of Commercial Products*. 4th ed. Williams and Wilkins, Baltimore.

Halder, M. R., and R. S. Shadbolt. 1975. Novel 4-hydroxy-coumarin anticoagulants active against resistant rats. *Nature* 253: 275–277.

Hamilton, K. C. 1981. Private communication. Department of Plant Sciences. University of Arizona. Tucson.

Hart, W. H. (Ed.). 1972. *Nematodes and Nematicides. Study Guide for Agricultural Pest Control Advisers*. Agricultural Publications, University of California, Berkeley, 53 pp.

Hawkins, D. E., F. W. Slife, and E. R. Swanson. 1977. Economic analysis of herbicide use in various crop sequences. *Illinois Agricultural Economics* 17(1): 8–13.

Hayes, W. J., Jr. 1982. *Pesticides Studied in Man*. Williams and Wilkins, Baltimore, 672 pp.

Hensley, J. R. 1981. A method for identification of triazine resistant and susceptible biotypes of several weeds. *Weed Science* 29(1): 70–73.

Holan, G. 1969. New halocyclopropane insecticides and the mode of action of DDT. *Nature* (London) 221: 1025.

Hollingworth, R. M. 1976. "The Biochemical and Physiological Basis of Selective Toxicity." In *Insecticide Biochemistry and Physiology*, C. F. Wilkinson (Ed.). Pp. 431–506. Plenum Press, New York.

Hollingworth, R. M., and L. L. Murdock. 1980. Formamidine pesticides: Octopamine-like actions in a firefly. *Science* 208: 74–76.

Holm, L. 1969. Weed problems in developing countries. *Weed Science* 17: 113–118.

Holm, L. 1971. The role of weeds in human affairs. *Weed Science* 9: 485–490.

Kenaga, E. E., and R. W. Morgan. 1978. *Commercial and Experimental Organic Insecticides*. Special Pub. 78-1. Entomological Society of America, College Park, Md. 79 pp.

Klingman, G. C., F. M. Ashton, and L. J. Noordhoff. 1975. *Weed Science. Principles and Practice*. Wiley, New York, 478 pp.

Koehler, P. G., and B. Kuhlman. 1979. *Household Pesticide Survey*. Extension Report No. 59. Florida Cooperative Extension Service. University of Florida, Gainesville. 50 pp.

Kohn, G. K. (Ed.). 1974. *Mechanism of Pesticide Action*. American Chemical Society, Washington. D.C. 180 pp.

LaBrecque, G. C., and C. N. Smith (Eds.). 1968. *Principles of Insect Chemosterilization*. Appleton-Century-Crofts, New York. 354 pp.

Lukens, R. J. 1971. *Chemistry of Fungicidal Action*. Molecular Biology Biochemistry and Biophysics Series, No. 10. Springer-Verlag, New York. 136 pp.

Matsumura, F. 1975. *Toxicology of Insecticides*. Plenum Press, New York. 503 pp.

McCain, A. H. 1970. *Chemicals for Plant Disease Control (Fungicides, Nematicides, Bactericides)*. Agricultural Extension Service, University of California, Berkeley. 213 pp.

McHenry, W. B., and R. F. Norris (Eds.). 1972. *Weed Control. Study Guide for Agricultural Pest Control Advisers*. Agricultural Publications, University of California, Berkeley. 64 pp.

Mead, A. R. 1979. "Economic Malacology with Particular Reference to *Achatina fulica*." In *Pulmonates*, V. Fretter and J. Peake (eds.). Vol 2B. Academic Press, London.

Meister, R. T. (Ed.). 1983. *Farm Chemicals Handbook*. Meister Publishing Co., Willoughby, Ohio.

Melnikov, N. N. 1971. *Chemistry of Pesticides*. Springer-Verlag, New York. 480 pp.

Menn, J. J., and M. Beroza (Eds.). 1972. *Insect Juvenile Hormones, Chemistry and Action*. Academic Press, New York. 341 pp.

Metcalfe, R. L., and J. J. McKelvey, Jr. 1976. *The Future for Insecticides*. Wiley, New York. 524 pp.

Moller, W. J. (Ed.). 1972. *Plant Diseases. Study Guide for Agricultural Pest Control Advisers*. Agricultural Publications, University of California, Berkeley. 232 pp.

Mullins, L. J. 1955. Structure toxicity in hexachlorocyclohexane isomers. *Science* 122 (July): 118–119.

National Academy of Sciences. 1975. *Contemporary Pest Control Practices and Prospects*. 2 vols. National Academy of Sciences, Washington, D.C.

O'Brien, R. D. 1967. *Insecticides: Action and Metabolism*. Academic Press, New York. 332 pp.

Olson, F. J. 1973. "The Screening of Candidate Molluscicides Against the Giant African Snail, *Achatina fulica* Bowdich." Unpublished master's thesis. Graduate College, University of Hawaii, Honolulu.

Orlob, G. B. 1973. Ancient and medieval plant pathology. *Pflanzenschutz-Nachrichten* 26(2): 62–281.

Pelczar, M. J., Jr., R. D. Reid and E. C. S. Chan. 1977. *Microbiology*. 4th ed. McGraw-Hill, New York. 952 pp.

Saleh, M. A., and J. E. Casida. 1978. Reductive dechlorination of the toxaphene component 2,2,5-endo, 6-exo, 8,9,10-heptachlorobornane in various chemical, photochemical, and metabolic systems. *Journal of Agricultural Food Chemistry* 26(3): 583–590.

Savage, E. P., T. J. Keefe, and H. W. Wheeler. 1979. *National Household Pesticide Usage Study, 1976–1977*. Epidemiologic Pesticide Studies Center, Colorado State University, Fort Collins. 126 pp.

Shorey, H. H. 1973. Behavioral responses to insect pheromones. *Annual Review of Entomology* 18: 349–380.

Smith, E. G., and P. A. Baker. 1976. *Wiswesser Line-Formula Chemical Notation (WLN)*. 3rd ed. CIMI, Cherry Hill, New Jersey.

Soloway, S. B. 1965. *Advances in Pest Control Research*. Vol. 6:85–126. Wiley, New York.

Staal, G. B. 1975. Insect growth regulators with juvenile hormone activity. *Annual Review of Entomology* 20: 417–460.

Storck, W. J. 1980. Pesticide profits belie mature market status. *Chemical Engineering News*, April 28, pp. 10–13.

Thomson, W. T. 1981. *Agricultural Chemicals. Book 3: Fumigants, Growth Regulators, Repellants, and Rodenticides*. Thomson Publications, Fresno, Calif. 182 pp.

Thomson, W. T. 1981. *Agricultural Chemicals. Book 2: Herbicides*. Thomson Publications, Fresno, Calif. 274 pp.

Torgeson, D. C. (Ed.). 1967. *Fungicides, an Advanced Treatise*. Vol. 1: *Agricultural and Industrial Applications, Environmental Interactions*. Academic Press, New York. 697 pp.

Torgeson, E. Y. (Ed.). 1969. *Fungicides, an Advanced Treatise*. Vol. 2: *Chemistry and Physiology*. Academic Press, New York. 742 pp.

United Nations. 1974. *The Population Debate, Dimensions and Perspectives*. Vol. 1. Papers of the World Population Conference, Bucharest.

University of California. 1973. *Plant Growth Regulators. Study Guide for Agricultural Pest Control Advisers*. Agricultural Publications, University of California, Berkeley. 79 pp.

U.S. Department of Agriculture. 1981. *Agricultural Statistics*. U.S. Government Printing Office, Washington, D.C.

U.S. Environmental Protection Agency. 1975. *Apply Pesticides Correctly: A Guide for Commercial Applicators*. U.S. Government Printing Office, Washington, D.C. 35 pp.

Van Valkenburg, W. 1973. *Pesticide Formulations*. Marcel Dekker, New York. 481 pp.

Ware, G. W. 1982. *Fundamentals of Pesticides—A Self-Instruction Guide*. Thomson Publications, Fresno, Calif. 257 pp.

Ware, G. W. 1980. *Complete Guide to Pest Control—With and Without Chemicals*. Thomson Publications, Fresno, Calif. 290 pp.

Washington Farmletter. 1979. Doane Agricultural Service, Inc., Washington, D.C. June 19.

Watson, T. F. 1975. "Practical Considerations in Use of Selective Insecticides Against Major Crop Pests." In *Pesticide Selectivity*, J. C. Street (Ed.). Pp. 47–66. Marcel Dekker, New York.

Watson, T. F., L. Moore, and G. W. Ware. 1976. *Practical Insect Pest Management*. W. H. Freeman and Company, San Francisco. 209 pp.

Weaver, R. J. 1972. *Plant Growth Substances in Agriculture*. W. H. Freeman and Company, San Francisco. 594 pp.

Weed Science Society of America. 1979. *Herbicide Handbook*, 4th ed. Champaign, Ill. 479 pp.

Wilkinson, C. F. 1976. *Insecticide Biochemistry and Physiology*. Plenum Press, New York. 768 pp.

Wistreich, G. A., and M. D. Lechtman. 1980. *Microbiology*. 3rd ed. Glencoe, Encino, Calif. 786 pp.

Wiswesser, W. J. (Ed.). 1976. *Pesticide Index*. 5th ed. Entomological Society of America, College Park, Md. 328 pp.

Wolfe, H. R., and W. F. Durham. 1966. "Safety in the Use of Pesticides." In *Proceedings, Second Eastern Washington Fertilizer and Pesticide Conference*. Washington State University, Pullman.

Worthing, D. R. (Ed.). 1979. *The Pesticide Manual—A World Compendium*. 6th ed. British Crop Protection Council, Glasshouse Crops Research Institute, Croydon, England. 655 pp.

Zbenden, G. 1968. "Mechanism of Toxic Drug Reaction." In *Pharmacologic Techniques in Drug Evaluation*, P. E. Esieyler and J. H. Mayer III (Eds.). Pp. 38–47. Yearbook Medical Publishers, Chicago, Ill.

Index